PALAEOHISTORIA

PALAEOHISTORIA

ACTA ET COMMUNICATIONES
INSTITUTI BIO-ARCHAEOLOGICI
UNIVERSITATIS GRONINGANAE

28

A.A.BALKEMA/ROTTERDAM/BROOKFIELD/1986

Editorial staff: Mette Bierma, J.W. Boersma, J.N. Lanting & H.R. Roelink
Address: Biologisch-Archaeologisch Instituut, Poststraat 6, 9712 ER Groningen, Netherlands

Published by
A.A.Balkema, P.O.Box 1675, 3000 BR Rotterdam, Netherlands
A.A.Balkema Publishers, Old Post Road, Brookfield, VT 05036, USA

ISSN 0552-9344
ISBN 90 6191 832 4

Printed in the Netherlands

CONTENTS

D. STAPERT
The Vermaning stones: some facts and arguments 1

A.L. BRINDLEY
Hunebed G2: Excavation and finds 27

A.L. BRINDLEY
The typochronology of TRB West Group pottery 93

J.J. BUTLER
Drouwen: end of a 'Nordic' rainbow? 133

W.A. CASPARIE
The two Iron Age wooden trackways XIV(Bou) and XV(Bou) in the raised bog of Southeast Drenthe (the Netherlands) 169

THE VERMANING STONES: SOME FACTS AND ARGUMENTS

Dick Stapert
Biologisch-Archaeologisch Instituut, Groningen, Netherlands

ABSTRACT: This paper concerns a large-scale forgery affair, commonly known as the 'Vermaning affair'. An attempt is made to recapitulate the most important facts and arguments that convinced the author that the artefacts in question are forgeries.

KEYWORDS: Vermaning stones, forgeries, Middle Palaeolithic, Late Palaeolithic, Neolithic, postdepositional surface modifications, boulderclay, bouldersand, artificial gloss, traces of modern grinding.

1. INTRODUCTION

From 1965 to 1986 at any rate, Mr. T. Vermaning (Assen; deceased in 1987) sold collections of artefacts to several museums in the Netherlands. In 1975 the present author had the misfortune to discover that many of these artefacts are recent fabrications.[1] Among these are 'Middle Palaeolithic' 'collections' or 'isolated objects' (from the 'sites' of Hoogersmilde, Hijken, Ravenswoud, Eemster, Leemdijk, Blauwmeer, Voshaar, and others), parts of one 'Upper Palaeolithic' collection (from 'Norgervaart'), parts of one 'Neolithic' collection (from 'Ravenswoud'), and perhaps also parts of one 'Mesolithic' collection (from 'Appelscha'). Several preliminary articles have been published, especially about the 'Middle Palaeolithic' collections (Stapert, 1975a; 1975b; 1980; see also Stapert, 1976a; 1976b); furthermore a report (unpublished) was prepared for use in lawsuits in 1977 and 1978 (Stapert, 1976c). One or possibly two isolated authentic Middle Palaeolithic artefacts from the same Mr. Vermaning were also described (Stapert, 1985a). It should be mentioned here that Vermaning has also sold many collections of authentic artefacts to museums and collectors, mostly dating from the Late Palaeolithic and younger phases of prehistory.

It is stressed that the main falsifiable argument is of a geological nature. It is based on the observation of weathering phenomena on natural flints present in the bouldersand (the presumed find layer of the 'Middle Palaeolithic' Vermaning stones). The Vermaning stones in question do not show any kind of natural postdepositional surface modifications. Additional arguments concern the presence on many Vermaning stones of artificial gloss, and flattened ridges between flake scars which were evidently produced by the application of an electrically driven grinding machine. Such ground ridges also occur on 'Late Palaeolithic' and 'Neolithic' artefacts sold by Vermaning, while a number of 'Neolithic' ground axes (from Vermaning's 'findspot' of Ravenswoud) must have been ground also with the help of a modern grinding machine.

2. THE GEOLOGICAL ARGUMENT

It can be said that the term 'patina' is not well defined in archaeology; elsewhere I have called it a dustbin-concept (Stapert, 1976a). Phenomena such as the following have been grouped under this term: white patina, yellow/brown patina, black patina (also called 'under-water patina' or 'marsh patina') and gloss patina. The first three of these are unambiguous: their presence or absence can be established without any trouble. Gloss patina is a more problematical phenomenon in this respect (see 7 for some reasons why natural gloss patina can be difficult to demonstrate; it apparently can be confused with artificially created gloss, or with a certain smoothness created by frequent handling). Whatever the genesis of gloss patina, mechanical (soil movements) or chemical or both, it can at least be said that flint surfaces with gloss patina are smoother than they would have been if no gloss patina had developed. Of course, I am discussing only natural postdepositional surface modifications here.

Apart from patinas, several other kinds of surface modifications on flint exist. In the northern Netherlands especially, windgloss is a recurrent phenomenon. In my experience it is nearly always possible to distinguish gloss patina from windgloss. On one and the same piece, windgloss is often variable in intensity, while in most cases it is associated with 'small pits' (described in Stapert, 1976a).

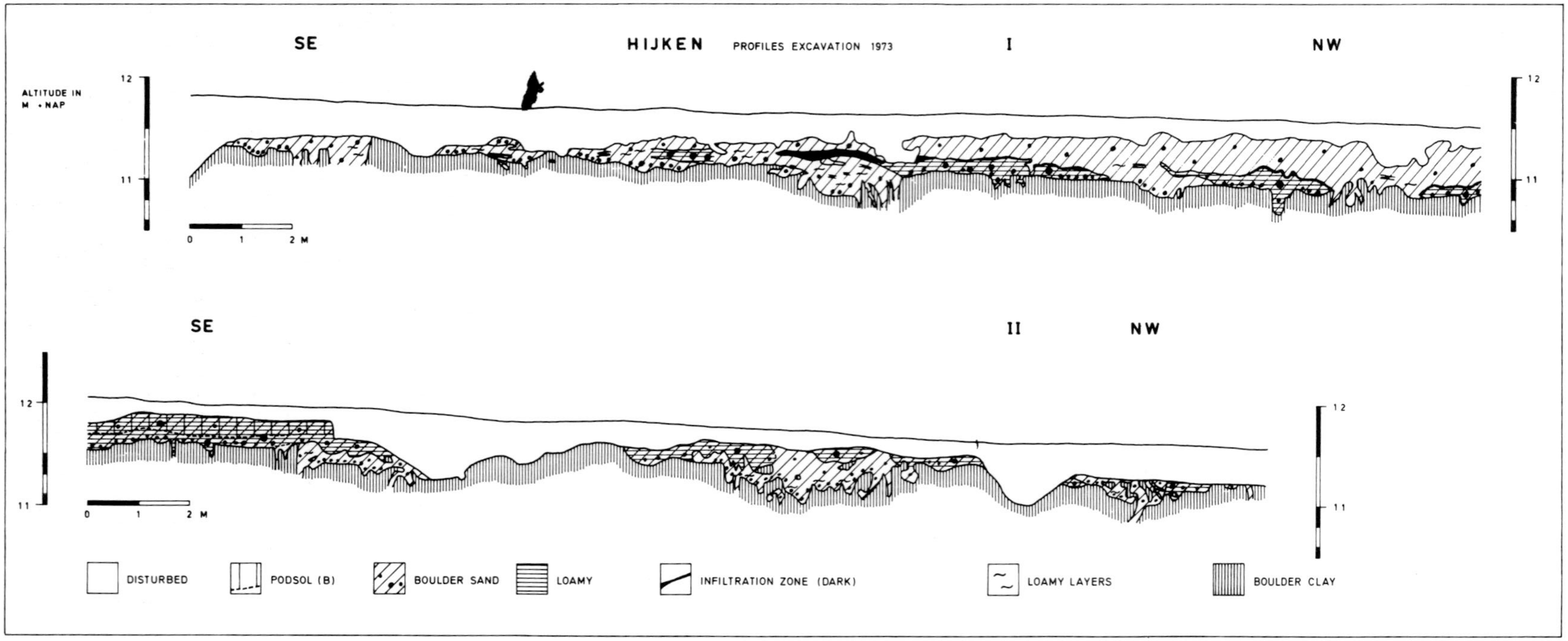

Fig. 1. Sections at the 'site' of Hijken (indicated by Vermaning), drawn during the B.A.I.-excavation in 1973. The sections are at right angles to the valley of the Vorrelveen. The vertical scale is exaggerated 2 x. The most northwestern part of section II is represented in figure 2. The stratigraphy here consists only of boulderclay and bouldersand; coversand is absent. The bouldersand is not an undifferentiated sediment, for some structure is visible. This consists of loamy and more compact layers in the bouldersand, which is otherwise fairly coarse-grained. The drawing schematically shows the following features: a. the stones in the bouldersand are most abundant at the base of this layer on top of the boulderclay. This is the case both in the coarser bouldersand and in the loamy layers where these lie immediately on top of the boulderclay. In some places one can even speak of a compact pebble layer; b. In general, larger stones occur in the loamy layers than in the coarser bouldersand; c. Contortions due to cryoturbation are especially clearly visible in the transition from boulderclay to bouldersand, though much distortion is also in evidence within the bouldersand. In some places, e.g. in the most northwestern part of section II, bands of boulderclay have been distorted together with bouldersand in a complex manner (fig. 2). The loamy layers possibly represent solifluction levels. Drawing by D. Stapert.

Other surface modifications on flint that are frequently encountered in the northern Netherlands are: frostsplitting; rounding of edges and ridges (in most cases not as a result of fluviatile rolling, but probably because of slow solution processes in the soil and/or as one of the results of the formation of windgloss); and scratches, 'pressure cones' and 'cryoturbation-retouch' as a result of cryoturbation and similar soil movements (Stapert, 1976a).

It is evident that in causing these various surface modifications several different processes are at work. Some of these alterations are the result of processes that are limited in time mainly to glacial periods, especially the phenomena produced by periglacial soil movements. Heavy windgloss also mainly arose during glacials, but it is known that windgloss can also be created in recent sanddrifts. Other modifications arose as a consequence of the flints lying at or near the surface for a long time.

Hijken and Hoogersmilde are locations indicated by Vermaning, situated on the boulderclay plateau of the province of Drenthe.[2] The stratigraphy at these sites can be summarized as follows, going from bottom to top: boulderclay (till), bouldersand (the alleged provenance of the 'Middle Palaeolithic' Vermaning stones; this deposit is also called 'cobble sands': e.g. Ter Wee, 1979), and locally a thin layer of coversand. According to Ter Wee (1966; 1979), mainly Late Glacial coversands are present on the higher parts of the plateau, as at the sites of Hijken and Hoogersmilde; only locally in depressions is older coversand found.

Bouldersand can be described as a lag deposit (weathering residue) originating from boulderclay, formed over a long period of time. Locally the bouldersand contains some coversand that was incorporated into it as a result of solifluction and similar processes. The number of stones per cubic metre of sediment is much higher in bouldersand than in boulderclay, implying that quite a thick layer of boulderclay has disappeared as a result of erosion – a reasonable estimate would be that in may places the topmost 1 to 2 m of the boulderclay was divested of its clay particles by the action of wind and water in the course of time, leaving a few dozen cm of bouldersand. Soils dating from the Eemian have disappeared as a result of later erosion (Ter Wee, 1979).

Especially the bouldersand and the top part of the boulderclay were heavily subjected to cryoturbation during the Weichselian, as could be observed during the excavation I carried out at the Hijken location in 1973 (Stapert, 1976a). Figures 1 and 2 give examples of this phenomenon seen in a profile and horizontally.

When flints occurring in boulderclay are compared with flints in the bouldersand on top of the boulderclay in the same site, a striking difference is apparent: the bouldersand flints are much more often heavily weathered than the boulderclay flints. Most bouldersand flints show phenomena such as windgloss, patinas, etc. In boulderclay severely weathered flints also occur, but these are scarce. Most boulderclay flints are not or hardly patinated (low gloss may be present), and not or hardly affected by windgloss, though many flints occurring in the uppermost 1 to 2 m of boulderclay are fragmented as a result of frostsplitting. The few heavily weathered flints in boulderclay presumably acquired their surface modifications when lying at the surface, before they were picked up by the ice-sheet. Some boulderclay flints (and, of course, bouldersand flints) show phenomena that were created during glacial transport, such as very coarse scratches and series of oblique pressure cones (Stapert, 1976a; see also fig. 4). On the majority of the bouldersand flints we see a wide range of surface alterations: a typical bouldersand flint exhibits, for example, windgloss, brown or white patina, scratches and rounding. Furthermore, bouldersand flints have on average become more fragmented as a result of frostsplitting than boulderclay flints. This has resulted in a gradual decrease in size of the flints. Larger pieces of flint show cracks due to frost action along which splitting has not yet taken place. Large, intact pieces of flint are virtually absent from bouldersand, and this has always been a problem for prehistoric flintworkers. It is known that frostsplitting of flints still took place after the end of the Upper Pleniglacial, for example during the Late Dryas Stadial (Stapert, 1982c); this resulted in new faces on flints that now will still be relatively fresh.

The conspicuous difference in weathering between the bouldersand and the boulderclay flints clearly means that the massive alteration of the bouldersand flints took place after deposition of the boulderclay, thus during the Eemian, the Weichselian and the Holocene. Surface modifications that arose during a glacial period – heavy windgloss, scratches etc. produced by cryoturbation – in most cases must have been created during the Weichselian. It is known that especially the Upper Pleniglacial of the Weichselian was a period characterized by extreme conditions: geologists speak of a 'polar desert' during this period (e.g. van der Hammen et al., 1967; Zagwijn, 1975; Kolstrup, 1980). Most ventifacts that are found in the Netherlands were formed during the Upper Pleniglacial (The Beuningen Gravel Bed: e.g. van der Hammen et al., 1967; van der Hammen & Wijmstra, 1971; Kolstrup, 1980; Meyer, 1986), and the same applies to most of the severe windgloss on flints encountered in the bouldersand; of course, during the Lower Pleniglacial and other stadials flints will also have acquired windgloss.

In western Europe the Middle Palaeolithic ends

Fig. 2. Cryoturbation at the transition between boulderclay and bouldersand, seen in a horizontal plane. B.A.I.-excavation at the 'site' of Hijken, northwestern part of section II (fig. 1). Photo by D. Stapert.

40,000 to 35,000 radiocarbon years ago, thus before the beginning of the Upper Pleniglacial. The 'finds' of Hoogersmilde and Hijken should, on the basis of their typology, date from the Eemian or the first half of the Weichselian. At the sites of Vermaning, geological circumstances do not deviate locally. Therefore, it cannot be understood how the Vermaning stones should have escaped all the various weathering processes that did leave their traces on the other flints in the bouldersand. In this connection it should be noted that many (natural) flints now occurring in the bouldersand came to be part of that deposit only in the course of the Weichselian – also during the Upper Pleniglacial. Middle Palaeolithic artefacts were, all of them, already in existence before the beginning of the Upper Pleniglacial, and have, therefore, been present in the bouldersand for a longer period, on average, than the other flints now to be found in that deposit. The 'Middle Palaeolithic' Vermaning stones do not show patinas, windgloss, or other natural surface modifications (see 7, however, for a contrasting opinion on this matter of Kars and others). Therefore, they cannot have been present in the bouldersand since the Eemian or the first half of the Weichselian. Moreover, assuming they are Middle Palaeolithic artefacts, there is no other layer, at the sites indicated by Vermaning, from which they could have derived.

Of course, geological circumstances do exist in which Middle Palaeolithic artefacts could have escaped weathering processes (creating patinas, windgloss etc.), for example in loam. Except within the boulderclay, no such circumstances exist at the sites of Vermaning. It can be excluded with certainty that the Vermaning stones come from the boulderclay, however. According to Vermaning, his 'Middle Palaeolithic' artefacts were found in very small concentrations, only a few metres across (van der Waals & Waterbolk, 1973)! This cannot be explained if they were incorporated in the boulderclay, given the genesis of this type of sediment (ground moraine).

My geological argument, presented above, can be summarized in the form of a syllogism as follows:

1. The 'Middle Palaeolithic' Vermaning stones, if they are genuine artefacts, must derive from the bouldersand.

2. Given their assumed age (on the basis of their typology), and the geological circumstances prevailing in the bouldersand at the sites indicated by Vermaning, they could not have escaped a series of

weathering processes in the course of time.

3. The Vermaning stones do not exhibit any of the surface modifications produced by such processes.

4. Therefore, the Vermaning stones cannot derive from the bouldersand at the sites indicated by Vermaning, which implies that they are forgeries.

Of course, it should be stressed that this argument is only valid within one specific geological context: the boulderclay plateau, outside valleys and other depressions. In other situations my argument might not apply, or only in a modified form. It is nonsense, in my opinion, to point to unpatinated finds elsewhere, for example deriving from loess, as evidence against my arguments (e.g. Franssen, 1978), because these finds do not come from geologically comparable situations.

At the end of this section it may be of interest to note that the argument I used to prove that the Vermaning stones are forgeries, is in itself not new. Similar reasoning has been used in other forgery affairs, from as early as the 19th century. As an example I should like to point to a famous publication by G. & A. de Mortillet in 1911 (this book is a revised edition of Préhistorique by G. de Mortillet, which first appeared in 1883), in which it was stated that it is necessary to prove the authenticity of Palaeolithic finds by establishing *caractères d'authenticité* such as patinas, rounding, gloss, etc. Of course, this type of argument should be adapted to the specific geological conditions prevailing at the sites under discussion, and this is what I have endeavoured to do in 1975.

3. 'APPLES AND PEARS'

The petrologist Kars has stated on several occasions (newspaper interview; pers. comm.), that "it is a comparison of apples and pears", when 'Middle Palaeolithic' Vermaning stones are compared with natural flints deriving from the bouldersand at the sites indicated by Vermaning. By this metaphor he suggests that something nonsensical is done, geologically speaking, when these two groups of flints are compared. According to Kars it is not permitted to compare these two groups of flints; in my opinion they should be compared.

This idea has been formulated earlier by Wouters and other amateur archaeologists defending the authenticity of the Vermaning stones. To Wouters c.s. their conviction implies, that other Middle Palaeolithic artefacts from the northern Netherlands, not found by Vermaning, which are indeed

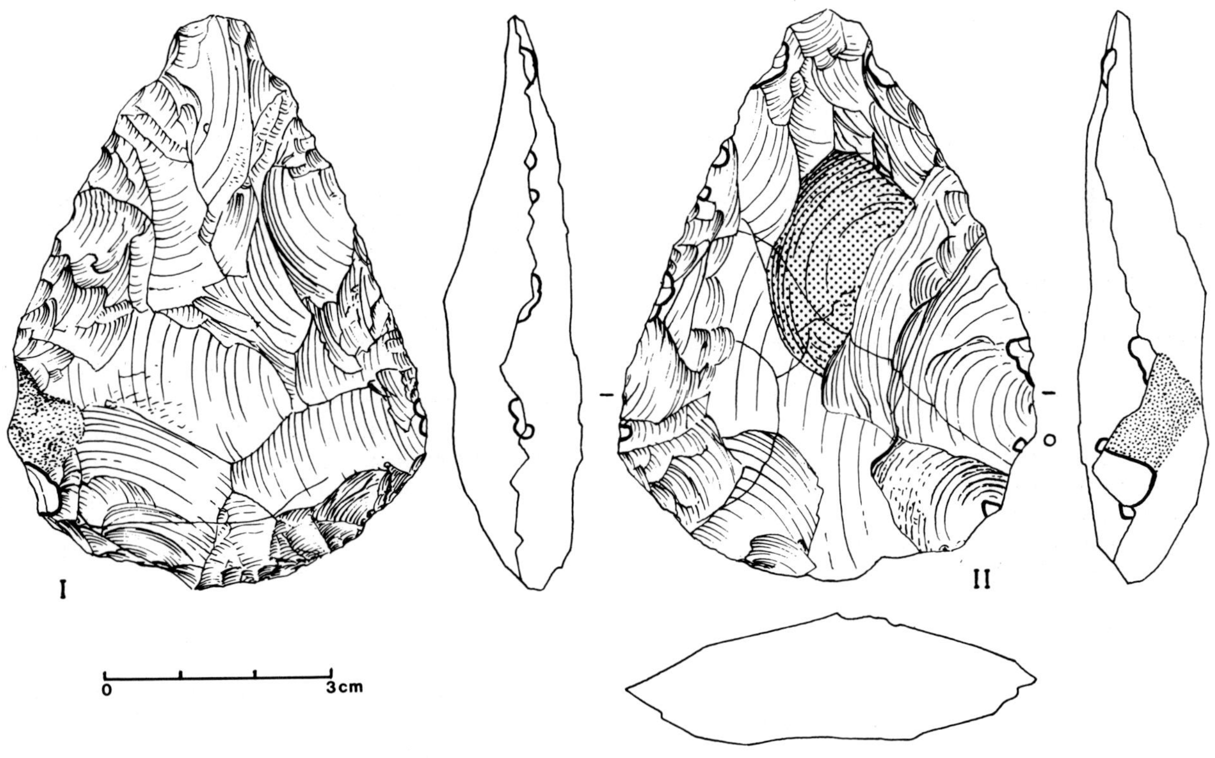

Fig. 3. The small triangular handaxe from Mander, manufactured out of a flake (Stapert, 1982a). Key: left white with a slightly thicker outline – (sub)recent damage; dense stippling – cortex; regular stippling – secondary frostsplit face; open circle – presumable position of the point of percussion of the flake. Drawing by H.R. Roelink, B.A.I.

patinated, covered in windgloss, etc. (see 4), must have been transported to this region by the ice-sheet during the Saalian. According to Wouters c.s., they had acquired their surface modifications before the arrival of the ice-sheet. The Vermaning stones, according to Wouters c.s., were left behind after the deposition of the boulderclay and therefore, still according to this theory, they could escape patination etc. (e.g. Wouters, 1980; Musch, 1980).

In itself the idea of glacial transport is worth considering. In the literature, several instances have been described of Palaeolithic artefacts incorporated in boulderclay (e.g. Baumann et al., 1983). In the Netherlands several sites are known that date from before the arrival of the Saalian ice-sheet (Rhenen: Stapert, 1981a; Belvédère: Roebroeks, 1984). Therefore, assuming that Middle Palaeolithic people also visited the northern Netherlands during the Early Saalian (of which we have no proof), the possibility exists that the ice-sheet picked up artefacts lying at the surface.

There are two remarks I want to make concerning this idea. First, I fail to understand what difference it would make to my argument presented above. Even if Wouters' theory should be correct, the Vermaning stones, if they are genuine artefacts, would have been left behind before the beginning of the Upper Pleniglacial. Therefore, my argument remains intact. Secondly, the hypothesis that the Middle Palaeolithic artefacts from the northern Netherlands that were not found by Vermaning were transported to this area by the Saalian ice-sheet seems, at least for some of these finds, very improbable from an archaeological point of view. On the basis of their typology, several handaxes, for example those from Anderen (Stapert, 1976b; 1976d), Anreep (Stapert, 1982b; Looyenga et al., 1984) and Mander (Stapert, 1982a), fit best in the Moustérien de tradition acheuléenne type A of Bordes (e.g. 1954/55; 1968), a tradition that is dated to the Early Glacial of the Weichselian – much later, therefore, than the period when the Saalian ice-sheet covered part of the Netherlands (see fig. 3 for the triangular handaxe of Mander).

In conclusion, it can be said that I agree with Kars that the Vermaning stones and natural flints from the bouldersand are 'apples and pears' when compared, meaning that they are different with respect to their surface modifications, but to my mind this difference is in need of a convincing explanation.

4. MIDDLE PALAEOLITHIC ARTEFACTS, NOT FOUND BY VERMANING

My argument, presented above, leads to the expectation that Middle Palaeolithic artefacts, when found in the same geological situation as that prevailing at the 'sites' of Vermaning (on the plateau, outside valleys or glacial basins) will show the same surface modifications, by and large, as can be observed on natural flints in the bouldersand. This is indeed the case. Apart from the handaxes of Anderen and Mander (mentioned above), we can also point to the handaxes of Wijnjeterp (Stapert, 1976b), Drouwen (Stapert, 1979) and Exloo (Stapert, 1976b), and to the points of Havelterberg and Ambt-Delden (Stapert, 1976b), as well as several dozens of other finds. It may be useful to note that we are discussing very clear phenomena here, such as white patina, brown patina and windgloss, which are present on these artefacts in, roughly speaking, the same degree and the same kinds of combinations as on natural flints occurring in the bouldersand. This also applies to finds that typologically belong to later phases of the Middle Palaeolithic, such as the small triangular handaxe (made out of a flake) from Mander. Of the handaxes from Wijnjeterp and Anderen colour photos were published, in order to give an impression of their surface modifications (in Stapert, 1976b; 1976d, respectively). It can be noted here that in Schleswig-Holstein (Western Germany), Middle Palaeolithic artefacts from surface sites show the same kinds of severe surface modifications as in the northern Netherlands (Hartz, 1986).

A circular argument could arise if I proposed the following: "in view of the existence of examples of Middle Palaeolithic tools from the northern Netherlands showing severe surface modifications, all other Middle Palaeolithic artefacts from this region have to look the same". It is possible in theory that these examples constitute a selection of some kind. Moreover, we have to face the problem that it is impossible to recognize isolated artefacts as Middle Palaeolithic solely on the basis of their typology. It is known that handaxe-like tools were also produced during the Neolithic (Stapert, 1981b), and there are more problems of the same kind. In fact, the mentioned examples of Middle Palaeolithic artefacts from the northern Netherlands have been identified as such only because they exhibit the same kind of surface modifications as seen on natural flints in the bouldersand. Again: this procedure is only valid within the geological context delimited above.

Taking this into account, I do not consider Middle Palaeolithic tools from this region with severe surface modifications as proof for my argument; they illustrate my argument, and prove that Middle Palaeolithic artefacts with such surface alterations exist in this region. My conclusion that the 'Middle Palaeolithic' Vermaning stones are forgeries cannot be falsified by finding unpatinated handaxes if it cannot be proved independently that these handaxes are indeed Middle Palaeolithic in age. It is evident, therefore, and I want to make this clear, that my argument is based only on the geological circumstances prevailing at the sites indicated by Vermaning. Basing myself on observations of surface modifications occurring on natural flints in bouldersand at these sites, I contend that Middle Palaeolithic artefacts from the bouldersand should show traces of the same weathering processes, because they could not have escaped all of these processes simultaneously. I at any rate fail to understand how they possibly could, seeing that they are manufactured from the same kinds of flint, and have been in exactly the same situation, for long periods in both cases. It is perhaps useful to state again that I am using this argument only for places on the plateau, outside valleys or glacial basins in which fine-grained sediments might be present (for some notes concerning the circumstances prevailing in glacial basins and valleys, see Stapert, 1985b; 1986).

I would gladly subject my conclusions to tests aimed at falsification (Popper, 1959). In my opinion, my conclusions can be falsified only by locating a spot on the boulderclay plateau where the flints in the bouldersand are not more frequently heavily altered than the flints in the boulderclay below. Personally, I have never come across such a situation, and until I am shown such a spot I maintain that my conclusions are not falsified. I repeat for the sake of clarity that we are concerned with very clear phenomena in this connection, such as windgloss and white or brown patinas.

The argument I used is independent of the existence of Middle Palaeolithic artefacts in the northern Netherlands showing severe surface alterations. It is certainly gratifying that they do exist, because they illustrate my argument. If they should not be known, my argument would remain intact, though devoid of examples. Similarly, it would be nice if we had more knowledge than we do concerning the precise nature and mode of origin of the various surface modifications that can be observed on flints in bouldersand, but my argument is in principle independent of such knowledge: it is only based on the empirical observation that these modifications took place on flints in bouldersand after the deposition of the boulderclay.[3]

5. ANOTHER GEOLOGICAL ARGUMENT

There are other arguments that independently lead to the same conclusion, viz. that the 'Middle Palaeolithic' Vermaning stones are forgeries. The main argument, presented above, is of a geological nature: it is based on the observation of weathering phenomena on flints in bouldersand. In this section I intend to discuss the consequences of a further characteristic of bouldersand. It is clear that this deposit was formed over a long period as an erosion residue of boulderclay, in many places under the influence of solifluction processes, among other factors. These soil movement processes are at work even on very gentle slopes during glacial periods (e.g. French, 1976). I observed possible indications of solifluction when studying sections on the 'site' of Hijken (fig. 1), during the excavation (unproductive, as far as the Middle Palaeolithic was concerned) that I carried out there in 1973 (Stapert, 1976a; 1976c). At the Middle Palaeolithic site of Mander it was observed that the several dozens of finds were scattered over a wide area, at least 250 m in diameter. This scattering is likely to be the result of solifluction over a long period, after the artefacts had been left behind (Stapert, 1982a).

'Concentrations' of Middle Palaeolithic artefacts in the bouldersand within sloping areas in the province of Drenthe, with diameters of 3.10 x 2.00 and 2.15 x 1.20 m (Concentrations A and B respectively of Vermaning's 'findspot' Hoogersmilde: van der Waals & Waterbolk, 1973), seem to me unthinkable for this reason, even with gentle slopes. In view of the genesis of bouldersand, original concentrations of artefacts would have become scattered considerably in the course of several tens of thousands of years after their deposition. Again, this argument is in principle independent of finding occurrences of scattered Middle Palaeolithic artefacts, as at Mander. Besides, although it is likely that the artefacts from Mander belong together as the residue of a single occupation, this cannot be proved with absolute certainty.

6. TYPES OF FORGERIES

If one should carefully read the existing literature on forgeries (e.g. Vayson de Pradenne, 1932), probably many different types of forgeries could be defined. For practical purposes, forgeries are divided into four major groups here:

1. Newly made objects.
2. 'Findspot forgeries'; these concern authentic artefacts that are falsely claimed to have been found at the site indicated by the finder or owner (for a probable example, see Lanting, 1977).
3. Recently modified, authentic artefacts; for example recently retouched prehistoric flakes.
4. 'Compositions'; for example the recently created combination of a mandible of giant deer with an authentic Tjonger point (Stapert, 1977).[4]

In the above I concluded that the 'Middle Palaeolithic' Vermaning stones are forgeries. In principle they could be forgeries of types 1 or 2. In my opinion there are good reasons for assigning them to type 1, i.e. newly made objects. Indications in this respect are the following: the presence of artificial gloss (see 7), the presence of traces of modern grinding (see 8), the improbable composition of the various collections in terms of the types of handaxes that are represented (Stapert, 1975b), and the technically crude workmanship of many artefacts. This last indication in my opinion is the strongest; it was also mentioned by McBurney (letter to Waterbolk, d.d. May 5th, 1969), and by Bosinski (1976) in his report for the lawsuits.

7. GLOSS

The petrologist Kars has claimed that the Vermaning stones "certainly possess patina" (newspaper interview; pers. comm.). This patina in his opinion mainly consists of a low, silky, gloss, but brown patinas are also present according to Kars. This opinion was formulated earlier by Wouters, during the court hearings, and by others (e.g. Franssen, 1978; Dijkstra, Musch & Wouters, 1986).

In 1975 I concluded that the Vermaning stones had been treated on purpose with substances that create artificial gloss (Stapert, 1975a; 1975c). I was able to wash away these gloss-producing substances by using water and soap, after which, as far as I could see, unpatinated surfaces remained (fig. 4). Together with Dr. G. Boom (Laboratory of Physical Metallurgy, University of Groningen) I tried to analyse these substances chemically, but due to the method used (EDAX) this attempt failed (Boom, 1976). However, we did establish the presence, apart from the artificial gloss, of many elements that were most probably deposits left behind by the hands of the many people who handled the Vermaning stones in the course of time: perspiration and skin-fats, etc. Moreover, the pieces had been marked with numbers and drawn. Draughtsmen commonly use crayon and other materials in their work. Attempts at refitting had been carried out with the larger 'Middle Palaeolithic' collections of Vermaning, involving the use of glue.

Furthermore, I have often stated that the mere fact that the pieces have been handled in many different ways for more than 10 years, has necessarily resulted in a certain surface modification that is not reversible. It is very well possible that this modification could be described as a 'silky gloss'. We know from use-wear analyses that repeated contact with meat and other soft materials produces irreversible gloss on flint (e.g. Keeley, 1980; Moss, 1983). Therefore, I am convinced that the Vermaning stones in the meantime have duly acquired a surface modification that cannot be

Fig. 4. Some refitted artefacts from the Hijken collection. The uppermost face is an old frostsplit face (dating from before the production of the artefacts), showing severe scratching (which probably resulted from glacial transport) and windgloss. Below this face, part of the ventral surface of a flake is visible (at the left), exhibiting gloss. The left half of this surface was cleaned by me using water and soap. This resulted in disappearance of the gloss, and a non-patinated surface remained. The Juridical Laboratory performed a similar test on this surface (using 'Vim'), with the same result (Witte, Verburgt & Groeneveld, 1976). Photo B.A.I.

removed completely. However, to my mind this has nothing to do with patinas. In my opinion the term patina, although not well defined, should only be used for natural surface modifications that can be clearly identified – even if their origin is not understood precisely.

In short, in my opinion the Vermaning stones show the following surface alterations that could perhaps be confused with gloss patina: a) artificial gloss (see also below); b) skin-fats and many other recent deposits as a result of repeated manipulation; c) a certain smoothness as a result of frequent handling. None of these I consider as natural surface modifications. I should like to add that a chemical analysis of the multifarious contaminations now present on the Vermaning stones seems to me a senseless undertaking after all these years.

Returning to the artificial gloss that could be easily washed away, the Juridical Laboratory analysed two Vermaning stones in this respect (report of August 22nd, 1975 by Witte, Verburgt & Groeneveld). A handaxe from Hijken was analysed by pyrolysis-gaschromatography, and

> ... it was concluded that a polyester-product, i.e. a modern artificial substance, was present. Therefore, an artificial substance has been sprayed on the handaxe.

Another citation from this report:

> A handaxe from the Hoogersmilde collection exhibited a gloss that was caused by a clearly visible fat-like substance, attached to the surface. An infra-red chemical analysis showed that this substance consisted of Vaseline. [Vaseline is a kind of petroleum jelly]

In itself, the presence of artificial gloss does not prove that the Vermaning stones are forgeries. It is possible that genuine artefacts were treated with gloss-creating substances, or worked with a modern grinding machine – examples of such behaviour do exist (for a possible instance, see Stapert, 1985a). Of course this proves swindling, but not necessarily that the objects in question were made recently, though in the case of the Vermaning stones this clearly seems to be the case.

8. TRACES OF GRINDING

Almost all the 'Middle Palaeolithic' Vermaning stones show a peculiar alteration of the ridges between flake scars, and sometimes of edges. In these cases the ridges appear to be somewhat rounded, but on close inspection it can be seen that in fact they mostly display 'facets' – the ridges have been truncated and present themselves as small level planes (for examples, see figs 5 and 6). There are several reasons why I interpreted these rounded or flattened ridges as the result of grinding with the help of a modern grinding machine (Stapert, 1976c; 1980). In the first place it can be noted that this is a partial phenomenon. On one and the same piece not all the ridges are altered: one finds very much flattened ridges close to unaltered ridges that have remained fresh and sharp. Furthermore, in several cases it can be seen that not only the most protruding ridges have been affected; these have sometimes remained sharp while less protruding ones have been levelled off, on the same piece. In a few cases within the flattened parts scratches can be observed that are strictly parallel. However, sets of parallel scratches occur more often just beside the flattened ridges.

Experiments I carried out with a corundum grindstone, electrically driven, showed that exactly the same kind of flattened ridges could be produced in this way (fig. 7). These experiments also showed that in most cases no scratches were created within the altered parts – in general these presented surfaces reminiscent of emery-paper, when seen through a stereomicroscope. Scratches sometimes occurred just outside the ridges, if for a moment the worked flint had accidentally slipped aside on the grindstone.

Perhaps more important than the results of these experiments is the observation that flattened ridges similar to those on the Vermaning stones do not at all occur on natural flints in the bouldersand at the sites indicated by Vermaning. I saw thousands of such flints from the bouldersand at the site of Hijken during the excavation I carried out there in 1973. Wouters and other defenders of the Vermaning stones (e.g. Dijkstra, Musch & Wouters, 1986), and also Kars (pers. comm.), explained the altered ridges on the Vermaning stones as being the result of cryoturbation or similar processes, but never offered an explanation for the fact that they are absent on natural flints in the bouldersand.

An interesting case in point is presented by the material of 'Ravenswoud'. This collection of 'Neolithic' artefacts was sold by Vermaning to the Fries Museum in Leeuwarden in 1968. Among other objects, this collection includes a number of ground axes and several so-called 'spearheads'. These 'spearheads' are unknown from other prehistoric collections from the Netherlands; they could be described as enlargements (scale approximately 4:1) of Neolithic arrowheads (figs 8 and 10). These grotesque 'spearheads' show exactly the same flattened ridges, in some cases accompanied by parallel scratches (figs 9 and 11), as the 'Middle Palaeolithic' Vermaning stones (Stapert, 1975a; 1976c). I also concluded, on the basis of observations with a stereomicroscope (figs 12-16), that a number of the polished axes from Ravenswoud had been ground with the help of a modern grinding machine. A few other ground axes from Ravenswoud are indeed genuine Neolithic artefacts, but have probably been

Fig. 5. A handaxe from the Hoogersmilde collection (1965/X-48). The length of the specimen is c. 10 cm. In many spots flattened ridges (between flake negatives) can be seen. Some clear examples have been marked. Note also that in several places unaffected ridges occur right beside extremely flattened parts. Photo B.A.I.

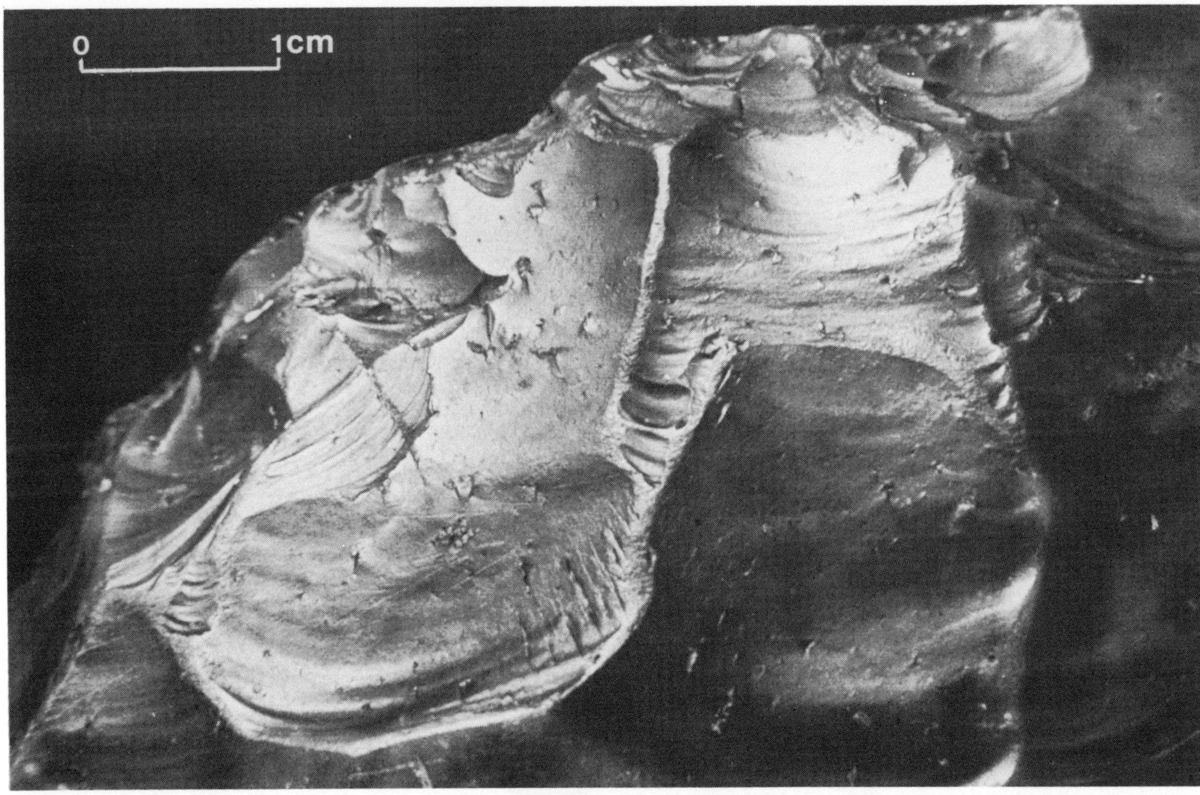

Fig. 6. Part of the surface of the 'Leemdijk handaxe', manufactured from brown flint. Flattened ridges are clearly in evidence; some scratches are also visible. In this case the altered parts are coloured red, presumably due to the circumstance that the flattened ridges were produced with the help of an electrically driven grinding machine (see text under 8). Photo B.A.I.

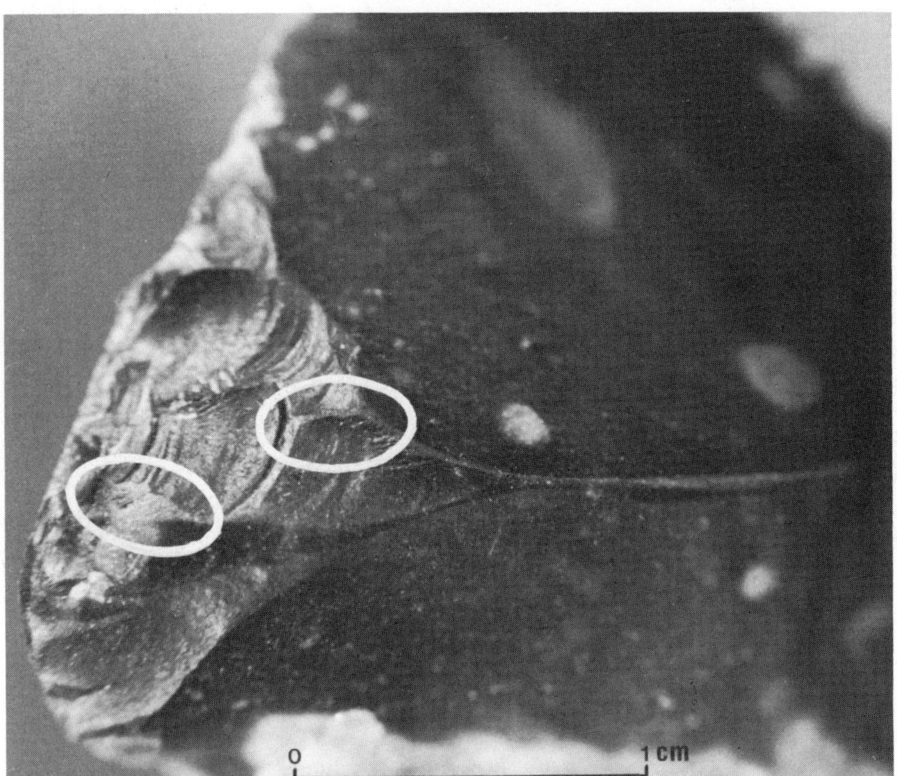

Fig. 7. Flattened ridges produced experimentally by using an electrically driven grinding machine on a freshly-struck piece of flint. These ground ridges are very similar to the flattened ridges occurring on the Vermaning stones. Some clear examples have been marked. Scale in cm. Photo B.A.I.

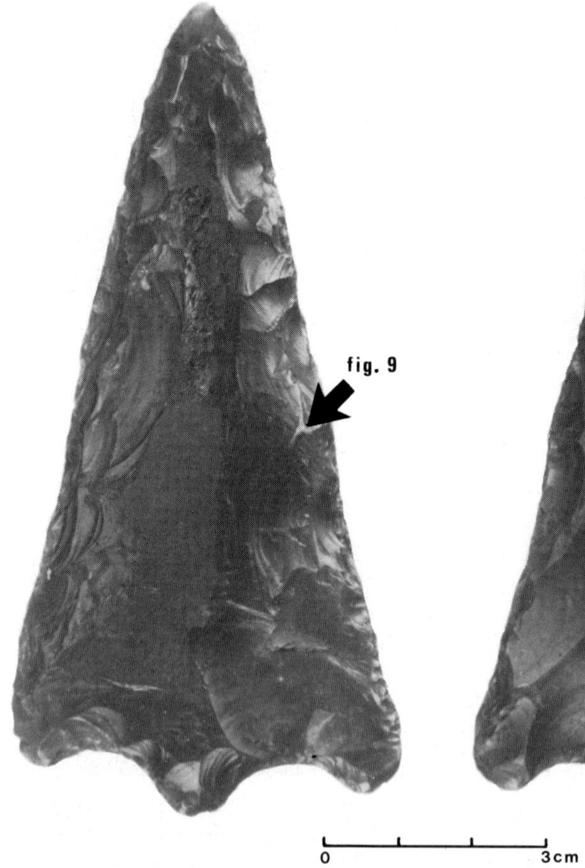

Fig. 8. One of the 'spearheads' of the 'Neolithic' collection of Ravenswoud (FM 1968-1-18). An arrow indicates the place where the photo shown in figure 11 was taken. Scale in cm. Photo B.A.I.

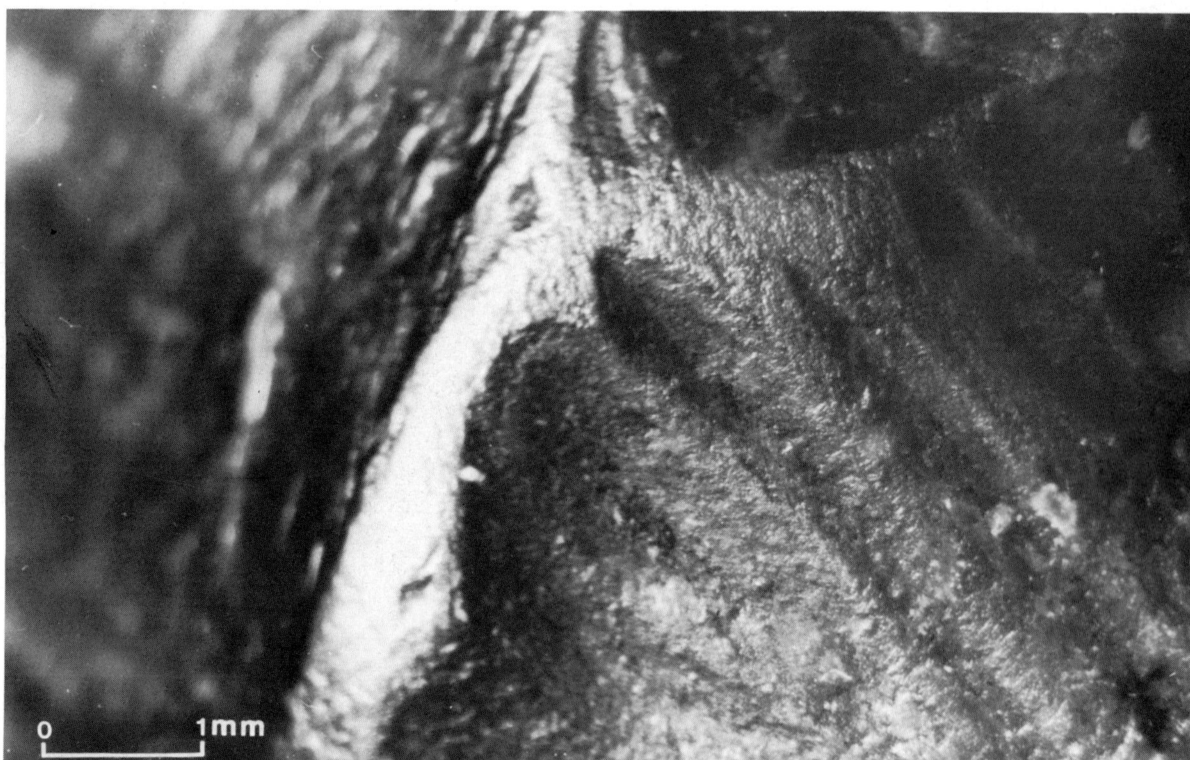

Fig. 9. Flattened ridges on one of the 'spearheads' of Ravenswoud (FM 1968-1-18, see fig. 8). Photo made by means of a stereomicroscope (Wild M5A) by D. Stapert.

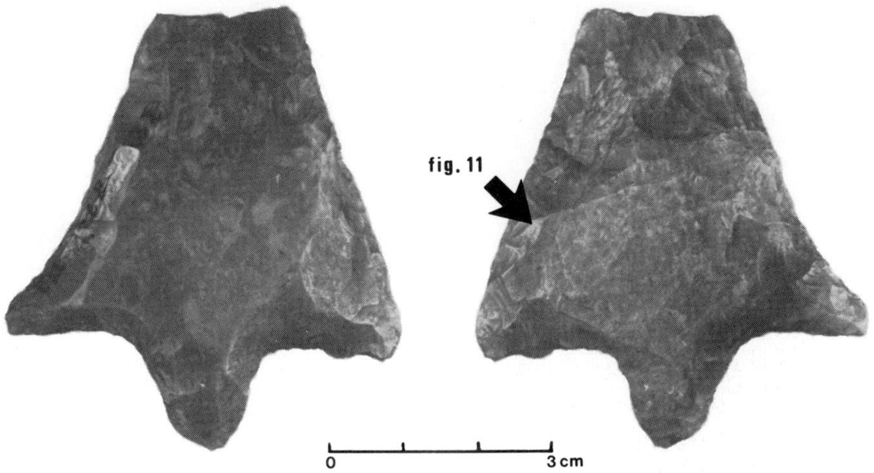

Fig. 10. One of the 'spearheads' of Ravenswoud (broken: FM 1968-1-22). An arrow indicates the place where the photo shown in figure 11 was taken. Scale in cm. Photo B.A.I.

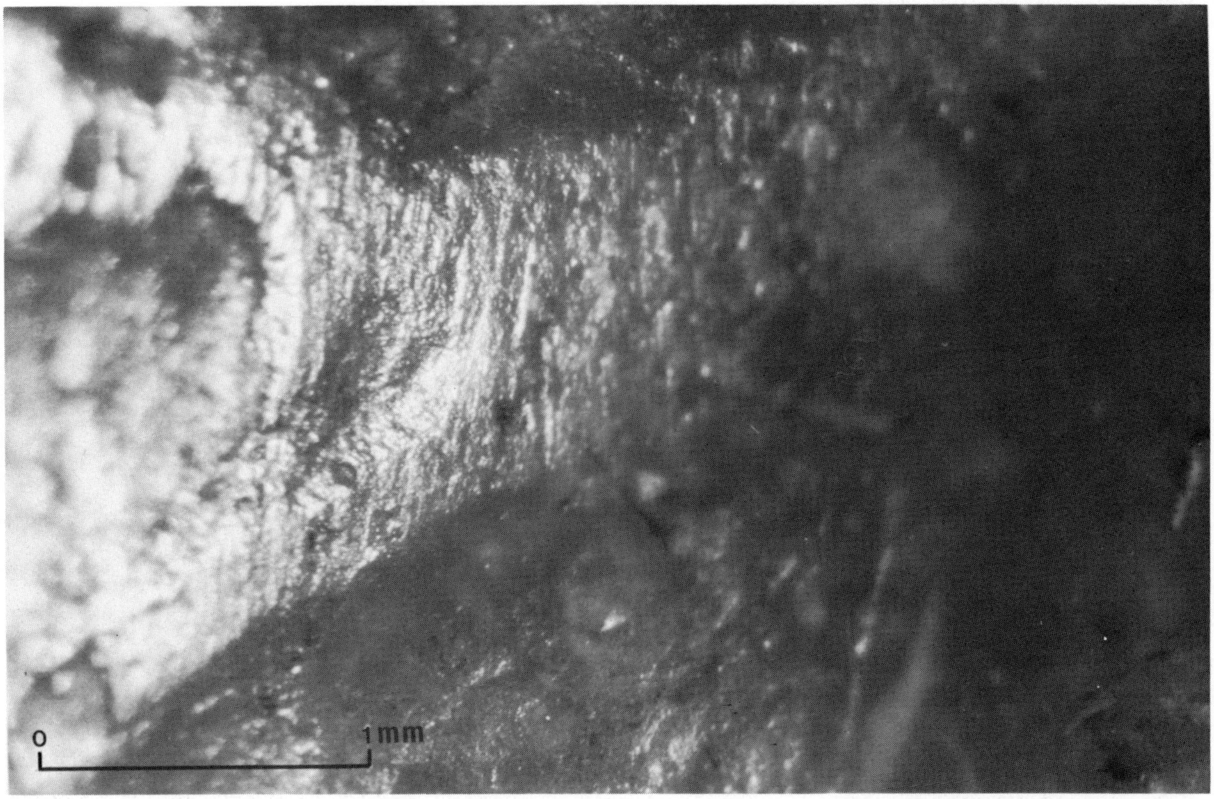

Fig. 11. Flattened ridges with parallel grinding scratches on one of the 'spearheads' from Ravenswoud (FM 1968-1-22, see fig. 10). Photo made by means of a stereomicroscope by D. Stapert.

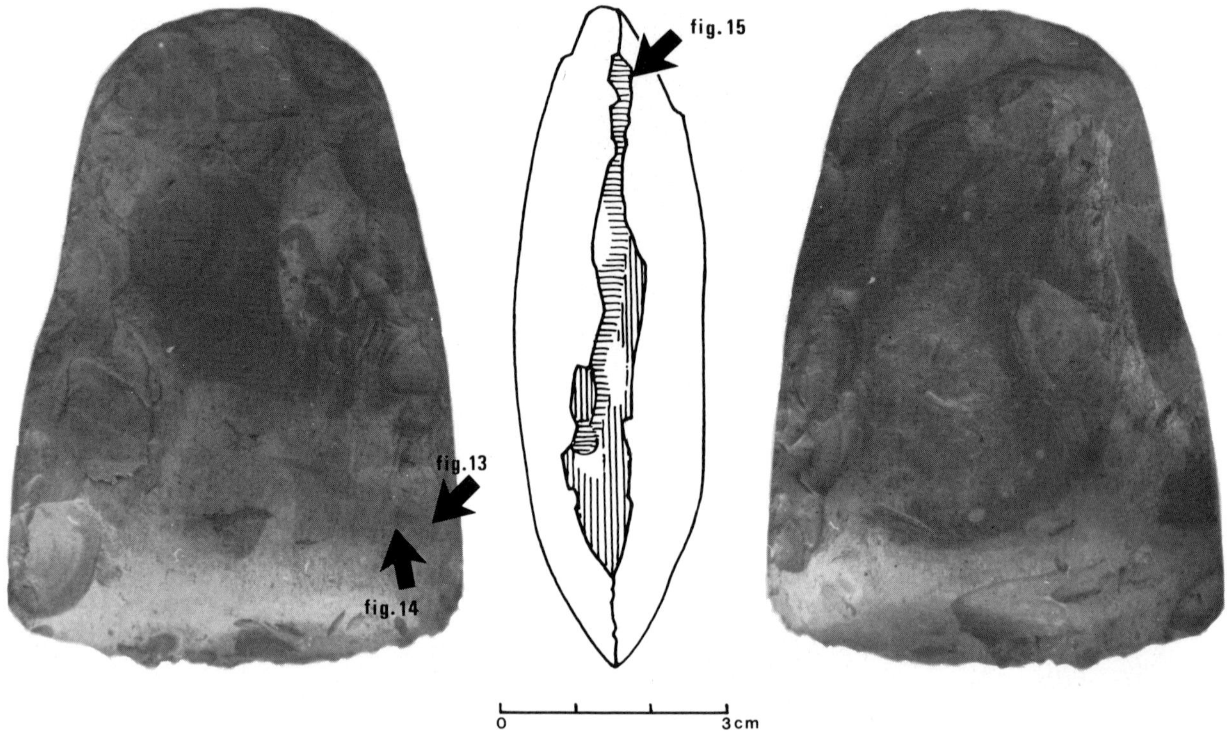

Fig. 12. One of the ground axes of Ravenswoud (FM 1968-1-8). The side-view shows the direction of grinding scratches present on that side. Arrows indicate the places where the photos shown on figures 13, 14, and 15 were taken. Scale in cm. Drawing by H.R. Roelink, photos B.A.I.

Fig. 13. Grinding scratches on one of the faces of an axe from Ravenswoud (FM 1968-1-8, see fig. 12). The scratches are very regular and strictly parallel, indicating that they were produced with the help of an electrically driven grinding machine. Photo taken by means of a stereomicroscope by D. Stapert.

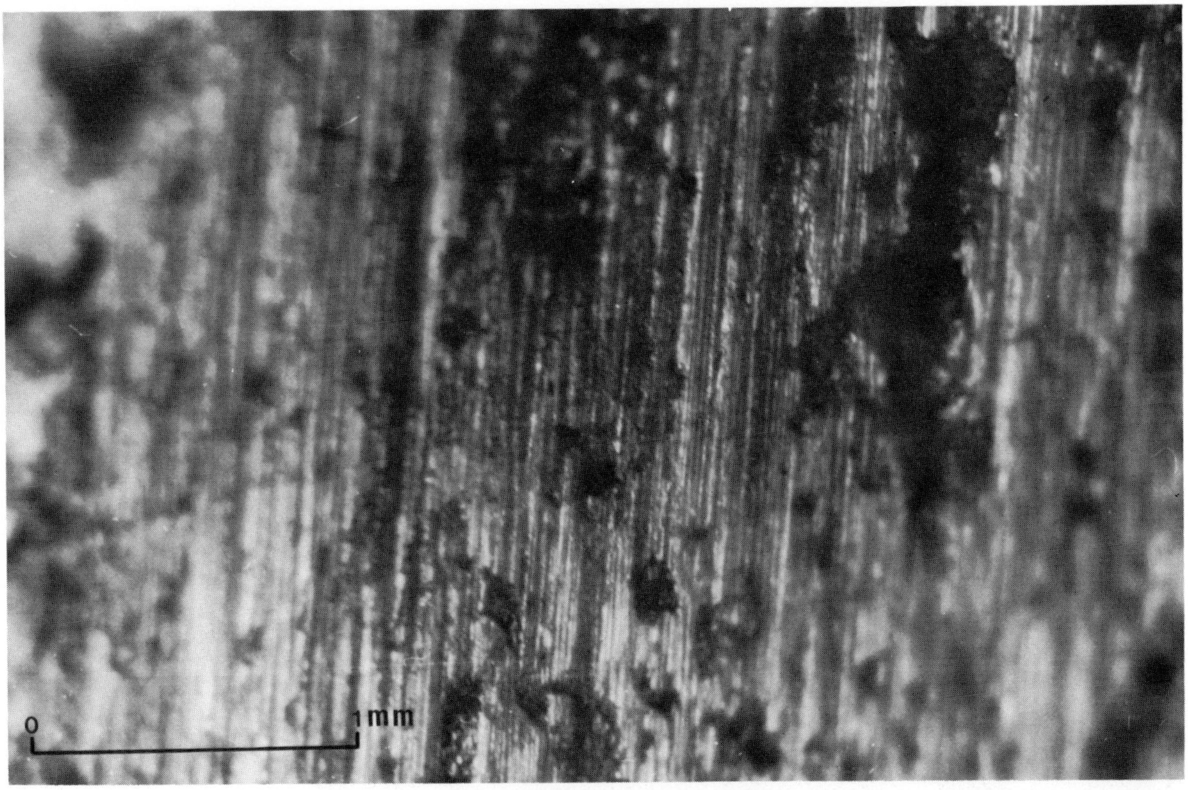

Fig. 14. Grinding scratches on one of the sides of an axe from Ravenswoud (FM 1968-1-8, see fig. 12). This photo is taken at a higher magnification than that shown in figure 13; it shows the same characteristics of grinding with an electrically driven grinding machine. Photo taken by means of a stereomicroscope by D. Stapert.

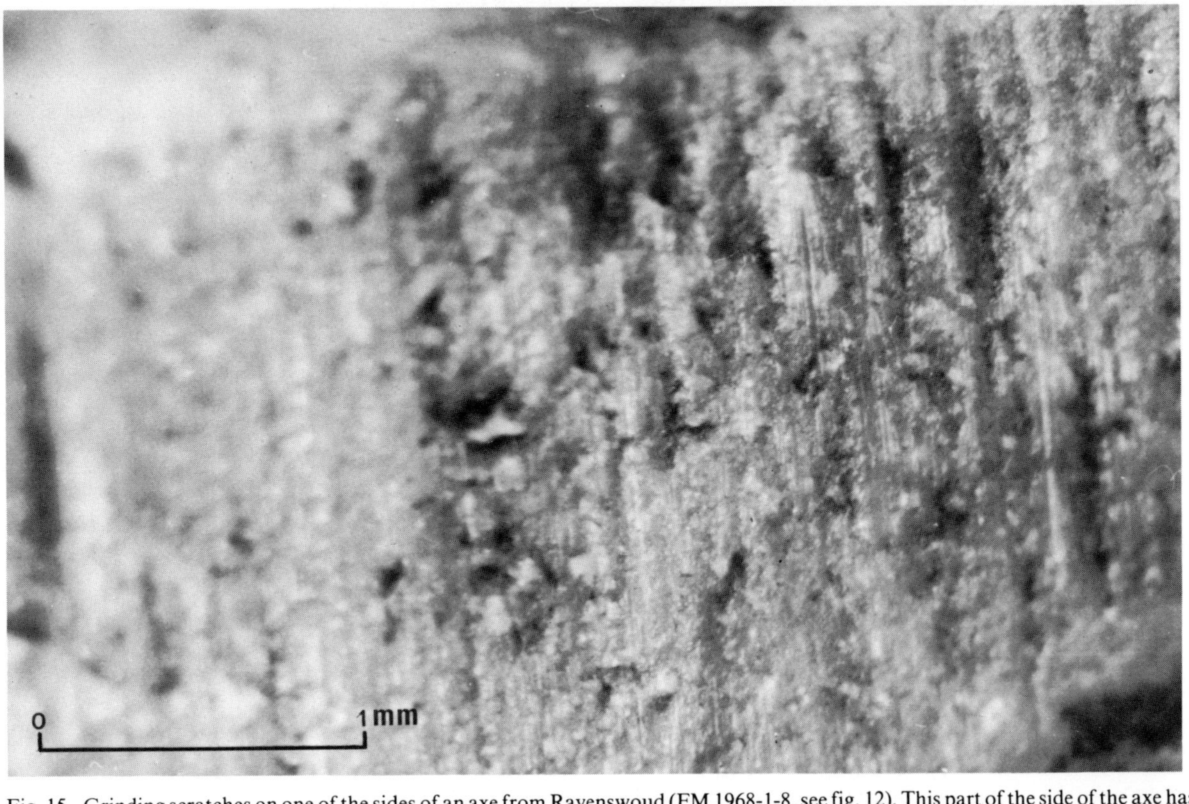

Fig. 15. Grinding scratches on one of the sides of an axe from Ravenswoud (FM 1968-1-8, see fig. 12). This part of the side of the axe has been ground *transversely*, i.e. at right angles to the longitudinal axis of the axe (see fig. 12). The scratches are similar to those on the faces of this axe shown in figures 13 and 14. Photo taken by means of a stereomicroscope by D. Stapert.

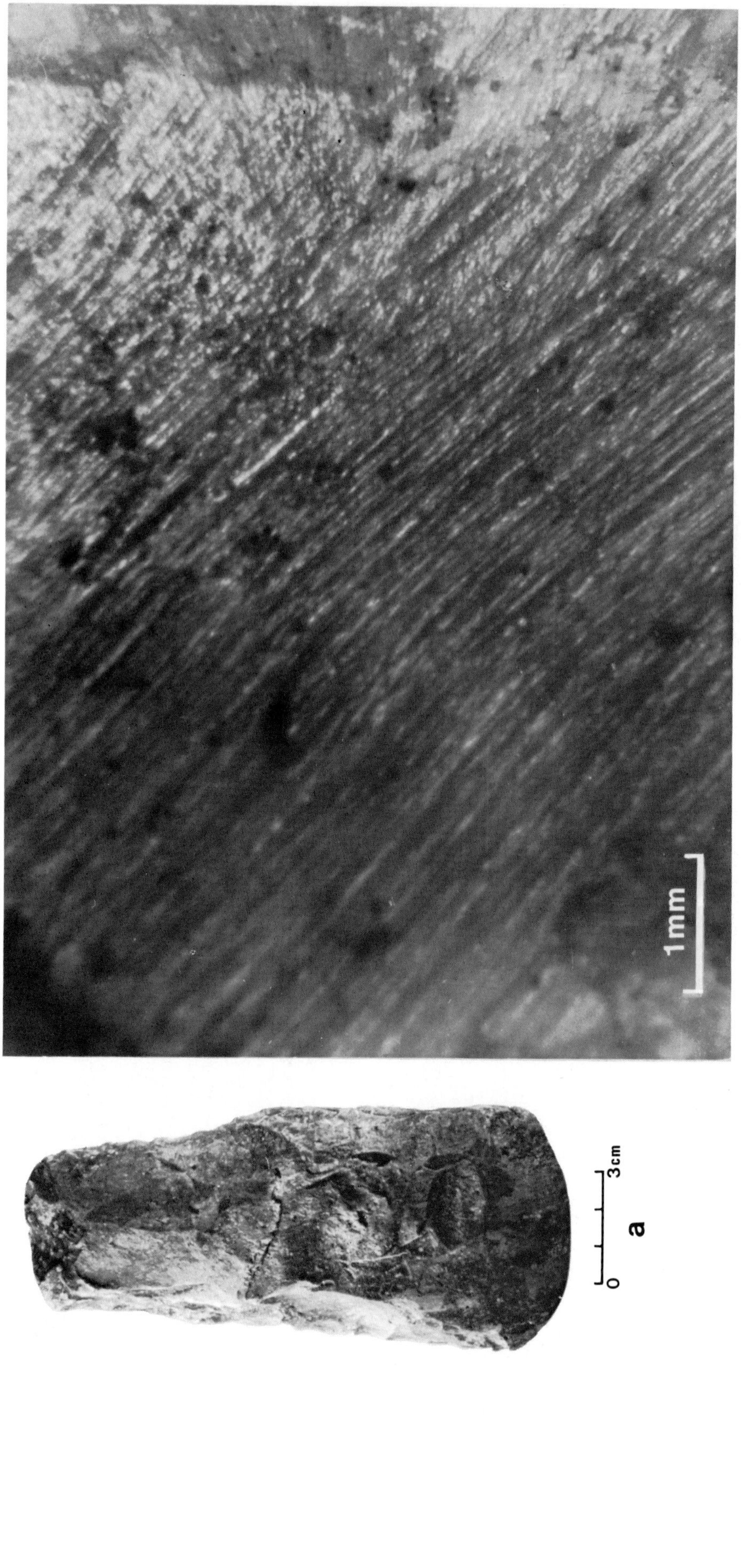

Fig. 16(a). One of the ground axes from Ravenswoud (FM 1968-1-5). Scale in cm. Photo B.A.I. (b) Scratches on one of the faces of this axe, showing the characteristics of grinding with the help of an electrically driven grinding machine. Photo taken by means of a stereomicroscope by D. Stapert.

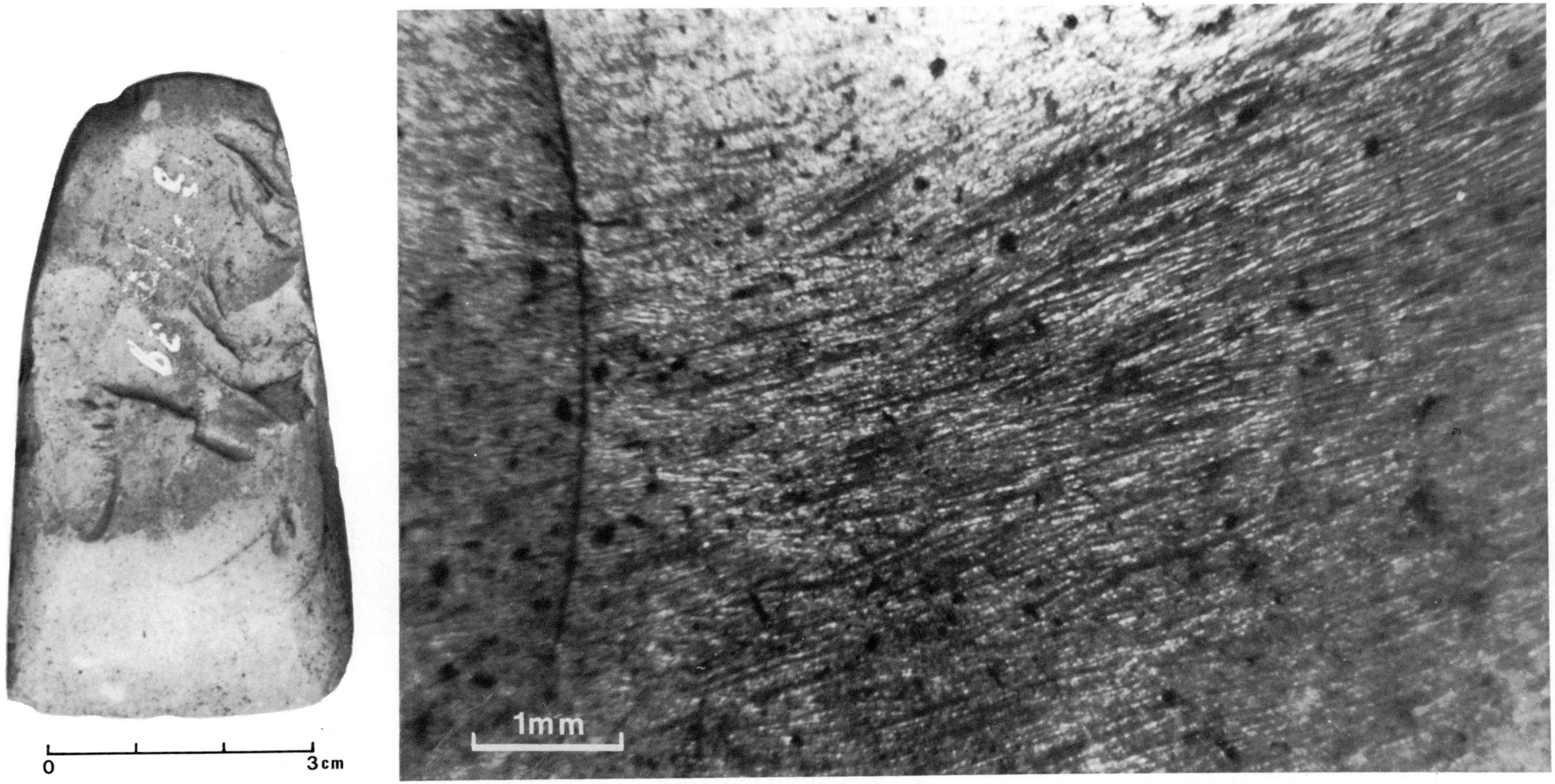

a b

Fig. 17(a). Neolithic polished axe from a grave near Havelte, excavated by the B.A.I. (B.A.I. collection 1969/VI-39). (b) Scratches on one of the faces of this axe. These scratches are irregular and not strictly parallel, therefore clearly produced by hand. Compare with figures 13-16. Photo taken by means of a stereomicroscope by D. Stapert.

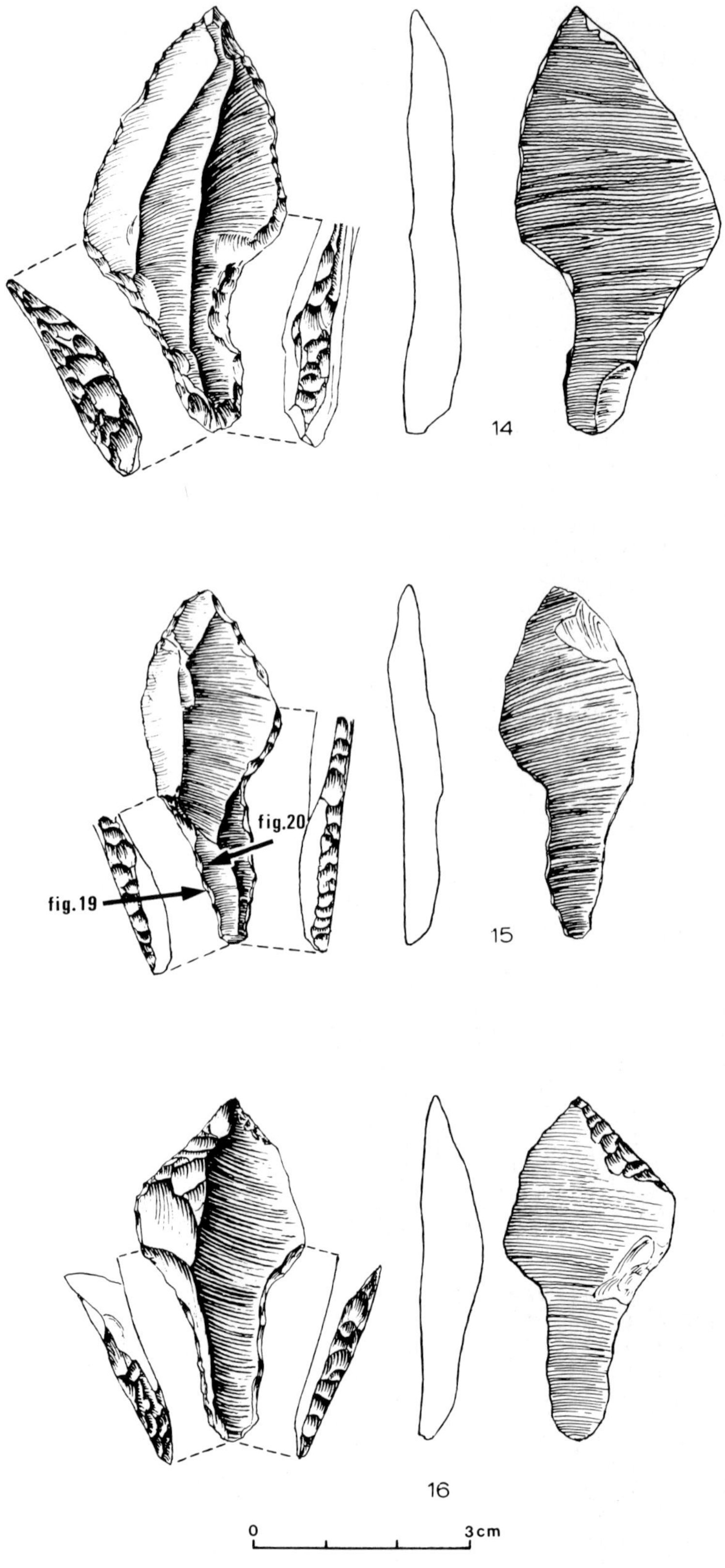

Fig. 18. The three tanged points from the *Federmesser* collection of 'Norgervaart' (from Paddayya, 1973). Arrows indicate the places on piece No. 15 where the photos shown in figures 19 and 20 were taken. Drawing B.A.I.

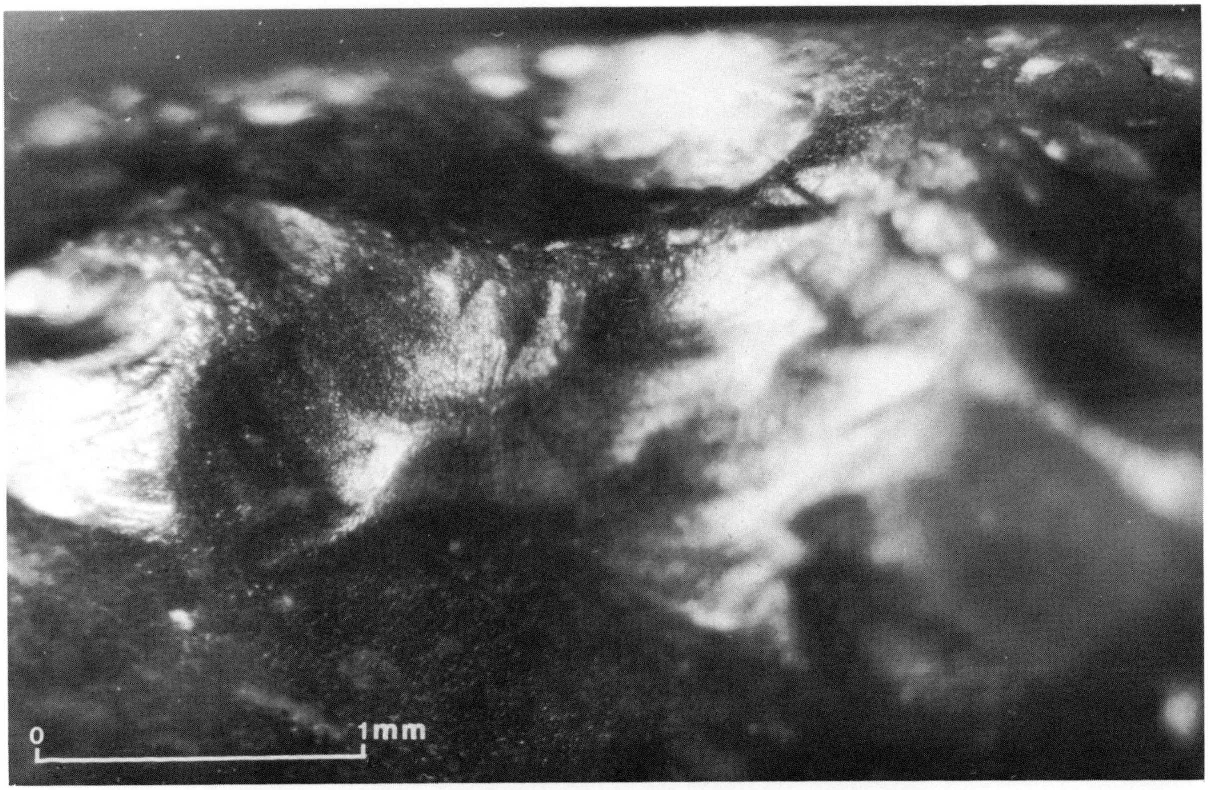

Fig. 19. Flattened ridges on one of the tanged points from Norgervaart (No. 15, see fig. 18). These flattened ridges are very similar to those occurring on 'Middle Palaeolithic' Vermaning stones and on the 'spearheads' of Ravenswoud. Photo taken by means of a stereomicroscope by D. Stapert.

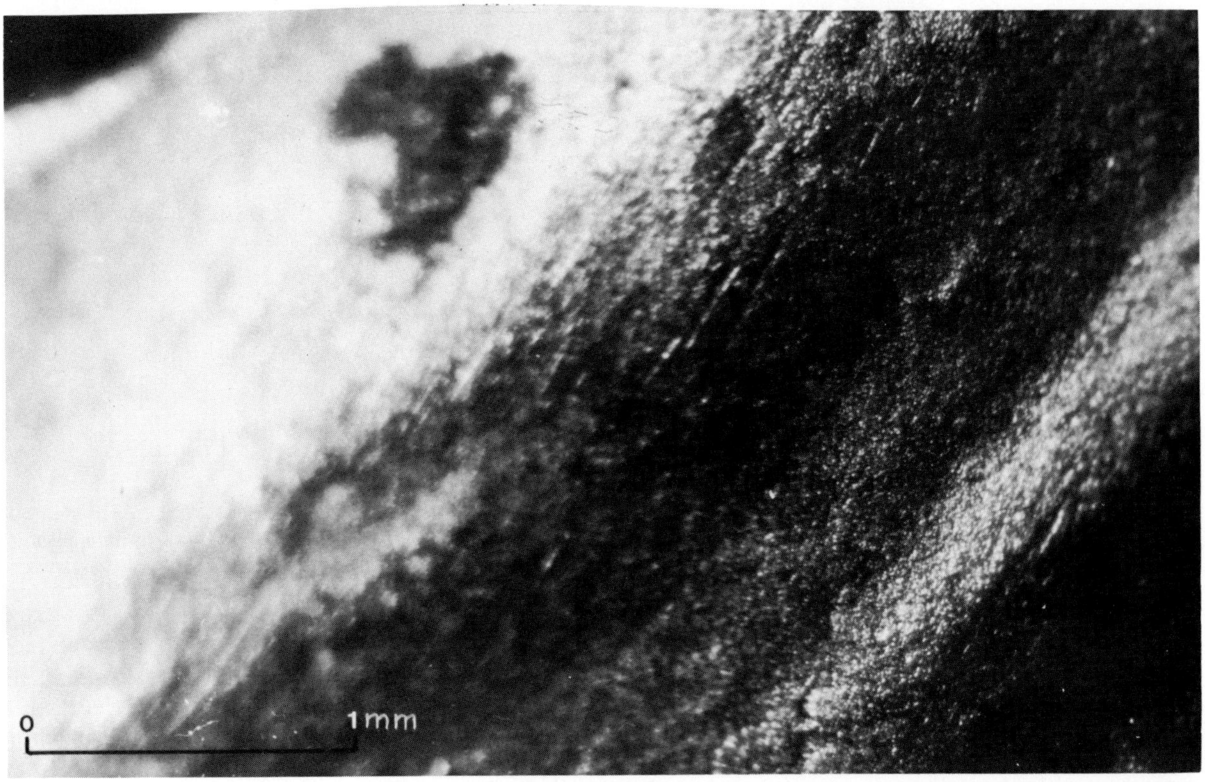

Fig. 20. Scratches occurring on one of the tanged points from Ravenswoud (No. 15, see fig. 18). These scratches are close to some flattened ridges, shown in figure 19. In several cases scratches of the same type occur near flattened ridges on 'Middle Palaeolithic' Vermaning stones. Photo taken by means of a stereomicroscope by D. Stapert.

Fig. 21(a). The 'handaxe from Blauwmeer'. Scale in cm. The rectangle shows the part that is represented in b. (b) Detail of one of the faces of this handaxe. Almost the whole surface of this handaxe is densely covered in scratches, presumably produced by the application of some hard type of emery-paper. Photos B.A.I.

imported from Denmark (van der Waals, pers. comm.).

Genuine Neolithic polished axes, when studied with the help of a stereomicroscope, show grinding scratches with the following characteristics: the scratches are never strictly parallel, they often taper towards one end, and they are variable in coarseness and depth (see for example fig. 17). The axes from Ravenswoud that in my opinion have been ground with an electrically driven grinding machine show scratches that are strictly parallel over relatively large distances, and very homogeneous in terms of coarseness and depth, as can be seen in the microscope photos. Furthermore, their ground

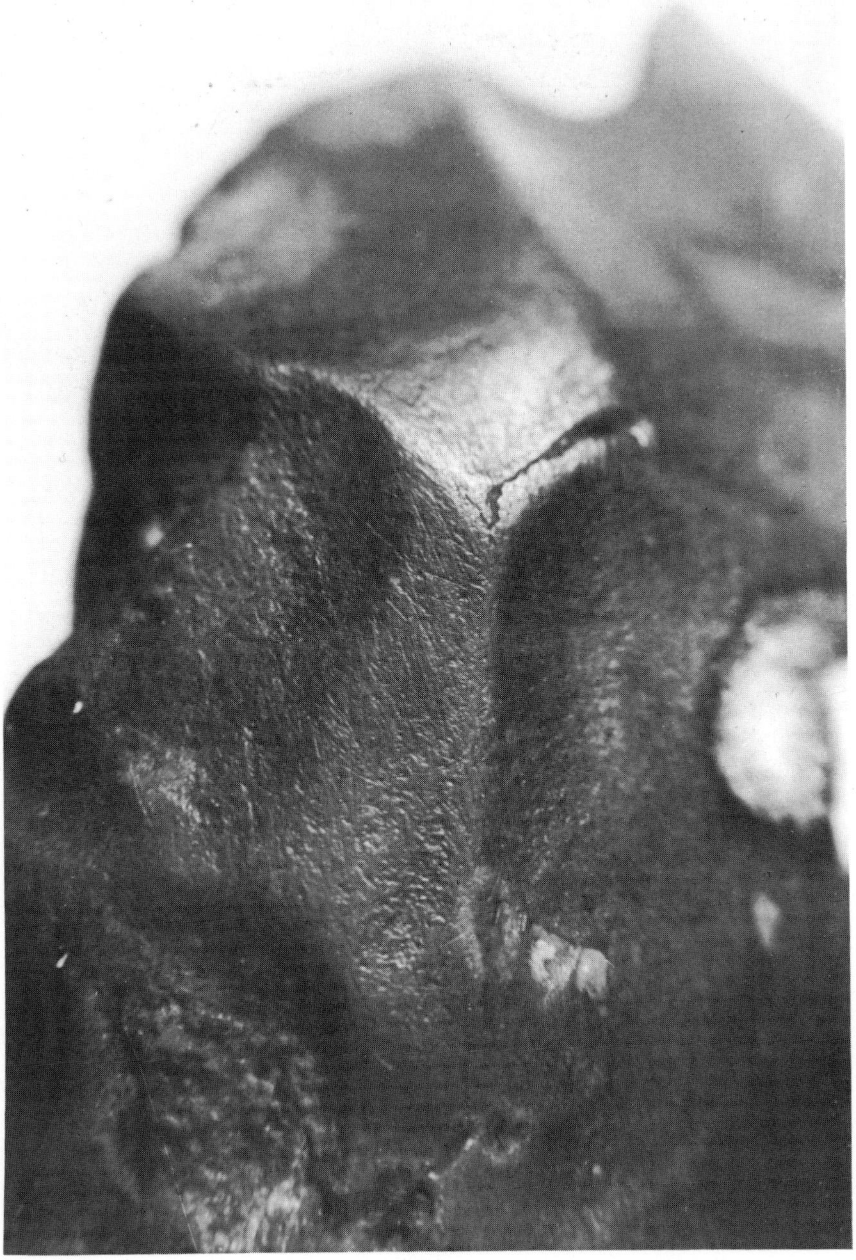

b

surfaces show different sets of scratches with different directions.

A 'Late Palaeolithic' (*Federmesser*) collection from 'Norgervaart', sold by Vermaning to the Provincial Museum of Drenthe in Assen, was published by Paddayya (1973). Especially striking within this collection are three rather broad tanged points (fig. 18), that are in general uncommon with *Federmesser* material. These are forgeries of type 3: recently retouched authentic blades. The retouches break through the patination (this is evident especially with No. 16, which possessed a brown patina prior to the retouching), and must therefore have been recently created. In 1986 I studied these

a

b

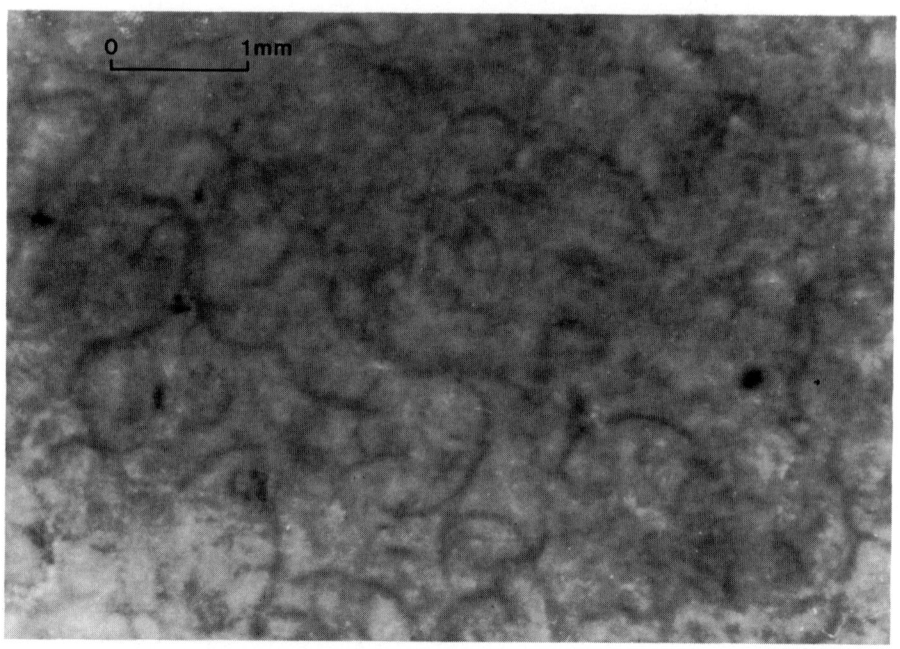

Fig. 22(a). One of the large rolled flakes in the museum 'It Bleekerhûs' in Drachten, said to come from 'Weperpolder'. See note 4. Photo Frisian Museum, Leeuwarden. (b) Many collision cones in the surface of this flake. Photo made by means of a stereomicroscope by D. Stapert.

pieces with the help of a stereomicroscope, and observed the same kind of flattened ridges and scratches (figs 19 and 20) as present on the 'Middle Palaeolithic' Vermaning stones and on the 'spear-heads' of Ravenswoud.

A special case is presented by the 'handaxe from Blauwmeer' (fig. 21), sold by Vermaning in 1973 (Stapert, 1976c). This piece does not show the flattened ridges that are so characteristic of most of the 'Middle Palaeolithic finds' of Vermaning. Almost the whole surface of the handaxe from Blauwmeer is densely covered in scratches (fig. 21b). In this case it is evident from the distribution of the scratches that they were not produced by an electrically driven grinding machine. It is very probable that this massive scratching was produced by the application of some very hard type of emery-paper, or suchlike. Apart from the myriads of scratches the piece does not show any kind of surface modification, let alone natural modifications. Flints having been suction dredged and collected at the site, did not show this massive scratching.

If my interpretation of the flattened ridges is not accepted, one has to offer another theory, which will also have to explain the fact that they are not present on natural flints in bouldersand. Furthermore, it must be taken into account that not only 'Middle Palaeolithic' pieces show these flattened ridges, but also 'Late Palaeolithic' and 'Neolithic' ones. It might also be a good idea to offer an explanation for their only seeming to occur on material sold by Vermaning.

The Juridical Laboratory carried out interesting research concerning these flattened ridges (Witte, Verburgt & Groeneveld, report of March 18th, 1976). There are a few brown 'Middle Palaeolithic' Vermaning stones: the handaxe from 'Leemdijk' (fig. 6), and a scraper from 'Hijken'. These pieces do not possess brown patina, but are manufactured out of brown flint. The Juridical Laboratory observed that in these cases the flattened ridges were coloured red, and was able to produce such red-coloured ridges experimentally by grinding brown flints with an electrically driven corundum grindstone. A few quotations from the report follow:

> When brown flints are pressed gently against the grindstone, then no red discoloration occurs. Only when strong pressure is applied, with the result that the flints start to spark, and therefore have become very hot at the ground surface, the typical red discoloration is produced. The flint starts to 'bleed', so to speak. ... We assume that the red discoloration is caused by the high temperature produced in the grinding process, possibly resulting in water being extracted irreversibly from built-in iron-hydroxides, so that red iron-oxides are formed.

In the report of the Juridical Laboratory also one ground axe from 'Ravenswoud' is described as possessing traces of 'undoubtedly' mechanical grinding (with the help of an electrically driven grinding machine) on one of the sides – the parallel grinding scratches there do not run longitudinally, as is normal on genuine Neolithic axes, but across (see also Witte, Verburgt & Groeneveld, 1975) (figs 12 and 15).

As stated above, I do not consider the presence of traces of modern grinding on the 'Middle Palaeolithic' pieces of Vermaning as proof of their being newly made, but only as one of the indications. The ground ridges and artificial gloss must have been meant to make the implements look 'weathered', and therefore older than they are.

9. SOME CONCLUDING REMARKS

In this article my main concern has been to formulate some facts and arguments that in my opinion are indispensable if a realistic understanding of the Vermaning affair is striven for, and to show that my basic argument is of a geological nature. Of course, many of the issues raised could be reported upon in greater detail; for some aspects this has already been done (Stapert, 1976a; 1976b; 1982a), and for some others only in Dutch (Stapert, 1976c). Some aspects which to my mind are less important I have not considered worth going into (for example the absurd arguments used by Professor D. Goosen (Enschede) during the court hearings: he explained the ground ridges as the result of mudstreams that according to him deposited the boulderclay). For an overview of the Vermaning history, reference can be made to Waterbolk (1980). To my mind, the arguments I presented above must lead to the conclusion that the Vermaning stones dealt with in this article are forgeries.

10. ACKNOWLEDGMENTS

Thanks are due to Xandra Bardet (Groningen) for improving my English text and making an excellent job of it, to H.R. Roelink (B.A.I.) for technical assistance with the illustrations, and to Dr. H. Kars (R.O.B., Amersfoort) for critically reading a first draft of the manuscript.

11. NOTES

1. Since the discovery of the forgeries in 1975, a number of amateur archaeologists (especially P. Dijkstra, C.J.H. Franssen, J.E. Musch and A.M. Wouters, but many others also) have opposed my conclusions; their objections are formulated in a spate of articles, which can be found in their journal *Archeologische Berichten*. Most journalists chose to take the side of Vermaning; some of them wrote books on the affair (Hulst, 1975; Vermeulen, 1980; Terpstra, 1986). Very few Dutch colleagues outside Groningen have made public statements on the Vermaning stones. One of these is the archaeo-petrologist Dr. H. Kars of the *Rijksdienst voor het Oudheidkundig Bodemonderzoek* in Amersfoort (see e.g. *Drentse en Asser Courant*: September 13th, 1986; *Nieuwsblad van het Noorden*: October 22nd, 1987); he made it clear that he is of the opinion that my arguments are erroneous. His opinion is shared by Mr. A. Bruijn of the R.O.B. Several other colleagues have supported my conclusions: the late Professor C.B.M. McBurney, the late Professor F. Bordes (for example in a letter to Wouters, dated February 5th, 1981, of which he sent a copy to the Biological Archaeological Institute), Professor G. Bosinski (Cologne), Professor H.T. Waterbolk (Groningen), Professor J.D. van der Waals, and the geologist Professor G.J. Boekschoten (*Vrije Universiteit*, Amsterdam).
2. The finds of Hoogersmilde and Hijken are manufactured out of northern moraine flints. Within the material of Eemster, of which I could only study a small part, flints of other sources seem to be present; one flake probably consists of

Meuse flint (opaque, dark-coloured with small white spots).

3. Kars (pers. comm.) is of the opinion that my arguments are based on an 'axioma' that is incorrect (in a newspaper interview he speaks about my 'dogma'). He suggests that the existence of the Vermaning stones, which lack windgloss, should be taken as an indication that the bouldersand is not everywhere on the plateau characterized by the circumstance that most flints in it show windgloss (and other modifications). However, my arguments are based on empirical observations made also at sites indicated by Vermaning himself, for example at Hijken (Stapert, 1976a). Therefore, Kars' reasoning is unclear to me. In 1987 I visited sections at Hijken and in a quarry near Emmen (Griendsveen), studied by Kars. Again, it was clear that at both sites bouldersand flints were much more often heavily weathered than boulderclay flints.
4. In the Museum 'It Bleekerhûs' in Drachten (province of Friesland) several 'Middle Palaeolithic finds' of Vermaning are exhibited since 1987. One of these finds is a handaxe, said to come from 'Ravenswoud' (prov. of Friesland). This handaxe is comparable in all essential aspects with the material from 'Hoogersmilde' and 'Hijken', discussed in this paper (no patinas or windgloss, but instead traces of working with the help of a modern grinding machine are present).
More interesting are 5 other finds, said to have been found by Vermaning at various findspots in Friesland (Fochtelo 2 x, Weperpolder, Ravenswoud, Haule). Most of these finds were in 1984 offered for sale to the Frisian Museum in Leeuwarden (in 1984 the collection consisted of 6 pieces: 1 pseudo-artefact, and 5 large flakes, said to come from: Ravenswoud 3 x, Appelscha/Fochtelo, Weperpolder). These pieces were described in an unpublished report (D. Stapert, 1984. Kort verslag betreffende zes vuurstenen, door Vermaning in 1984 aan het Fries Museum te koop aangeboden. Groningen, unpublished).
The material in the Museum in Drachten comprises 3 or 4 large white-patinated flakes, which are heavily rounded due to rolling (fig. 22a shows one of these flakes). These artefacts possess myriads of large collision cones (up to 1.5 mm in diameter), that cover the whole surface (fig. 22b). In all probability these flakes derive from a gravelly beach, where they must have been repeatedly subjected to surf action. Such finds are known, for example, from the eastern coast of Denmark, where many Neolithic finds are to be found. It can be excluded that such finds could be collected on the boulderclay plateau of Friesland, which is where they are said to have been found by Vermaning. It is obvious, therefore, that we are dealing here with findspot forgeries, a category not represented by other Vermaning stones, as far as I am aware.

12. REFERENCES

BAUMANN, E., D. MANIA, V. TOEPFER & L. EISSMANN, 1983. *Die paläolithische Neufunde von Markkleeberg bei Leipzig*. Berlin.

BOOM, G., 1975. Onderzoek van vuurstenen, zowel erkend authentieke artefacten en natuurlijke stukken, als stukken uit collecties afkomstig van of toegeschreven aan Tj. Vermaning. State University of Groningen, unpublished.

BORDES, F., 1954/1955. Les gisements du Pech-de-l'Azé (Dordogne). I. Le Mousterien de tradition acheuléenne. *l'Anthropologie* 58, pp. 401-432; 59, pp. 1-38.

BORDES, F., 1968. *Le Paléolithique dans le monde*. Paris.

BOSINSKI, G., 1977. Report for use during the lawsuits concerning the 'Middle Palaeolithic' Vermaning-stones. University of Cologne, unpublished.

DIJKSTRA. P., J.E. MUSCH & A.M. WOUTERS, 1986. De Jong-Archeuléen vindplaats 'Eemster'. *Archeologische Berichten* 17, pp. 7-176.

FRANSSEN, C.J.H., 1978. De kenmerken van stenen artefacten en steencomposities in het proces Vermaning. *Archeologische Berichten* 2, pp. 2-42.

FRENCH, H.M., 1976. *The periglacial environment*. London/-New York.

HAMMEN, T. VAN DER & T.A. WIJMSTRA (eds.), 1971. The Upper Quaternary of the Dinkel Valley. *Mededelingen Rijks Geologische Dienst*. (N.S.) 22, pp. 55-213.

HARTZ, S., 1986. Paläolithische Funde aus dem Altmoränengebiet Nordfrieslands. *Offa* 43, pp. 105-147.

HULST, T., 1975. *Tjerk Vermaning, steen des aanstoots*. Groningen.

KEELEY, L.H., 1980. *Experimental determination of stone tool uses: a microwear analysis*. Chicago/London.

KOLSTRUP, E., 1980. Climate and stratigraphy in northwestern Europe between 30.000. and 13.000 B.P., with special reference to The Netherlands. *Mededelingen Rijks Geologische Dienst* (N.S.) 32, pp. 181-253.

LANTING, J.N., 1977. De zg. TRB-vondsten van Herpen (N.-Br.) (met een bijdrage van D. Stapert). *Helinium* 17, pp. 249-260.

LOOYENGA, T., P. SCHUDDEBEURS & D. STAPERT, 1984. Een vuistbijl van helleflint, gevonden bij Assen. Met opmerkingen over het gesteente en het woord helleflint. *Grondboor en Hamer* 38 (1), pp. 2-14.

MEYER, H.H., 1986. Steinsohlen - ihre Genese und Altersstellung nach neueren Forschungsbefunden. *Eiszeitalter und Gegenwart* 36, pp. 61-73.

MORTILLET, G & A., 1911. *La Préhistoire*. Paris.

MOSS, E.H., 1983. *The functional analysis of flint implements. Pincevent and Pont d'Ambon: two case studies from the French Final Palaeolithic* (= B.A.R. Intern. Series 177). Oxford.

MUSCH J.E., 1980. Overzicht van de archeologische ontdekkingen in Noord-Nederland, periode: 1978-1980. *Archeologische Berichten* 8, pp. 107-116.

PADDAYYA, K., 1973. A *Federmesser* site with tanged points at Norgervaart, province of Drenthe (Netherlands). *Palaeohistoria* 15, pp. 167-213.

POPPER, K.R., 1959. *The logic of scientific discovery*. London.

ROEBROEKS, W., 1984. The Middle Palaeolithic site Maastricht-Belvédère (Southern-Limburg, the Netherlands). *Helinium* 24, pp. 3-17.

STAPERT, D., 1975a. Voorlopig rapport over de steentijdvondsten van Tj. Vermaning (bijgewerkte versie, zoals uitgereikt op de persconferentie d.d. 18 maart 1975). *Westerheem* 24, pp. 70-75.

STAPERT, D., 1975b. Preliminary notes on the 'Early and Middle Palaeolithic' finds of Mr. T. Vermaning. *Palaeohistoria* 17, pp. 7-8.

STAPERT, D., 1976a. Some natural surface modifications on flint in the Netherlands. *Palaeohistoria* 18, pp. 7-41.

STAPERT, D., 1976b. Middle Palaeolithic finds from the Northern Netherlands. *Palaeohistoria* 18, pp. 43-72.

STAPERT, D., 1976c. De 'steentijdvondsten' van Tj. Vermaning. Bewijsvoering betreffende valsheid. State University of Groningen, unpublished.

STAPERT, D., 1976d. De vuistbijl van Anderen (gem. Anloo) en natuurlijke oppervlakteveranderingen bij vuursteen. *Nieuwe Drentse Volksalmanak* 93, pp. 145-167.

STAPERT, D., 1977. The combination of 'the mandibula of a giant deer and a Tjonger point having been shot into it', from Roermond, is of recent date. *Helinium* 17, pp. 235-248.

STAPERT, D., 1979. The handaxe from Drouwen (province of Drenthe, the Netherlands), and the Upper Acheulian. *Palaeohistoria* 21, pp. 127-142.

STAPERT, D., 1980. Het Middenpaleolithicum in Noord-Nederland. In: M. Chamalaun & H.T. Waterbolk (eds.), *Voltooid verleden tijd?*. Amsterdam, pp. 63-82.

STAPERT, D., 1981a. Archaeological research in the Kwintelooijen Pit, municipality of Rhenen, the Netherlands. *Mededelingen Rijks Geologische Dienst* 35, pp. 204-222.

STAPERT, D., 1981b. Hand-axes in Southern Limburg (the

Netherlands) - How old? In: F.H.G. Engelen (ed.), *Third International Symposium on flint, Maastricht 1979* (= Staringia 6). Heerlen, pp. 107-113.

STAPERT, D., 1982a. A Middle Palaeolithic artefact scatter, and a few younger finds, from near Mander NW of Ootmarsum (province of Overijssel, the Netherlands). *Palaeohistoria* 24, pp. 1-33.

STAPERT, D., 1982b. A handaxe made of *hälleflinta* found at Anreep, near Assen (province of Drenthe, the Netherlands). *Palaeohistoria* 24, pp. 41-51.

STAPERT, D., 1982c. A site of the Hamburg tradition with a constructed fire-place at Oldeholtwolde (prov. of Friesland, the Netherlands). *Palaeohistoria* 24, pp. 53-89.

STAPERT, D., 1985a. Twee vuistbijlen, in 1984 verworven door het Drents Museum, en het *Moustérien de tradition acheuléene* (MTA) in noordelijk Nederland. *Nieuwe Drentse Volksalmanak* 102, pp. 97-124.

STAPERT, D., 1985b. Oud hout uit een zuigkolk bij Zwolle (met een bijdrage van W.A. Casparie). *Grondboor en Hamer* 39, pp. 130-141.

STAPERT, D., 1986. The handaxe from Elahuizen (province of Friesland, the Netherlands); an exercise in interpreting secondary surface modifications. *Helinium* 26, pp. 3-19.

TERPSTRA, P., 1986. *Tjerk Vermaning. Een leven tussen stenen.* Drachten/Leeuwarden.

VAYSON DE PRADENNE, A., 1932. *Les fraudes en archéologie préhistorique. Avec quelques examples de comparaison en archéologie générale et sciences naturelles.* Paris.

VERMEULEN, F., 1980. *De onderste steen. Recht en wetenschap in de knel. De zaak Vermaning.* Doetinchem.

WAALS, J.D. VAN DER & H.T. WATERBOLK, 1973. The Middle Palaeolithic finds from Hogersmilde. *Palaeohistoria* 15, pp. 35-120.

WATERBOLK, H.T., 1980. Het onderzoek van het oudere paleolithicum van Noord-Nederland. Een persoonlijke terugblik over de jaren 1939-1979. State University of Groningen, unpublished.

WEE, M.W. TER, 1966. *Toelichtingen bij de geologische kaart van Nederland, 1:50.000; blad Emmen west (17W) en Emmen oost (170).* R.G.D., Haarlem.

WITTE, A.H., J.W. VERBURGT & E.R. GROENEVELD, 1975. Rapport betreffende het onderzoek van vuurstenen artefacten afkomstig van Tj. Vermaning. Juridical Laboratory, Rijswijk, unpublished.

WITTE, A.H., J.W. VERBURGT & E.R. GROENEVELD, 1976. Rapport betreffende het onderzoek van vuurstenen artefacten in de zaak contra Tj. Vermaning, verdacht van het recent bewerken van deze stenen. Juridical Laboratory, Rijswijk, unpublished.

WOUTERS, A.M., 1980. De Leemdijkbijl, de bewogen geschiedenis van een Drentse vuistbijl. II. *Archeologische Berichten* 7, pp. 107-116.

ZAGWIJN, W.H., 1975. Indelingen van het Kwartair op grond van veranderingen in vegetatie en klimaat. In: W.H. Zagwijn & C.J. van Staalduinen (eds.), *Toelichtingen bij de geologische overzichtskaarten van Nederland.* R.G.D., Haarlem, pp. 109-114.

HUNEBED G2: EXCAVATION AND FINDS

A.L. Brindley
Archaeological Survey OPW, Earl Court, Adelaide Road, Dublin, Ireland

ABSTRACT: The remains of a levelled *hunebed* with seven pairs of sidestones were excavated during the winter 1969-1970. Almost all structural parts had been removed or destroyed, but the contents of the chamber remained in situ, although fragmented. A large number of (mainly reconstructable) pots, many flint artefacts (arrowheads, *bikkels*, axes) and amber beads were found. In addition, sherds of an EGK amphora, three Bell Beakers and two Bell Beaker knives indicate later activity. A sherd of a *kogelpot* dates the destruction of the *hunebed* to the late 10th or 11th century AD.

KEYWORDS: Northern Netherlands, Neolithic, megalithic monument, TRB pottery, typochronology.

1. INTRODUCTION

The *hunebed* G2 is one of two destroyed megalithic tombs located on the Glimmer Es (municipality of Haren), about 9 km south-east of Groningen, the Netherlands. Both were excavated by J.N. Lanting on behalf of the Biologisch-Archaeologisch Instituut, Groningen, who subsequently published the groundplans and a small selection of the finds (Lanting, 1974-75). In the intervening period, the flint has been examined by P.H. Deckers (Groningen) as part of doctorate research into Neolithic flint working and a definitive report of the excavation and finds from G3 has been published recently (Brindley, 1983).

2. THE DESTROYED MEGALITHIC MONUMENT

2.1. Discovery and excavation

In 1966 J.E. Musch, an amateur archaeologist, discovered two concentrations of small fragments of stone (chiefly crushed granite), sherds of TRB pottery, flint and fragments of burnt bone in the otherwise stoneless arable of the Glimmer Es (Bakker, 1982-83: p. 183). He identified these correctly as the sites of two destroyed megalithic tombs and duly informed the B.A.I. and the provincial archaeologist for Groningen. In 1969 it was decided to investigate both sites. The remains of the larger *hunebed*, G2, were excavated in the winter of 1969-70 and those of the smaller *hunebed*, G3, in the autumn of 1971.

Hunebed G2 was situated near the highest part of the Glimmer Es (maximum height c. 5.5 m above N.A.P.), in cadastral parcel Haren, Section H4, No. 2834, which at the time was owned by Mr. Kloots and worked by Mr. Scholtens, both of Noordlaren (fig. 1). The excavation took place between 3rd November 1969 and 26th February 1970. Because of hard frost, no work was possible between 9th December and 24th February. The excavation was carried out by K. Klaassens, J.H. Zwier, J.E.

NOTE ON THE FIGURES

The following conventions were adopted for the illustration of the reconstructed pottery:
- solid lines indicate certainty of reconstructed profile;
- broken lines indicate proposed reconstruction of profile;
- solid and heavy lines indicate certainty of reconstruction of decoration;
- fine lines or broken lines indicate proposed reconstruction of decoration;
- relevant sherds are shown on proposed reconstructions;
- sections with applied features (handles, lugs, bosses, cordons, footrings) shown by solid lines indicate certainty of location and reconstruction.

The drawings of the pottery are reproduced at the scale 1:3. The numbers refer to the catalogue. In figures 2 and 3 finds of pottery outside the chamber are indicated by catalogue numbers, as well.

The drawings of the flint and amber objects are reproduced at full scale, with the exception of the axes and axe fragments, which are shown at the scale 2:3. The numbers are the find numbers. Most of these numbers are indicated in figure 4, the remaining ones refer to either stray finds collected before, during and after the excavation (No. 1) or to material found during cleaning of areas larger than 1 x 1 m squares (Nos 20, 21, 37, 59, 60, 61, 67 and 68).

The majority of the pottery was drawn by S.W. Jager. Additional material was drawn by H.R. Roelink, who was also responsible for the illustrations of the flint, for tables 15 and 16, and for numbering and mounting all the figures. The flint axes were drawn by J.M. Smit; the plans are the work of J.H. Zwier.

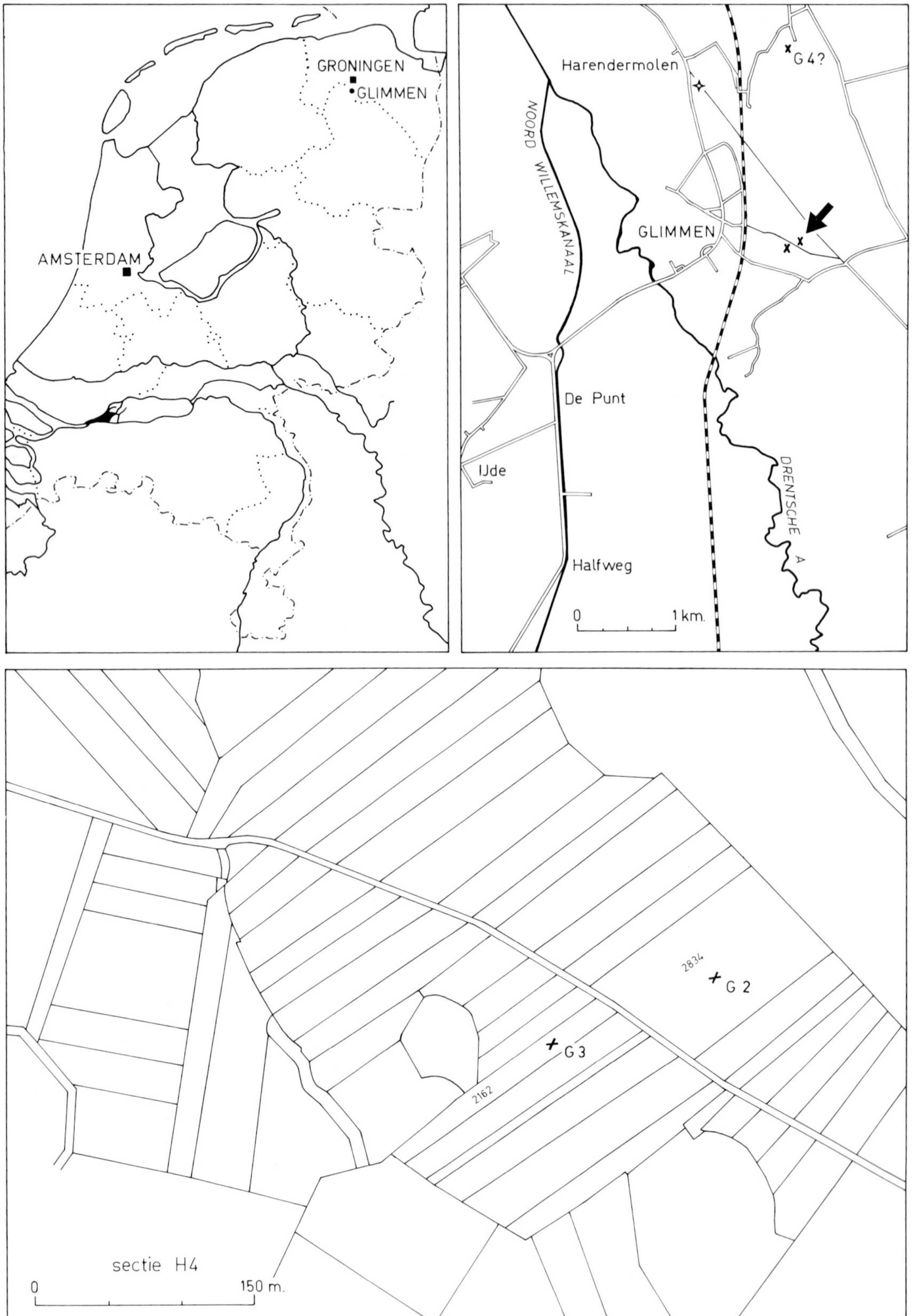

Fig. 1. Location of *hunebed* G2.

Musch, J.E. Heeres, A.E. Lanting and J.N. Lanting.

The remains of *hunebed* G2 appeared immediately under the ploughsoil as an elongated, subrectangular pit measuring c. 14 x 5 m, filled with brownish sand, crushed granite, field stones and pieces of larger stones (chamber debris). The surviving grave goods were all found in this material. They were collected by square metre in order to record any possible information about their original place of deposition. The entire layer of chamber debris was sieved through a 6 x 6 mm mesh mounted on a hand driven potato sorter to ensure the collection of even very small artefacts. This proved very effective. Only one tiny amber bead and a fragment of a second were found in the sand under the sieve afterwards. The topsoil was not sieved and therefore a small amount of pottery and flint was not collected. In fact some flint was found afterwards in the redeposited topsoil by the amateur archaeologist E. Ameling (see below).

It was originally intended to excavate the remains of G2 in such a way that profiles through the layer of chamber debris could be examined. Therefore the excavation pit was divided into eight parts by placing a line over the long axis, one over the short axis and two lines parallel to but three metres distant from the last mentioned line. After excavating the first (upper) layer of chamber material in four alternating parts, it became clear that little information could be extracted from the profiles. It was therefore decided to excavate by level. The plan and square numbers (fig. 4) still indicate the original intention.

An approximately oval shaped cutting measuring 22 x 19 m was laid out around the remains of the destroyed chamber. The ploughsoil above the layer of chamber debris was only 0.20 m thick. Along the edges of the cutting this increased to 0.30-0.35 m. The subsoil consisted of stratified sand, the upper 0.30-0.40 m of which had lost its stratification, due to the activities of burrowing animals and the penetration of roots. Originally a podsol profile had been present, but with the exception of the lowest part of the iron pan layer, most of this had been removed by ploughing. Neolithic pits could not be traced in the structureless sand. However, some deeper pits were recognizable in the stratified sand, not because of their colour, but as disturbances in the stratification.

Four levels in the chamber area were planned during the excavation (figs 2 and 3).

Level 1. Immediately below the ploughsoil. Only the outline of the large pit was recorded.

Level 2. At the level of the upper surface of the paved floor, about 15-20 cm below Level 1. Extraction pits with their black, humic filling were clearly visible at this level.

Level 3. Immediately under the floor, c. 16 cm below Level 2. The extraction pits were much smaller at this depth. Field stones became visible in apparently undisturbed yellow soil around the edges of some of these pits. These were the original packing stones along the edges of the sockets of the orthostats. These sockets were not visible, however, in the unstratified coversand.

Level 4. C. 10 cm below the floor. Most of the extraction pits had disappeared at this level. In some cases, however, a field stone appeared below the extraction pits, in apparently undisturbed yellow sand. These stones were also packing stones in the orthostat sockets.

2.2. The destruction of the chamber

As stated above, the remains of the destroyed chamber appeared immediately below the ploughsoil as an elongated pit filled with dark soil, crushed granite, small pieces of stone and cobbles. It is a well-established fact that the chambers of megalithic tombs of the TRB culture were constructed in pits. The sockets for the orthostats were dug into the base of this pit and lined with a packing of field stones. The gaps between the orthostats were filled with dry stone walling of small stone slabs. The floor of the pit within the chamber was paved with cobbles and then covered with a layer of crushed granite fragments. The chamber was surrounded with a packing of larger stones. Between these stones, the remainder of the foundation pit was filled with a mixture of sand and crushed granite.

Immediately below the ploughsoil, no difference was visible between the areas on either side of the orthostats (i.e. inside and outside the chamber), nor were the extraction pits of the orthostats recognizable. This was largely due to the fact that the top part of the layer of chamber debris contained a lot of dark topsoil, but also because it was impossible to clean this level adequately. The presence of this dark soil was due to infiltration and ploughing. When the chamber debris had been removed down to the level of the paved floor (small parts of which had survived), the original positions of the sidestones became visible as recent pits with a filling of dark, humic soil. These extraction pits cut through the layer of brownish sand, stones, and stone fragments; it was only due to plough damage and humic infiltration that they could not be recognized immediately below the ploughsoil. However, the profiles which were visible briefly during the excavation of the first layer (see 2.1) showed quite clearly that these extraction pits were younger. The orthostats must have been left in situ until most of the stone paving and the packing stones around the chamber had been removed. The removal of the orthostats and the backfilling of their extraction pits must have been the last stage in the destruction and levelling of this megalithic tomb.

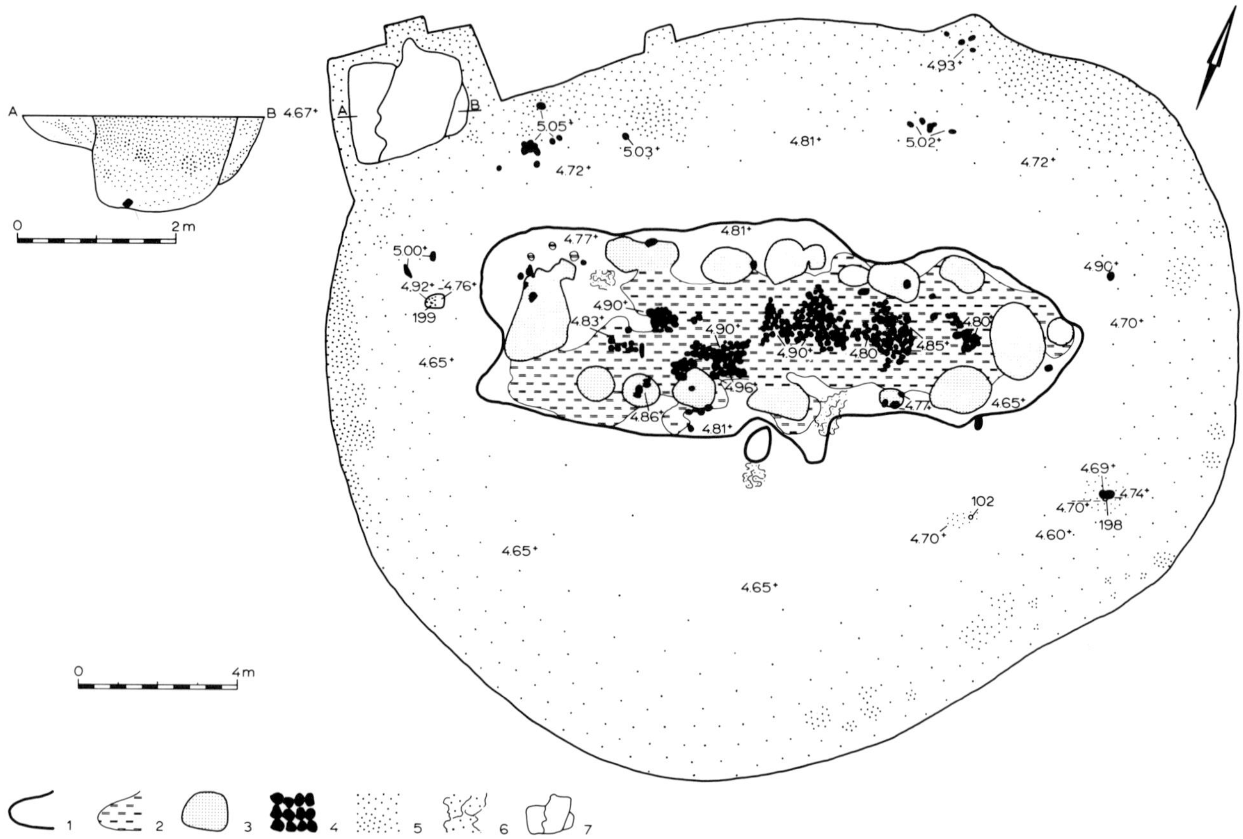

Fig. 2. Groundplan: Levels 1 and 2 combined. Legend: 1. level 1, outline of chamber debris; 2. level 2, chamber debris; 3. extraction pits; 4. floor and packing stones in situ; 5. iron pan infiltration at edge of mound; 6. iron pan with infiltration veins in chamber area; 7. Medieval pit.

A large pit in the extention of the NW side of the excavation cutting is apparently connected with the destruction of the chamber. In fact this consists of two pits, as a large part of the filling of the pit was dug out and backfilled again later. The colour of the filling and the sharpness of the contours indicated that the pits were relatively young and probably of medieval origin. Some field stones and a large fragment of a boulder were found in the filling. The pit may have been dug to bury an orthostat or capstone, or fragments of large stones which were dug out later.

As at G3 (Brindley, 1983) a fragment of a *kogelpot* with simple rim profile (No. 370) which can be dated to the late 10th or 11th century AD, was found amongst the chamber debris. The presence of this sherd almost certainly indicates that G2 was destroyed during this period. The jug (No. 363), described by Lanting (1974-75: p. 172) as medieval, is more probably of TRB origin.

2.3. The chamber

It was stated in the first publication of this excavation that G2 had six pairs of side stones (Lanting, 1974-75). However, on the published groundplan (Lanting, 1974-75: fig. 2), the correspondingly numbered extraction pits do not occur opposite each other, with the exceptions of Z5/Z5' and Z6/Z6'. Since this is unusual for a *hunebed*, the reconstruction must be incorrect. A re-examination of the field drawings by J.N. Lanting (who has been carrying out continuing research into Dutch megalithic tombs in recent years) has shown that G2 had originally seven pairs of side stones. The following reinterpretations can now be put forward.

a) The extention on the northeast side of the extraction pit of Sl 1 appears to be the extraction pit of a side stone. The small pits with a gritty filling north of this extention at Level 2 may indicate the positions of packing stones in the socket of this side stone. A field stone was discovered at Level 3 beneath the extention; this was possibly a packing stone at the bottom of this socket. This 'new' side stone should now be numbered Z1 and the pits Z1-Z6 of the first publication (Lanting, 1974-75: fig. 2) should now be renumbered Z2-Z7.

b) Between the extraction pits Z4' and Z5' in

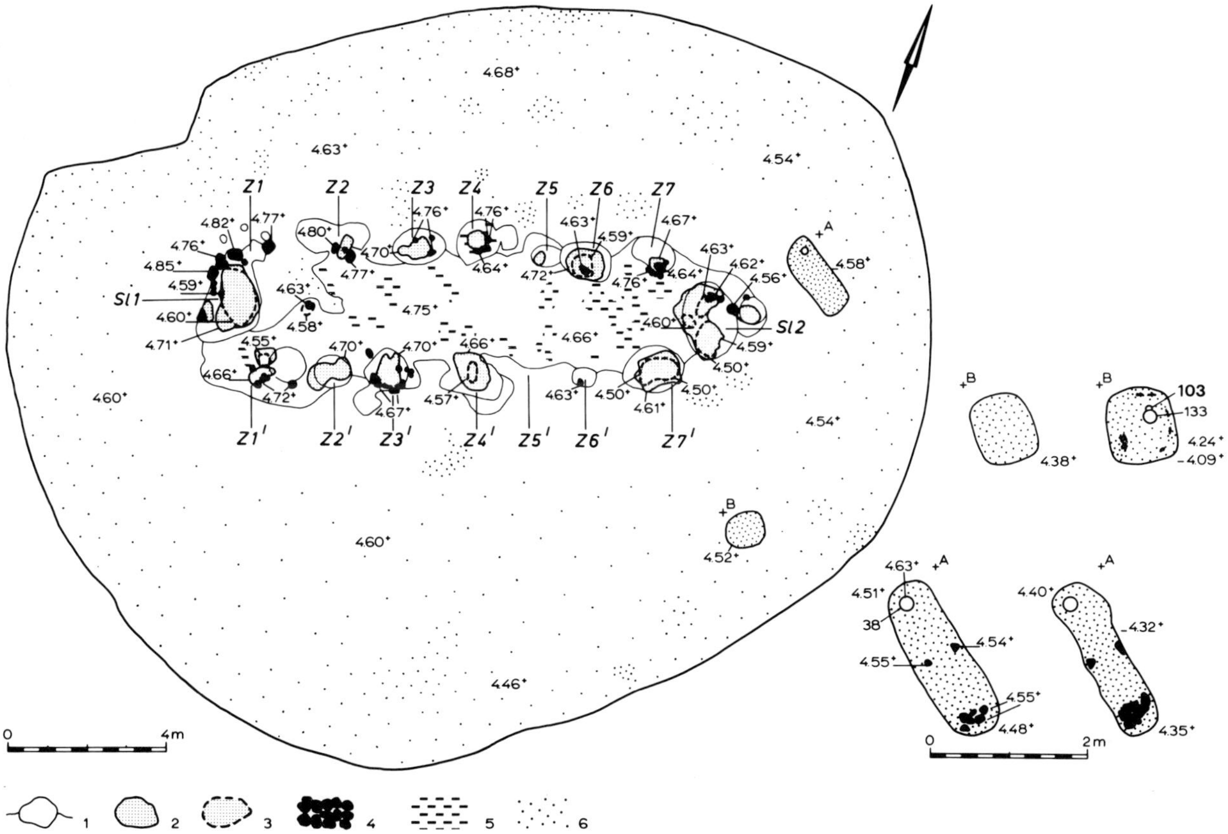

Fig. 3. Groundplan: levels 2, 3 and 4 combined. Legend: 1. level 2, outline of chamber debris and extraction pits; 2. level 3, extraction pits; 3. level 4, extraction pits; 4. level 4, packing stones in situ; 5. level 3, traces of chamber debris; 6. iron pan at edge of mound.

figure 2 of the earlier publication (see above), is a large gap of about 2 metres. This is far larger than the normal gap between two side stones but it is also too large for an entrance gap. The most likely explanation is that a side stone with a very shallow foundation pit stood here originally, and that by Level 2 traces of this stone had disappeared. It is possible, however, that the brown patch with infiltration veins recorded at Level 2 may be connected with this side stone. This new side stone should now be numbered Z5'. The extraction pits Z5' and Z6' of the earlier publication (Lanting, 1974-75: fig. 2) should be renumbered as Z6' and Z7'.

c) Only in the case of Zl' did the pit filled with dark humic soil at Level 2 not correspond fully with the traces of an extraction pit and packing stones in situ at Level 3 and 4. This means that Zl' stood closer to Sl 1 than was indicated in the earlier publication (Lanting, 1974-75: fig. 2).

Following these changes the chamber now appears to have been very regular with the correspondingly numbered extraction pits lying opposite each other. A new ground plan is published here on which the extraction pits have been renumbered (fig. 3).

The small pit with a similar dark humic filling to the extraction pits east of Sl 2 may have been due to the removal of a large boulder in the stone packing around the chamber.

The orientation of the chamber was ENE-WSW, its internal measurements were 11 x 2-2.2 m. Large parts of the paved floor of field stones had been broken up, but some substantial parts remained intact. The surface of the floor was uncovered c. 0.35-0.40 m below the surface of the field. This must correspond more-or-less to the original depth of the chamber below the surface at the time of its construction.

A *hunebed* orientated ENE-WSW would have had its entrance near the centre of its south side. There is evidence to suggest that the entrance to G2 occurred between Z3' and Z4', that is, west of centre. Between Z3' and Z4' the paved chamber floor was defined by a strip of slightly larger stones which projected above the level of the rest of the chamber floor. It is possible that these stones were originally laid along a sill stone which disappeared

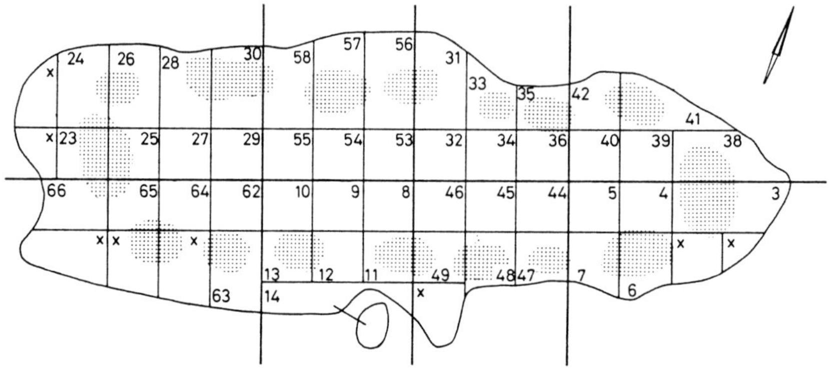

Fig. 4. Layout of excavation squares with find numbers. Positions of orthostats shaded, x indicates no finds.

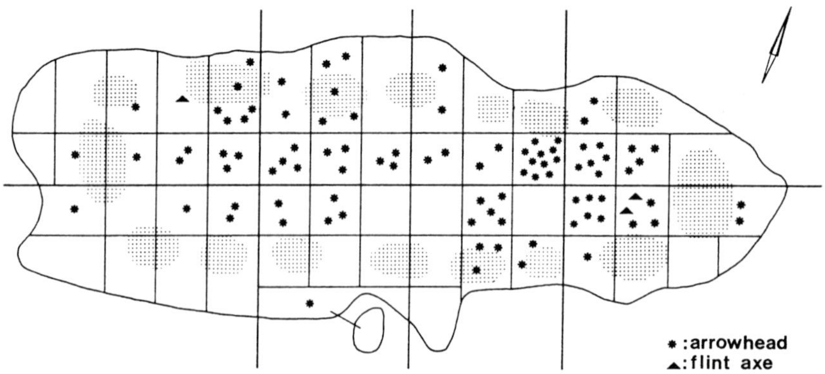

Fig. 5. Distribution of trapezoidal arrowheads and flint axes.

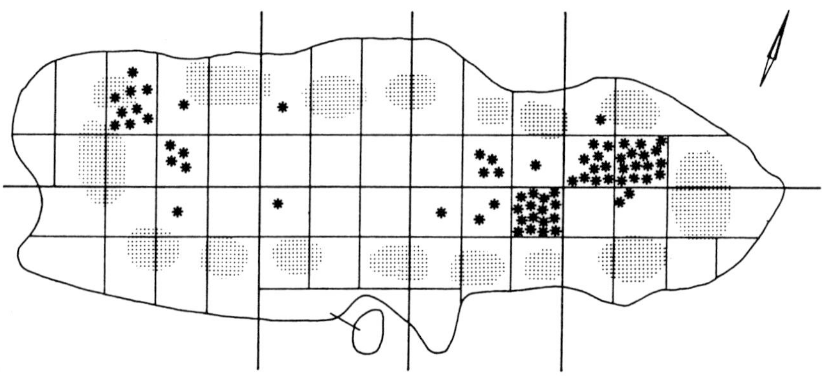

Fig. 6. Distribution of amber beads.

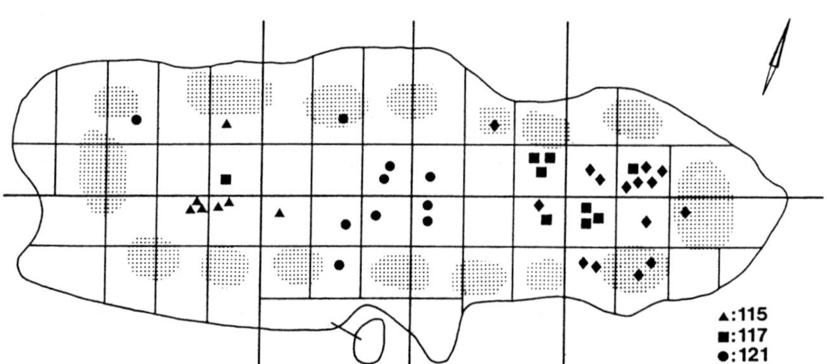

Fig. 7. Distribution of sherds of lugged bowls, Horizon 3.

during the destruction of the chamber. An oval shaped 'recent' pit filled with brownish sand and stone grit at Level 1 and a subrectangular patch of iron pan infiltration at Level 2 were noticed southwest of Z4' in the area where a portal stone would have stood if the entrance to the chamber had been between Z3' and Z4' The 'recent' pit may be due to the removal of the portal stone and the infiltration may have been caused by the former presence of the stone, as in the case of side stone Z5'. The absence of a socket is not surprising; in general, portal stones are smaller than side stones and do not have such deep sockets. This explains why no traces of the corresponding portal stone were recognized west of the entrance.

2.4. The mound

As with all megalithic tombs in the Netherlands, G2 was originally surrounded by a mound of earth. This mound was also levelled when the chamber was demolished. No podsol profile had yet developed at the time the chamber and mound were constructed; this only took place subsequently. Therefore the former mound was traceable by the absence of iron pan infiltration immediately around the chamber and its presence near the edges of the cutting. Unfortunately, only the lower part of the iron pan layer had escaped being ploughed away and in places only faint traces were still present. Nevertheless, an approximate outline of the shape and size of the mound could be made out. It was apparently an elongated subrectangular-to-oval shaped mound c. 19 m long and c. 10 m wide (fig. 2).

There is some evidence that the mound was originally defined by a kerb. Not all *hunebed*s had kerbs but they are most commonly associated with *hunebed*s with two pairs of portal stones. Concentrations of field stones were found in several places along the north and west sides of the chamber, more-or-less along the edge of the iron pan infiltration. These stones are possibly packing stones from kerb sockets. The sockets of kerb stones are generally very shallow and would therefore be amongst the first features to disappear. This may also explain why traces of kerb and portal features are almost never found during excavations of destroyed megaliths in the Netherlands. The concentrations of field stones along the edges of the former mound of G2 may be an exception to this general rule. The shape and size of the mound, and especially the distance of c. 2 m or possibly slightly more between the chamber and the edge of the mound can also be taken as indicative of the presence a kerb originally.

2.5. Features outside the chamber

Several features associated with pottery were uncovered outside the chamber area (figs 2 and 3).

a) Northeast of Sl 2 a burial pit, orientated NW-SE, became visible when the stratified sand was uncovered (length 2.1 m, width 0.8 m, depth below present ground surface 0.7 m). Eight field stones were found at the bottom of the pit, six close together near the southeast end and one each close to the middle of the long sides. A large funnel beaker (No. 38) was found near the NW end of the pit, in an inverted position. A small piece of flint without traces of either working or use was found in the pit fill. This should be considered as a stray find. Although no trace of a corpse silhouette was found, the absence of any cremated bone indicates that this was an inhumation burial. The function of the field stones is unclear. These would be explained as coffin supports in a Bronze Age grave but it is not known if they fulfilled the same purpose in TRB burials. The dimensions of the pit and the position of the stones suggest extended burial, but again, absolute certainty is not possible. Nor is the stratigraphic relationship between this burial and the megalithic tomb clear; the burial occurred near the edge of, but clearly beneath, the former mound. It could, therefore, be earlier than the mound but this cannot be proven. Its position under the mound and presumably just within the kerb are not necessarily arguments against the digging of the grave pit after the construction of the mound covering the megalithic chamber as Lanting (1974-75: p. 177) thought.

b) Southeast of the extraction pit of Sl 2 a small undecorated bowl with a footring (No. 198) was found beneath two field stones, close to the edge of the former mound. No traces of a pit were visible in the structureless sand.

c) Southeast of Z7' and outside the mound a subrectangular pit (0.90 x 0.90 m) was found. The base of the pit was c. 0.9 m below the present surface of the field. The outline of the pit was only visible in the stratified sand. It contained two vessels. A faint circle of small fragments of charcoal occurred at the level of the uppermost part of these pots. This charcoal was collected and dated: 4380±40 BP (GrN-6156). The pottery consisted of a straight walled pail with four lugs close to the rim and a five or six footed base (part of the base has decayed; No. 133). This was accompanied by a single lugged beaker with a thumb sized horizontal lug (No. 103). The function of this pit is unclear. No cremated bone or traces of a corpse silhouette were found. The dimensions of the pit were only large enough for a child burial. It is possible, of course, that this is not a burial pit but one for an offering.

d) South of Z7' an undecorated poorly made beaker with an uneven base (No. 102) was found in a pale grey patch. The findspot was well within the edge of the mound but it is clear in this case that the pot must have been buried after the construction of the mound.

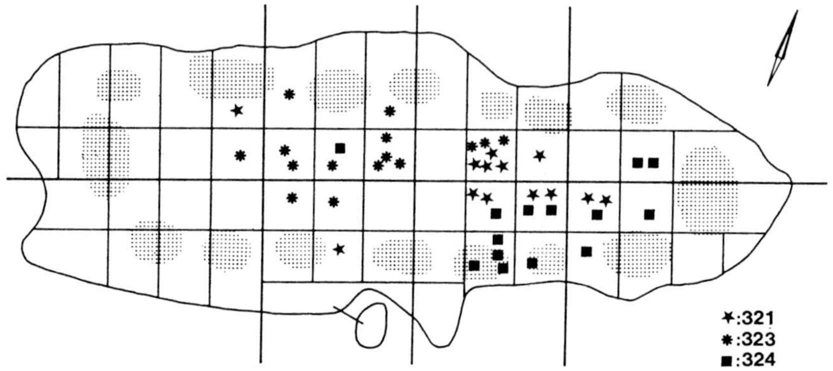

Fig. 8. Distribution of sherds of tureen-amphorae, Horizon 5.

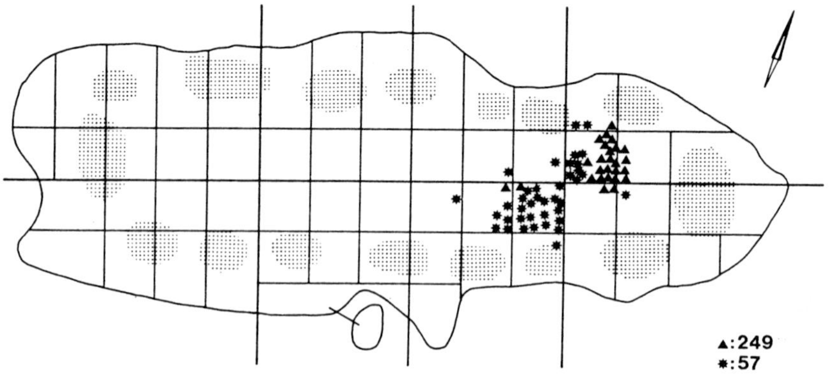

Fig. 9. Distribution of sherds of service set, Horizon 3.

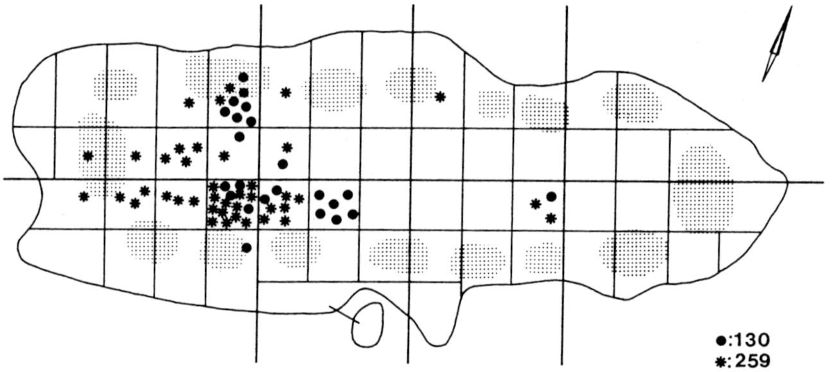

Fig. 10. Distribution of sherds of service set, Horizon 3.

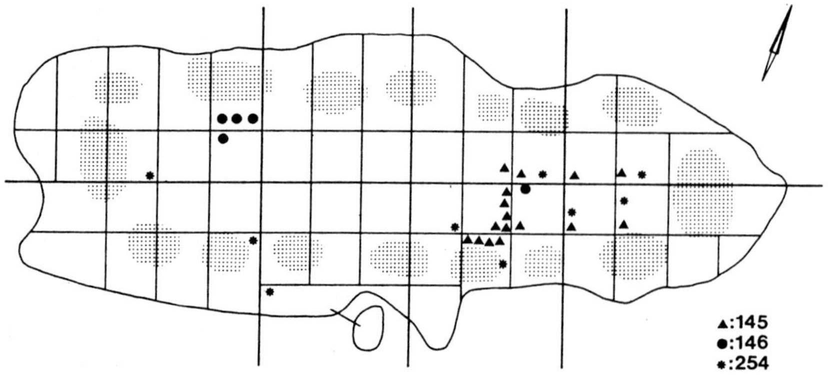

Fig. 11. Distribution of sherds of service set, Horizon 4.

e) 2.5-3 m west of Sl 1 a small, undecorated bowl with an uneven base (No. 199) was found in a grey patch with a few pieces of charcoal which was therefore visible in the structureless sand. This pit was dug close to the edge of the mound.

2.6. Distribution of material in the chamber (figs 5-14)

Although the structure of the *hunebed* was almost completely destroyed (all traces of the mound and orthostats having disappeared and only a small portion of the floor remaining in situ), the contents of the chamber survived not only relatively intact but more-or-less in their original position. In many cases, the distribution of sherds of a single pot was fairly well nucleated. These nuclei represent the original position of the pot, not a disturbed position. This is shown by the distribution of sherds of various 'service sets' (see figs 9-13). The term 'service set' was introduced by Schlicht in discussions of the pottery from Emmeln 2 (Schlicht, 1968: p. 21). The term refers to a group of two or more pots, either of the same types or of different types, with identical fabric and decoration, indicating that all were made by one potter on one occasion and fired together. In some cases, it has not been possible to distinguish sherds of different pots in a 'service set'. These figures show that the pots from a single 'service set' were deposited side-by-side, almost certainly in a single act. Similarly, the grouping of the amber beads into well-defined and distinct areas, suggests the original positions of three separate necklaces.

Figure 7 shows the distribution of several bowls of Horizon 3. The sherds indicate the original location of the individual bowls, the distribution of which show that no particular part of the chamber was preferred. A larger number of sherds and other items were found in the east end of the chamber, but finds from all horizons and of all types were spread equally over the whole chamber. The distribution of arrowheads (fig. 5) reflects this overall distribution of finds.

3. FINDS

All finds are lodged in the Groninger Museum, Groningen, inventory numbers 1969/X (excavation B.A.I.) and 1987/II.1 (Ameling collection).

3.1. Flint

994 pieces of flint were found before and during the excavation. E. Ameling, an amateur archaeologist, later collected a further 280 pieces from the redeposited and unsieved topsoil. These were only recently brought to the attention of, and presented

Table 1. Service sets.

Tureens	Funnel beaker	Pail	Bowl
276, 277			
270, 271			
(272, 273)			
249	57		
259		130	
248			122
254			145, 146
—			139, 140

to, the B.A.I. There is a curious difference between the material collected by the B.A.I. and by Ameling. In the B.A.I. collection, at least 154 pieces are clearly recognizable as tools and another c. 50 pieces show traces of working or use. Of the 280 pieces in Ameling's collection, the vast majority are unworked pieces and flakes. Only six tools are present. Furthermore, the large number of burnt pieces (75) is surprising. Amongst these, eight pieces of a large axe are recognizable. It is possible that the other pieces also belong to this axe.

As stated previously, during the excavation the finds were collected by sieving with a potato sorter and recorded by square metre. With the exception of the axes, all the flint material in the B.A.I. collection was included by Deckers within the context of his research on Neolithic flint working. Ameling's collection was only brought to his attention after Deckers finished his work on the G2 flint material. Therefore only the tools in this collection will be included in his publication.

A detailed description of the tools is given by Deckers (forthcoming). Here it is sufficient to present the tools according to traditional categories. The assemblage includes (figs 17-25):

- 119 transverse arrowheads. The small arrowhead with retouch along one side which was found in square 3, resembles a Mesolithic A-point, but is probably an unusually narrow TRB arrowhead. As in other *hunebeds*, unworked flakes with the proportions of transverse arrowheads were found at G2. These probably fulfilled the same function. These unworked pieces are not included in the numbers given or illustrated;
- 11 scrapers;
- 1 fragment of a bi-facially worked artefact (not illustrated);
- 4 blades with retouch, three of which show sickle gloss;
- 21 *bikkels*, one of which is made from a piece of a polished flint axe. Some of these *bikkels* seem to have been used as strike-a-lights. Recently a *bikkel* and an iron ore nodule were found together in a flat grave with Horizon 4 pottery near Heek-Ammerter Mark, Kr. Borken (information W. Finke). Other *bikkels* may have been used as retouching tools or borers;

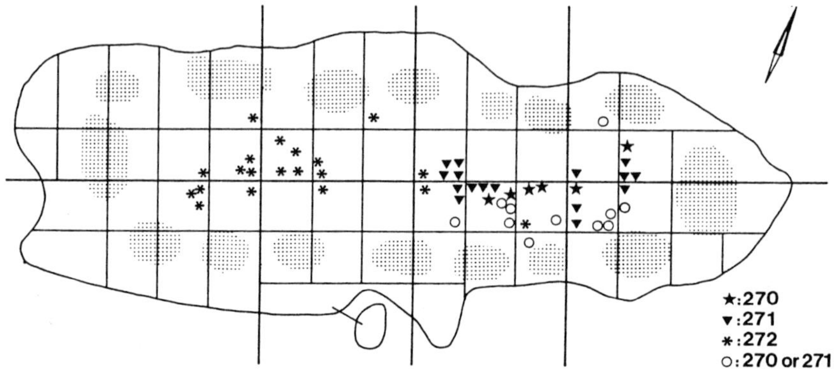

Fig. 12. Distribution of sherds of service set, Horizon 4, and similar but not identical tureen (No. 272).

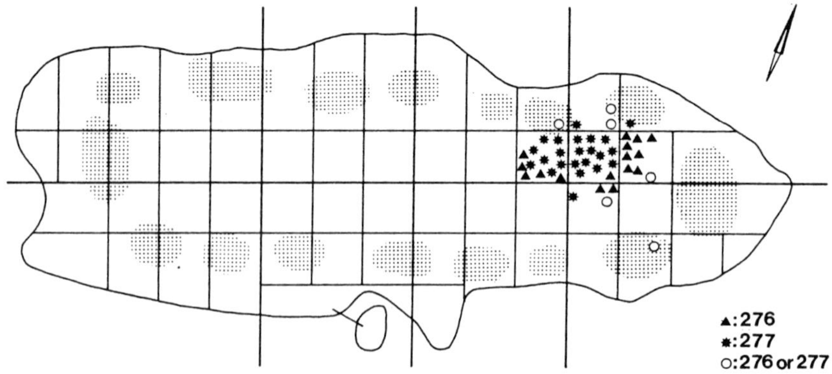

Fig. 13. Distribution of sherds of service set, Horizon 4.

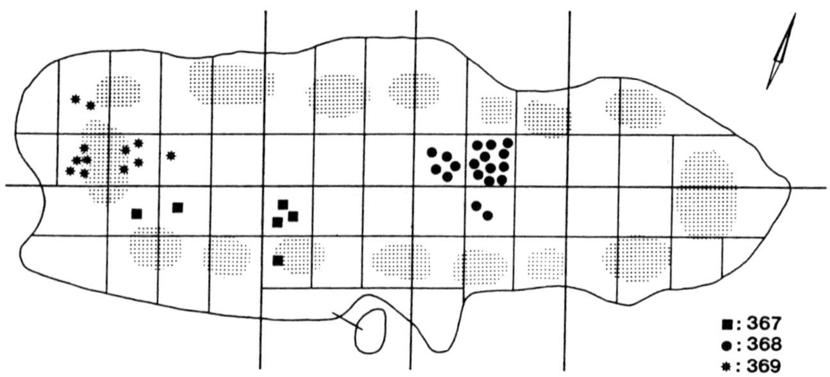

Fig. 14. Distribution of sherds of Bell Beakers.

- 4 flint axes. Two of these were found together in square 4 and are probably an original association. The small axe (No. 1) has a badly damaged cutting edge and was probably used as a hammer stone at some stage. In addition to these, there are two large pieces of flint axes with traces of secondary working. The pieces of burnt axe in the Ameling collection are not included in this discussion;
- 2 oval flint knives with marginal retouch which belong to the Bell Beaker phase of activity (Lanting, 1973: p. 232).

3.2. Amber

71 pieces of amber were recovered. This large number is attributable chiefly to the use of a 6 mm mesh sieve during the excavation. Apart from a single hemispherical unperforated piece found in the ploughsoil, all the pieces are either complete, or portions of, beads. The beads are of two main forms, tubular and disc, the latter being the more common. Most of the beads were found in three concentrations (fig. 26), two in the eastern half of the chamber, and the other (about 12 beads) in the

western half. The hemispherical unperforated piece is possibly an unfinished amber bead of the type found in Beaker graves.

3.3. Bone

No unburnt bone was preserved. *Hunebed* inventories from a.o. Hilter, Kr. Osnabrück (Schlüter, 1985: Fig. 126-129) and Wechte, Kr. Steinfurt (Knöll, 1983: Tafel 62) indicate the general range of bone objects deposited in megalithic tombs, but the soil conditions in this area are not suitable for the preservation of unburnt bone. It is very likely that bone points and tusk/teeth pendants originally formed part of the grave gifts.

3.4. Burials

A very small amount of cremated bone was recovered. Since it is unlikely that any was removed prior to the discovery and excavation of the tomb, and as no significant amount was lost during the excavation, this must mean that the chamber never contained large amounts of cremated bone. No estimate of the original quantity of unburnt bone is possible, but as several pots (often 5 or 6) are commonly found in what seem to have been originally single burials in flat graves, there is no reason to believe that there should have been an interment with every gift; it is also clear from the pottery itself, that groups of pottery (the so-called service sets) were deposited as single acts. Furthermore, there is no evidence one way or the other for burials without gifts or with perishable gifts, both of which conceivably took place.

3.5. TRB pottery

3.5.1. *General remarks*

The following sections are devoted to a descriptive account of the TRB pottery. About 8500 sherds of TRB pottery were found in the chamber debris and in the ploughsoil before and during the excavation. This material was sorted over a period of five months. The assemblage originally consisted of approximately four hundred pots representing all but the earliest stages of West Group TRB pottery. About 360 pots are reconstructable to a large extent. An additional forty can be recognized by distinctive rim, shoulder or decorated sherds, as not belonging to any other identifiable pot. The bases have not been included in this figure. Although examined exhaustively, 78 fairly complete bases could not be definitely associated with body sherds. Six more-or-less complete pots were also found outside the chamber (see 2.5).

The pottery is described by individual type based chiefly on shape (funnel beaker, pail, etc., see fig. 15) rather than by horizon. Two problems were encountered in doing this – one being the absence of terminology for various fairly specific and well-represented types. For these it was found necessary to introduce new terminology in the form of two new terms and to modify an existing one (see fig. 15):

- funnel beaker cup,
- shouldered vases, with or without lugs,
- tureen-amphorae/amphorae types 1 and 2.

This procedure was found to be fairly successful but a second problem was occasionally encountered – where several pots exist with similar shape, but are apparently otherwise unrelated. The vast majority of the material treated under the heading 'Lugged bowls and pails' belongs to the earlier TRB phases (i.e. Horizon 3). Pail No. 133, however, belongs to a much later stage of the sequence. An alternative approach, by established groups or phases, was also considered but rejected. Recent work by Voss (1982), Taayke (1985: pp. 131-132), and Brindley (1986) has shown that some modifications to the existing typochronology (Bakker, 1979) are necessary.

3.5.2. *Catalogue*

3.5.2.1. *Contents*

Funnel beakers	1-102
Funnel beaker cups	103-110
Lugged beakers	111-112
Lugged bowls and pails	113-133
Non-lugged decorated bowls	134-175
Non-lugged undecorated bowls	176-212
Necked bowls	213-235
Cordoned bowls	236-237
Jugs	238-241
Tureens	242-281
Amphorae	282-325
Undecorated lugged/ non-lugged shouldered vases	326-348
Collared flasks	349-359
Biberons	360-361
Miscellaneous	362-366
Total	366
Decorated rims	26
Body sherds	14
Bases	78

3.5.2.2. *Descriptions*

Funnel beakers

1. Almost complete. Flaring neck and rounded shoulder. Vertical lines.
2. Incomplete. Body with rounded profile. Vertical lines.
3. Incomplete. Body with rounded profile. Vertical lines.
4. Incomplete. Flaring rim and rounded body. Fine vertical lines on body.
5. Almost complete profile. Fairly straight, plain neck and short shoulder. Vertical lines. Poorly fired.
6. Incomplete. Plain neck. Vertical lines on body.

	TYPE	HORIZON
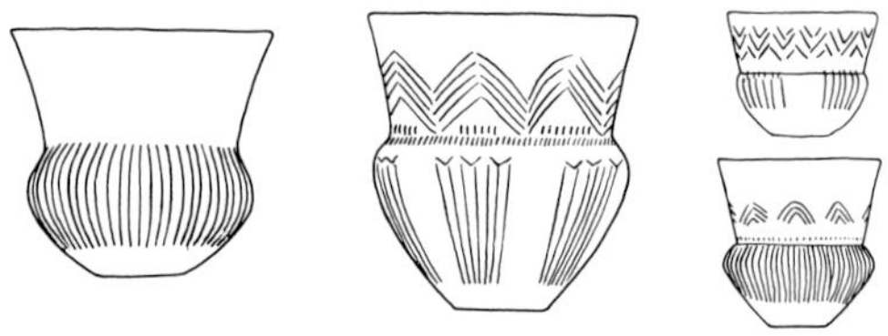	FUNNEL BEAKER	1–5
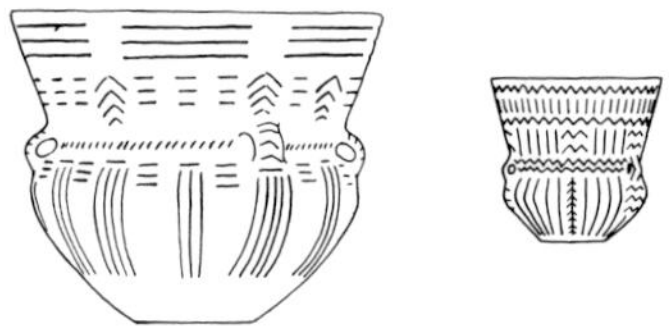	LUGGED BEAKER	1–4
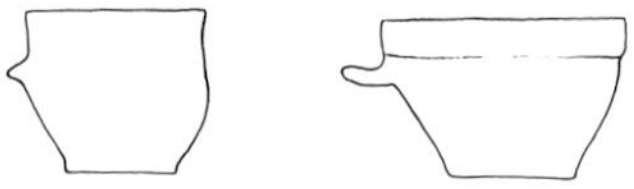	FUNNEL BEAKER CUP	5–7
	JUG	1–2
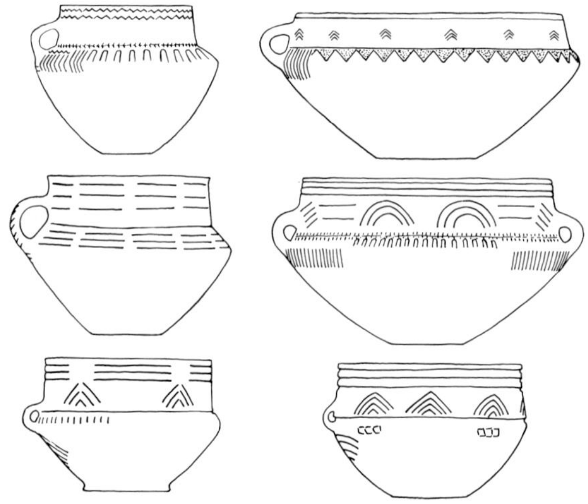	TUREEN	3–4
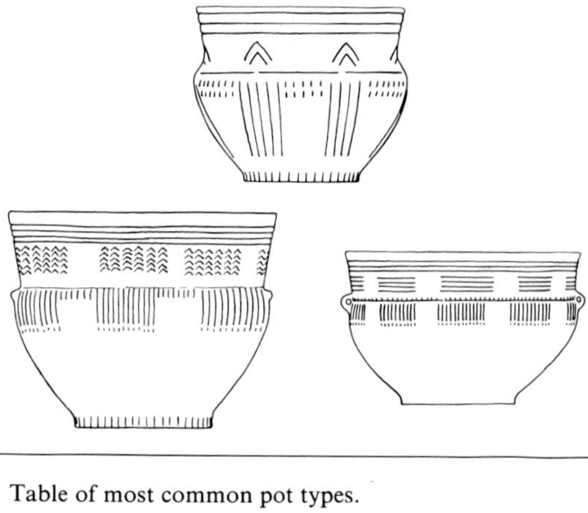	TUREEN–AMPHORA	4–5

Fig. 15. Table of most common pot types.

	TYPE	HORIZON
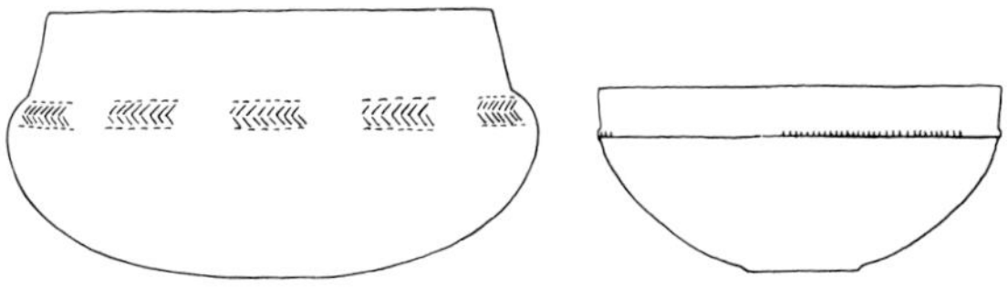	NECKED BOWL	7
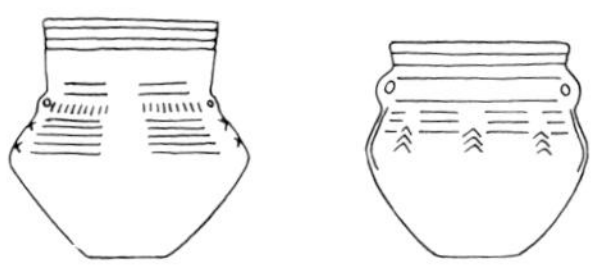	AMPHORA TYPE 1	3–4
	AMPHORA TYPE 2	4–5
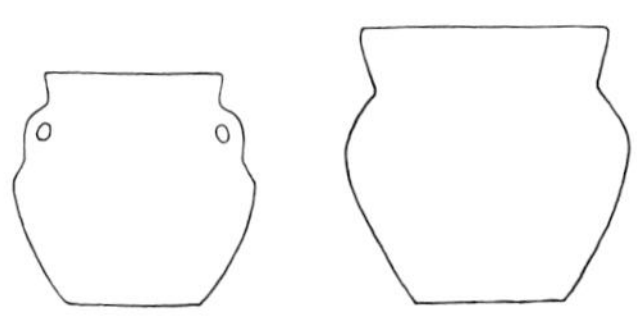	LUGGED OR NON-LUGGED SHOULDERED VASE	2–7
	LUGGED BOWL	1–3
	PAIL	2–4
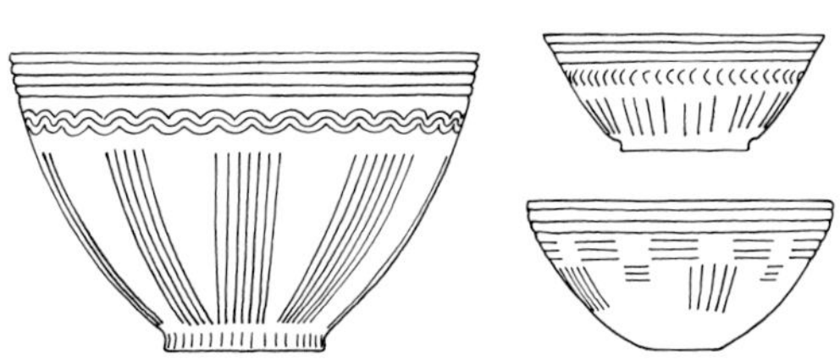	NON-LUGGED BOWL	decorated 3–5 undecorated 2–7
	COLLARED FLASK	1–7
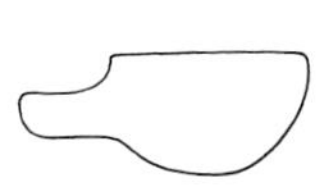	BIBERON	2–4

Fig. 15 (cont.).

DECORATION	
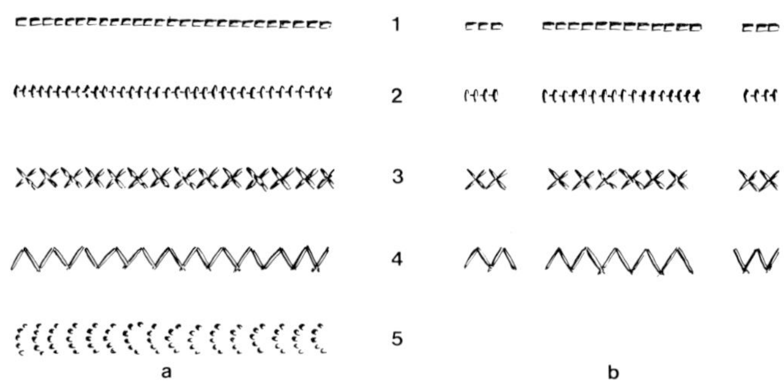	1. tiefstich 2. tvaerstik 3. incised lozenge 4. zigzag 5. C motif a. continuous horizontal line b. discontinuous horizontal line
	horizontal band of verticals (upper zone) a. continuous b. discontinuous
	large multiple zigzag multiple zigzag close set multiple zigzag
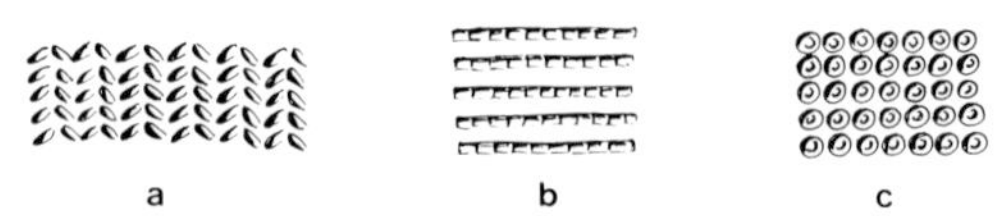	horizontal blocks a. zigzag b. tiefstich c. hollow stamp
	vertical blocks
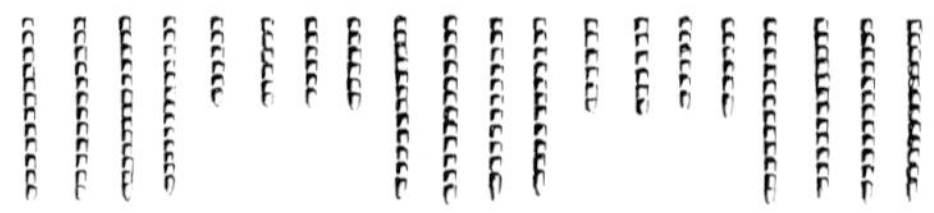	alternating long and short vertical blocks
	continuous verticals

Fig. 16. Table of main decorative elements, layout and applied features.

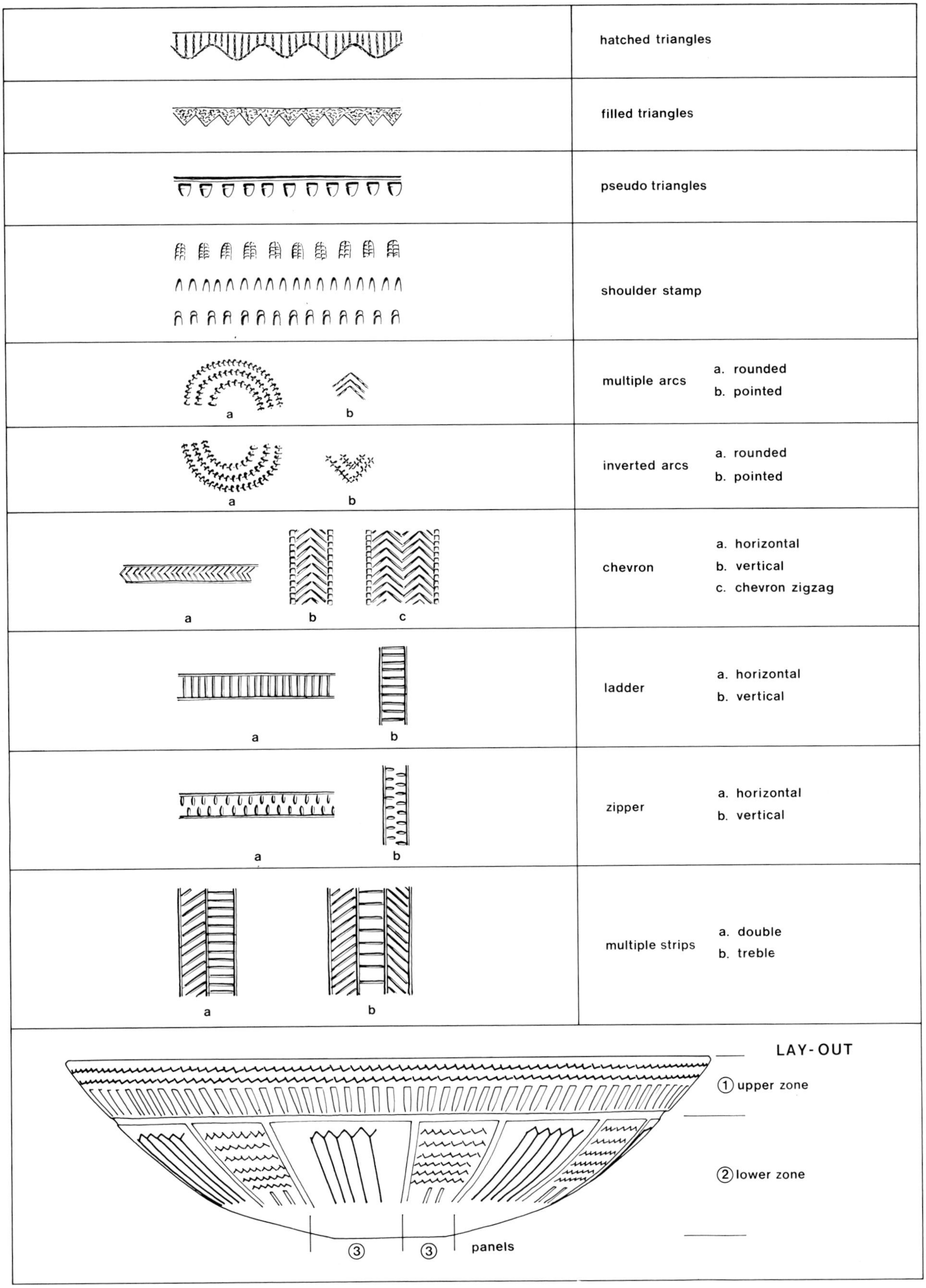
hatched triangles
filled triangles
pseudo triangles
shoulder stamp
multiple arcs
a. rounded
b. pointed
inverted arcs
a. rounded
b. pointed
chevron
a. horizontal
b. vertical
c. chevron zigzag
ladder
a. horizontal
b. vertical
zipper
a. horizontal
b. vertical
multiple strips
a. double
b. treble
LAY-OUT
① upper zone
② lower zone
③ ③ panels

Fig. 16 (cont.).

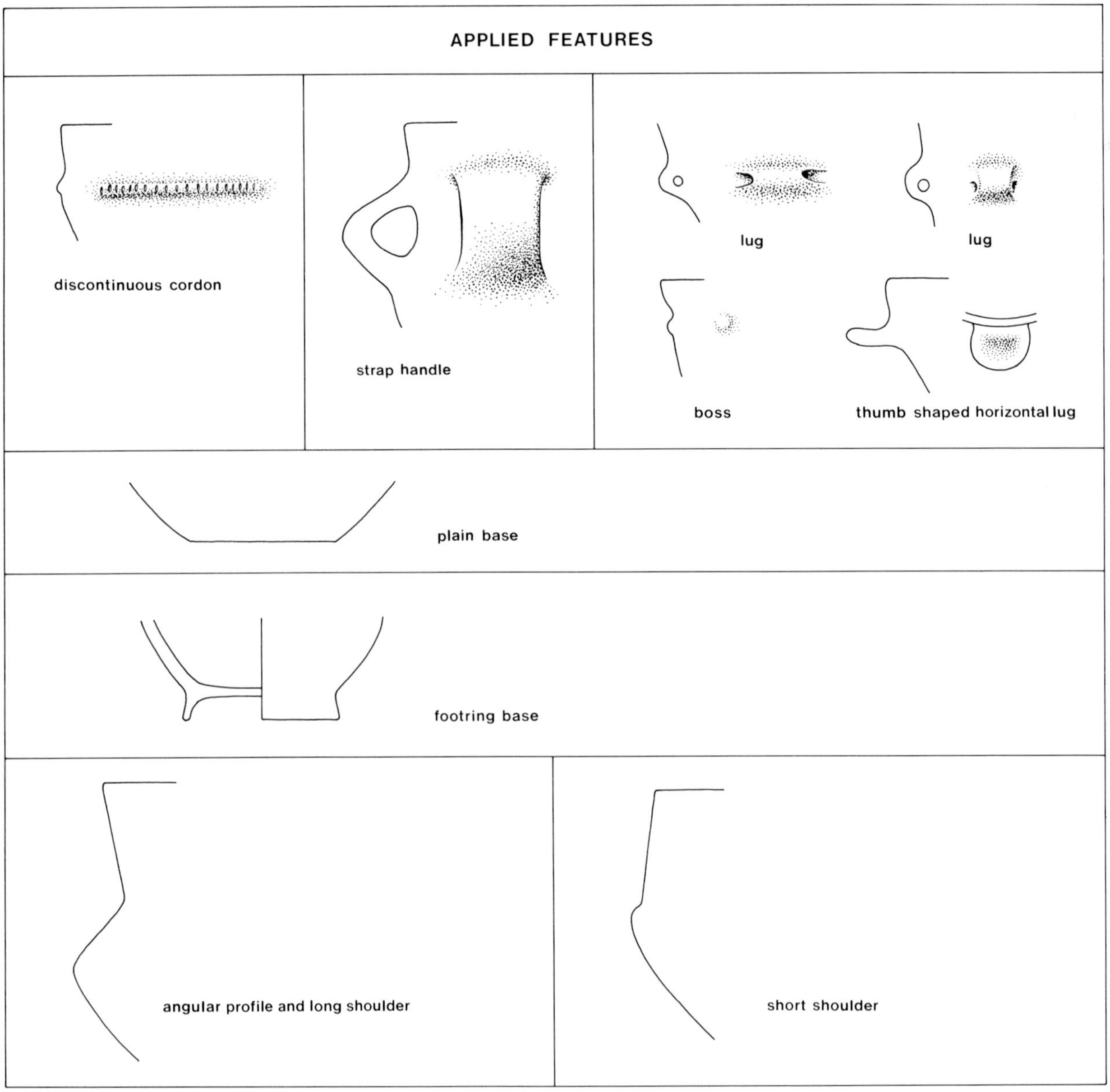
APPLIED FEATURES
discontinuous cordon
strap handle
lug
lug
boss
thumb shaped horizontal lug
plain base
footring base
angular profile and long shoulder
short shoulder

Fig. 16 (cont.).

7. Incomplete. Slightly flaring neck and almost unmarked shoulder. Verticals on body.

8. Incomplete. Cylindrical neck and rounded belly. Vertical lines on body.

9. Incomplete. Flaring rim. Verticals on body.

10. Incomplete. Rounded belly. Lightly scratched vertical lines on body.

11. Incomplete. Longish shoulder. Vertical lines.

12. Complete profile. Flaring neck and fairly short, rounded shoulder. 2 lines of *tvaerstik* at base of neck, vertical *Tiefstich* lines on body.

13. Fragmentary. Plain neck, vertical lines on body.

14. Incomplete. Plain neck. Vertical lines on body.

15. Complete. Plain neck. Vertical lines on body.

16. Almost complete. Flaring rim and short, rounded shoulder. Vertical lines on body.

17. Incomplete. Body sherds only. Vertical lines on body.

18. Incomplete. Short rounded shoulder. Stab marks at base of neck. Verticals on body.

19. Complete profile. Small, thin-walled beaker. Verticals on small body.

20. Almost complete. Long flaring neck and fairly short shoulder. Vertical lines on body.

21. Almost complete. Flaring neck and small, rounded shoulder. Vertical lines on body.

22. Incomplete. Rounded shoulder. Vertical lines on body.

23. Almost complete profile. Flaring neck and small, rounded shoulder. 2 *tvaerstik* lines at base of neck, vertical *Tiefstich* on body.

24. Incomplete. Similar to No. 23 but smaller.

25. Incomplete. Plain neck. Vertical lines on body.

26. Almost complete profile. Flaring neck and short, rounded shoulder. 3 lines on shoulder, vertical blocks on body.

27. Incomplete. Flaring neck and very slight shoulder. Vertical lines on body.

28. Fragmentary. Flaring neck. Small shoulder. Vertical lines on body.

29. Fragmentary. Vertical lines on body.

30. Incomplete. Flaring undecorated neck, verticals on body.

31a. Fragmentary. *Tvaerstik* line at base of neck, vertical lines below.

31b. Fragmentary. Not part of 31a or 31c.

31c. Fragmentary. Not part of 31a or 31b.

32. Fragmentary. Short shoulder. Vertical lines.

33. Fragmentary. Flaring neck, horizontal line at base of neck, vertical lines on body.

34. Incomplete. Body sherds only. Vertical lines.

35. Incomplete. Body sherds only. Vertical lines.

36. Incomplete. Cylindrical neck and sloping shoulder. Alternating blocks of vertical lines and stabs on body.

37. Complete profile. Cylindrical neck and almost imperceptable shoulder. Vertical lines on body in blocks alternating in length.

38. Complete. From the burial pit northeast of Sl 2. Plain neck, alternating panels of vertical lines and short fringe on body.

39. Almost complete profile. Flaring neck and short shoulder. Vertical blocks of alternating length on body.

40. Incomplete. Thick walled with rounded shoulder. Two lines at base of neck. Vertical blocks on body.

41. Incomplete. Body sherds only. Vertical blocks.

42. Fragment. Vertical blocks on body.

43. Incomplete. Short, flaring neck and gently curving profile. Fringe along shoulder.

44. Almost complete profile. Small, thick-walled beaker with rounded shoulder. Panels of vertical lines on body.

45. Incomplete. Long-neck, short shoulder. Large, double zigzag on neck, double line of very small zigzag on shoulder, vertical lines on body.

46. Fragment. Portion of large multiple zigzag. See No. 48.

47. Incomplete. Flaring neck and slight, rounded shoulder. Large treble zigzag on neck, vertical lines on body.

48. Complete profile. Flaring neck and small, rounded shoulder. Four large zigzags on neck, stab marks at base of neck, long tapering blocks on body. Horizon 5.

49. Incomplete. Short necked, wide funnel beaker with double zigzag below rim and fringe on small, rounded shoulder.

50. Incomplete. Flaring neck. Double zigzag below rim, 3 lines at base of neck and lines on body.

51. Almost complete profile. Slightly flaring neck, and very slight shoulder. Double zigzag below rim, line of small *tvaerstik* at base of neck. Horizon 5?

52. Incomplete. Short, cylindrical neck. Zigzag below rim, fringe on shoulder. Horizon 5?

53. Incomplete. Neck sherds only. Zigzag at base of neck, vertical lines on body.

54. Incomplete. Small, wide beaker with short, flaring neck. Zigzag at base of neck, vertical lines on body.

55. Complete profile. Flaring neck and short sloping shoulder. Double zigzag at base of neck, verticals on body.

56. Almost complete profile. Relatively straight neck, and very slight shoulder. Zigzag at base of neck, verticals on body.

57. Complete profile. Flaring neck and high well marked short shoulder. Double zigzag on lower neck. Verticals on body. Service set with tureen No. 249. Horizon 3.

58. Complete profile. Flaring neck and short, rounded shoulder. Zigzag at base of neck, verticals on body.

59. Complete profile. Flaring rim and slight rounded shoulder. Double zigzag on neck, broad *Tiefstich* line at base of neck and *Tiefstich* verticals on body.

60. Fragmentary. Body sherds only. Zigzag (?) at base of neck. Vertical lines on body.

61. Complete profile. Thick walled; sharp shoulder, slightly flaring neck. Zigzag at base of neck, line and fringe on shoulder, vertical blocks on body.

62. Incomplete. Flaring neck and short, rounded shoulder. Zigzag at base of neck, verticals on body.

63. Incomplete. Short shoulder; zigzag at base of neck, horizontal line on shoulder with blocks of pendant *Tiefstich* and *tvaerstik*.

64. Incomplete. Small, rather wide funnel beaker. Multiple zigzag related pattern on neck, vertical panels on body. Amphora-like funnel beaker. Horizon 5.

65. Fragmentary. Rim with multiple arcs on neck. Fine, well-finished thin fabric. Horizon 4.

66. Almost complete profile. Flaring neck and short, sharp shoulder. Multiple arcs on neck, verticals on body. Horizon 4.

67. Almost complete profile. Slightly flaring neck and very slight shoulder. Double arcs on neck. Horizontal line at base of neck, vertical blocks on body. Horizon 4.

68. Long, flaring neck, long shoulder and sharp angles between neck, shoulder and body. Plain neck, stab marks at junction of neck and shoulder, neatly executed lines of stab marks along shoulder, and vertical *Tiefstich* lines on body.

69. Incomplete. Very flaring neck and short, round shoulder. Single line of deeply impressed *Tiefstich* along base of neck. See Emmeln No. 114.

70. Incomplete. Flaring neck. Stab marks along shoulder, vertical lines on body. Decoration is very faint and may represent guidelines for ornament never executed.

71. Complete profile. Slightly flaring neck, and slight shoulder. Very faint lines on body may represent guidelines for ornament never executed.

72. Incomplete. Weak profile. Line of small pits below rim.

73. Incomplete. Slightly conical neck and weak shoulder.

74. Incomplete. Angular profile, vertical neck and flaring rim tip.

75. Incomplete. Undecorated neck.

76. Fragment. Flaring neck.

77. Incomplete. Slightly flaring neck, d. 17.9 cm.

78. Incomplete. Short, horizontal shoulder and slightly rounded body.

79. Incomplete. Neck, d. 13.5 cm.

80. Incomplete. Slightly flaring, straight neck and short,

slightly rounded shoulder.

81. Incomplete. Body sherds only. Short ledge-like shoulder.

82. Almost complete. Thick-walled beaker with weak profile.

83. Incomplete. Unaccentuated profile.

84. Incomplete. Slightly conical neck and weak shoulder.

85. Incomplete. Straight neck and weak shoulder.

86. Complete profile. Wide, short necked, thick-based funnel beaker with conical neck and weak shoulder.

87. Incomplete. Wide, short necked funnel beaker.

88. Fragmentary. Neck sherds only.

89. Fragmentary. Neck sherds only.

90. Almost complete. Crudely made funnel beaker. Outer surface is covered with vertical scratch marks.

91. Small funnel beaker with weak profile.

92. Incomplete. Neck sherds only.

93. Fragmentary. Cylindrical neck with thin, protruding ledge rim.

94. Incomplete. Small, thin-walled funnel beaker body.

95. Fragmentary. Thin protruding ledge rim.

96. Fragmentary. Thin, protruding ledge rim.

97. Incomplete. Very large, thick-walled, well-finished beaker with cylindrical neck and rounded shoulder, neck d. 22 cm.

98. Incomplete. Large, poorly fired funnel beaker, neck d. 16 cm.

99. Incomplete. Large funnel beaker of well baked coarse fabric, neck d. 19.6 cm.

100. Incomplete. Thick-walled beaker. Well finished. Neck d. c. 15.7 cm.

101. Incomplete. Thick-walled beaker. Well finished.

102. Complete. Slack profiled funnel beaker. Found outside the chamber.

Funnel beaker cups

103. Complete. Slightly 'S'-shaped cup with narrow, horizontal, unpierced lug.

104. Almost complete. Horizontal groove marking junction of neck and body. Narrow, horizontal, unpierced lug.

105. Complete profile. Conical bowl with short neck and vestigal shoulder visible in the slight increase in thickness of body wall.

106. Incomplete. Thin-walled with very slight, short neck and very slight almost unmarked shoulder. Narrow, horizontal, unpierced lug.

107. Incomplete. Relatively flaring very short neck.

108. Almost complete profile. Fairly thick-walled conical bowl with black, burnished leathery outer surface. Slight inward curve at base of neck. Negative of narrow, horizontal, unpierced lug.

109. Incomplete. Fairly thick-walled conical bowl with black, burnished leathery outer surface. Slightly flaring neck. Similar to No. 108 except in thickness and shape of rim tip.

110. Sherds of three narrow, horizontal, unpierced lugs from vessels similar to Nos 103-109.

Lugged beakers

111. Incomplete; neck sherds only. Line of pseudo-*tvaerstik* below rim, band of verticals in upper zone, lower zone filled with defined panels of alternating motifs, multiple zigzag, verticals and zipper. Between upper and lower zones and at base of the neck are straight grooved lines. Four small lugs span the junction of the neck and shoulder. Horizon 3.

112. Incomplete. Four small lugs at base of neck. *Tiefstich.* Horizontal blocks below rim, alternating horizontal blocks and multiple arcs on neck, line at junction of neck and body, and alternating long, vertical and short horizontal blocks on body. Horizon 4.

Lugged bowls

113. Almost complete profile. Horizontally pierced, decorated lugs within upper decorative zone. Upper zone defined by zigzag but otherwise 'empty'. Lower zone includes panels with zigzag and verticals. Internal rim decoration of single zigzag line is unusual, but not chronologically significant. Horizon 2.

114. Two rim sherds showing verticals between lines of zigzag. Horizon 3.

115. Almost complete profile. Horizontally pierced lugs within upper decorative zone. Upper band of verticals defined by single lines of zigzag. Lower zone of alternating broad panels of verticals and narrow panels of zigzag motif. Horizon 3.

116. Incomplete. Horizontal 'sausage lug' at junction of upper and lower decorative zones. Line of *hartstempel* (Bakker, 1979: p. 179) motif below rim; upper zone of vertical lines separated from lower zone of vertical lines by line of *hartstempel* motif. Horizon 3.

117. Almost complete profile. Horizontally pierced lugs within upper decorative zone. Upper zone of incipient groups of verticals between lines of *hartstempel* motif. Lower zone also shows incipient spacing of verticals within the panels and zigzags. Some inlay visible. Horizon 3.

118. Incomplete. Horizontally pierced, decorated lug at junction of upper and lower zones. Upper zone of verticals between horizontal lines of 'X' (crude skating) motif. Lower zone of panels of verticals interspersed with groups of zigzag motif. Lug is decorated with vertical chevron and ladder. Horizon 3.

119. Complete profile. No horizontal line below rim; zone of vertical lines above zigzag line. Lower zone of alternating undefined panels of verticals and zigzags. Late Horizon 2, or, more likely, early Horizon 3 – incomplete upper zone; panel decoration compares well with that on handle of late Horizon 2 jug No. 238.

120. Almost complete profile. Unperforated lug within upper zone. Zigzag line below rim; upper zone of verticals separated from lower zone by small gap. Lower zone of alternating groups of long and short verticals. Absence of zigzag in lower zone is unusual. Horizon 3.

121. Complete profile. Double line of zigzag below rim. Upper zone of verticals. Upper and lower zones and panels in lower zone separated by broad *Tiefstich* line. Lower zone of alternating panels of multiple zigzags and verticals. Verticals are pendant from a single line of zigzag giving a 'Y' effect. Late Horizon 3 – vertical lines are well spaced within panel.

122. Almost complete profile. Unperforated, decorated lugs with upper decorative zone. Double zigzag line below rim; upper zone of verticals. Upper and lower zones and panels of lower zone separated by broad line. Lower zone of panels of verticals and zigzags. Horizon 3.

123. Incomplete. Single line of zigzag related motif below rim. Upper zone of verticals separated from lower zone by broad, straight line also used to define panels of zigzag from verticals making up lower zone decoration. Horizon 3.

124. Fragmentary. Double zigzag line below rim. Upper zone of verticals. Upper and lower zones are probably separated by zigzag line. Lower zone includes some zipper motif. Apart from the lugged beaker, No. 111, this is the only example of zipper motif in the Glimmen assemblage. It is not certain that the rim and body sherds represent a single pot. Horizon 3?

125. Incomplete. Band of vertical lines below rim. May have had zigzag line beneath. Horizon 3?

126. Incomplete. Maggot zigzag above zone of vertical lines. This is the only example of this decorative technique in the Glimmen assemblage. Maggot impressions belong chiefly to Horizons 1 and 2. The apparently relatively narrow zigzag panels together with the maggot impressions may indicate late Horizon 2.

127. Incomplete. Horizontal, decorated unpierced lugs on lower borderline of upper zone. Decoration is executed almost entirely in *tvaerstik*. 3 horizontal lines below rim and above very narrow band of short vertical stabs above fourth line. Lower zone consists of widely spaced panels of 'Y' motif, those below the lugs being bordered by single vertical *tvaerstik* lines. Lugs are decorated with simple vertical lines. Horizon 4.

128. Almost complete profile. Bowl with small, low, unperforated lugs. *Tvaerstik.* 3 lines below rim, zone of short,

verticals above zone of narrow spineless herringbone interrupted by 4 lugs. Below this, decoration breaks into vertical panels of chevron and *tvaerstik* below the lugs and short horizontal panels in between. Extensive traces of paste inlay preserved. Horizon 4.

Pails

129. Incomplete. Horizontally perforated lugs astride upper and lower zones. Upper and lower zones and panels of lower zone are separated by broad straight lines. Line of skating motif below rim. Upper zone is filled with verticals, lower with alternating panels of verticals and zigzag. Vertical line on lug. Horizon 3.

130. Complete profile. Horizontally perforated lug at junction of upper and lower zones. *Tiefstich* and *tvaerstik*. Upper and lower zones are separated by broad straight line. Similar lines divide both zones into panels. Horizontal lines of *tvaerstik*; remainder of decoration, *Tiefstich*. Upper zone of band of verticals, lower zone with alternating panels of vertical lines beneath two horizontal lines of *tvaerstik* and zigzags. Lugs are decorated with single vertical lines. Service set with tureen No. 259. Horizon 3.

131. Fragmentary. Vertical lines. Very worn.

132. Incomplete. Fairly straight-walled bowl with at least 2 horizontally pierced plugs. *Tvaerstik*. 3 lines below rim, empty zone, followed by at least 3 more lines. Lugs, of which one is preserved and portion of second, occur along the lower band of lines and are decorated with simple vertical lines. Chevron panel below lugs. Horizon 4.

133. Complete. Straight-sided pail with four horizontal pinched out lugs and six pressed out feet. One or two horizontal *Tiefstich* lines below rim; panels of close-set vertical *Tiefstich* ending in 'fringe'. Horizon 6.

Non-lugged decorated bowls

134. Complete profile. Small, fairly straight walled bowl. *Tiefstich*. Zigzag line below rim, verticals covering entire body. Traces of paste inlay visible. Horizon 3.

135. Complete profile. Fairly straight walled bowl. *Tiefstich*. Upper and lower zones defined by two horizontal lines, zones filled with verticals. Horizon 3.

136. Almost complete profile. *Tiefstich* and *tvaerstik*. 2 discontinuous lines of *tvaerstik* below rim, upper zone of grouped vertical lines, separated from lower zone of grouped verticals by horizontal line. Late Horizon 3?

137. Incomplete. *Tiefstich*. Two discontinuous lines below rim, alternating panels of verticals, crude zigzags and crude zipper (?) motif. Late Horizon 3?

138. Incomplete, Large, coarse bowl. *Tiefstich*. 3 discontinuous lines below rim, alternating panels of single, pointed arcs and verticals above three more horizontal lines from which hang pendant groups of 3 verticals alternating with paired verticals. Late Horizon 3?

139, 140. Two similar bowls, one larger than the other. *Tiefstich* and *tvaerstik*. 3 lines of *tvaerstik* below rim, zone of short vertical lines above panels of longer vertical lines. Horizon 4.

141. Complete profile. Bowl with footring. Coarse *Tiefstich*. 3 lines below rim, line of 'C' motif, vertical panels terminating at base. Traces of paste inlay visible. Horizon 4.

142. Almost complete profile. *Tvaerstik*. 4 lines below rim, panels of four verticals on body, finger nail impressions around base. Horizon 4.

143. Complete profile. Small bowl. *Tiefstich*. 2 lines below rim, pendant panels of three and four verticals. Small footring. Horizon 4?

144. Incomplete. *Tiefstich*. 4 lines below rim, vertical panels on body. Horizon 4?

145. Almost complete profile. Bowl with footring (now missing). *Tvaerstik*. 3 lines below rim, followed by two discontinuous lines. Panels of verticals not unlike 'Y' motif on body. Traces of paste inlay visible. Service set with Nos 146 and 254. Horizon 4.

146. Complete profile. Small bowl with footring. Decoration similar to No. 145. Service set with Nos 145 and 254. Horizon 4.

147. Incomplete. *Tiefstich*. 3 lines below rim, alternating panels of verticals and horizontals. Horizon 4?

148. Almost complete profile. Small, crude bowl. *Tiefstich*. Single line below rim followed by alternating panels of two long horizontal and three short vertical lines. Horizon 4?

149. Incomplete. *Tiefstich* and *tvaerstik*. 3 lines of *tvaerstik* below rim and above alternating panels of 'Y' motif and 2 lines of horizontal *tvaerstik*. Below the latter are panels of 'Y' motif. Horizon 4.

150. Incomplete. *Tiefstich*. Decoration similar to above but without lower zone. Horizon 4.

151. Complete profile. Bucket-shaped bowl. *Tiefstich*. Panels of horizontal lines below rim, vertical panels on body. Horizon 4.

152. Almost complete profile. *Tvaerstik*. 3 lines below rim, followed by zone of horizontal blocks, with lower zone of alternating long and short blocks. Traces of paste inlay visible. Horizon 4.

153. Incomplete. Fairly straight-walled bowl. *Tvaerstik*. 4 continuous and 2 discontinuous lines below rim; large, double line of zigzag with pendant spaced blocks of verticals, possibly terminating at base. Horizon 4.

154. Almost complete profile. *Tiefstich*. 3 lines below rim, 3 horizontal blocks immediately below these, followed by 2 zones of alternating long and short, vertical blocks. Some paste inlay visible. Horizon 4.

155. Incomplete. *Tvaerstik*. 3 lines below rim, 2 zones of horizontal blocks. Horizon 4.

156. Incomplete. *Tiefstich* and *tvaerstik*. 2-3 lines below rim, alternating verticals and 'M' type motif. Horizon 4.

157. Fragmentary. Deeply impressed *tvaerstik*. Badly worn. Horizon 4.

158. Incomplete. Bowl with thick walls. Deeply impressed *tvaerstik* lines below rim. No sherds of lower part of body. Horizon 4.

159. Fragmentary. *Tvaerstik*. Badly worn. Possibly large zigzag.

160. Incomplete. *Tvaerstik*. 3 lines below rim, line of lozenge motif, alternating blocks of wide inverted multiple arcs and horizontal lines. Horizon 4.

161. Complete profile. Bowl with footring. *Tvaerstik*. Horizontal blocks below rim, discontinuous line of lozenge motif, more horizontal blocks above vertical blocks which terminate close to base. Horizon 4.

162. Fragmentary. *Tvaerstik*. 4 lines beneath rim, alternating vertical and horizontal blocks below. Trace of paste inlay visible. Late Horizon 4.

163. Fragmentary. *Tiefstich* lines below rim.

164. Fragmentary. Bowl with *tvaerstik* below rim.

165. Fragmentary. Thick-walled bowl with 2 lines of *tvaerstik* below rim. Band of verticals?

166. Complete profile. Large bowl with footring. Broad *Tiefstich*. 4 lines below rim, panels of zigzag above and separated from panels of verticals terminating at base of horizontal zigzag. Groups of short verticals on footring. Late Horizon 4 or early Horizon 5.

167. Almost complete profile. Large bowl. *Tiefstich* and *tvaerstik*. 4 lines below rim, with line of skating motif immediately below. 3 lines of large zigzag, above vertical blocks of alternating *Tiefstich* and *tvaerstik*. No sherds of base present but it is very likely that this bowl had a footring. Late Horizon 4 or early Horizon 5.

168. Complete profile. Large bowl. *Tiefstich* and *tvaerstik*. 3 lines *Tiefstich* below rim, triple line of zigzag above vertical blocks of alternating lines of *Tiefstich* and *tvaerstik*. Vertical stabs on footring. Late Horizon 4 or early Horizon 5.

169. Fragmentary. *Tiefstich*. 3 lines below rim; blocks below. Horizon 5.

170. Fragmentary. *Tiefstich*. Horizontal lines under rim. Various motifs. Similar to No. 169. Horizon 5?

171. Almost complete profile. *Tiefstich*. At least 3 lines below rim and above 4 zigzag lines; blocks of verticals on lower body. Rather thick base. Horizon 5.
172. Almost complete profile. Bowl with incurving rim. *Tiefstich*. 3 lines below rim followed by double line of zigzag above alternating panels of verticals and pointed arcs. Horizon 5.
173. Very small bowl with faint line below rim.
174. Incomplete. Horizontal line with groups of short, pendant stabs at intervals. Horizon 7.
175. Incomplete. Large, thick walled shallow dish. Smoothed walls. Horizontal line with fringe below rim. Horizon 7.

Non-lugged undecorated bowls
176. Incomplete bowl with low footring. Fabric and general quality suggest Horizon 5/6.
177. Incomplete. Undecorated bowl.
178. Incomplete. Undecorated bowl.
179. Incomplete. Broad hemispherical bowl of heavy fabric.
180. Complete profile of thick-walled bowl. Original perforation in the centre of base.
181. Almost complete profile of bowl of friable, gritty fabric.
182. Lower part and base of large dish/bowl.
183. Incomplete. Bowl similar to No. 181 but less friable.
184. Complete profile of bowl. Similar to No. 181 but less friable.
185. Complete profile of bowl. Similar to No. 181 but less friable.
186. Complete profile of finely made hemispherical bowl. Black fabric with few visible grits.
187. Incomplete profile of well-finished bowl. Orange fabric with small grits.
188. Almost complete bowl. Orange fabric with black core. Uneven, although smoothed, surface.
189. Almost complete profile. Bowl.
190. Incomplete small, strongly built bowl. Mottled grey and pink hard fabric with small grits.
191. Incomplete bowl. Similar to No. 190 but smaller.
192. Incomplete small bowl. Black-grey fabric with pink, crushed granite grit.
193. Incomplete. Bowl of fairly hard, beige fabric with grey core and small grits.
194. Incomplete bowl. Friable black fabric with medium-sized grits. Slight weak shoulder.
195. Incomplete bowl. Yellow fabric with black core and medium-sized grits.
196. Incomplete thin walled, uneven surfaced bowl with medium-sized grits.
197. Almost complete profile of heavy-walled bowl. Similar to No. 202 but with sandy fabric.
198. Complete. Single find from outside the chamber.
199. Complete. Single find from outside the chamber.
200. Open thick-walled dish. Well smoothed fabric.
201. Base of open thick-walled dish.
202. Incomplete bowl. Similar to No. 196 but of reddish, slightly thicker fabric.
203. Incomplete small bowl.
204. Almost complete profile of small, crude bowl.
205. Incomplete very small bowl with inturned rim.
206. Incomplete small bowl.
207. Incomplete small bowl.
208. Incomplete small, smoothly finished bowl.
209. Incomplete bowl.
210. Complete profile of tall bowl. Fine, thin fabric.
211. Incomplete small bowl.
212. Incomplete small bowl.

Necked bowls
213. Incomplete. Short neck and slight, rounded shoulder. Blocks of small stabs on shoulder. Horizon 7.
214a, b. Two shoulder sherds, from one or two shouldered bowls with blocks of small stabs on shoulder. Horizon 7.
215, 216. Incomplete. Two bowls with fairly long neck and rounded shoulder. Short blocks of vertical zigzag on shoulder. On No. 216 these are relatively deeply impressed, on No. 215 lightly scored. Horizon 7.
217. Fragmentary. Slashed cordon. Horizon 7.
218. Incomplete. Long neck and slashed cordon at base of neck. Horizon 7.
219. Incomplete. Long neck and slight, rounded shoulder, emphasised at intervals by low, slashed cordon. Horizon 7.
220. Incomplete. Fairly short neck and body curving in from base of neck; this junction is emphasised by low, slashed cordon. Slightly coarse fabric in comparison with some of the bowls in this series. Horizon 7.
221. Incomplete. Fairly short neck and small ledge-like shoulder emphasised at intervals by small, slashed cordons. Horizon 7.
222. Incomplete. Short neck and slightly rounded shoulder. Irregular line at base of neck. Small, low, slashed cordons at intervals on shoulder. Horizon 7.
223. Incomplete. Short neck, body curving in from base of neck; this junction is marked at intervals by a low, slashed cordon. Horizon 7.
224. Incomplete. Fairly short neck and very slight shoulder, emphasised at intervals by low, slashed cordon. Horizon 7.
225. Incomplete. Neck of medium length, slight shoulder emphasised by hyphenated line with pendant fringe. Horizon 7.
226. Incomplete. Short neck and small, ledge-like shoulder emphasised at intervals by slashed cordons. Horizon 7.
227. Incomplete. Short, slightly curved neck and small but clearly profiled shoulder with short lengths of slashed cordon most of which has fallen off. Horizon 7.
228. Fragmentary. Short neck, slight shoulder; junction of which is marked by grooved line. Horizon 7.
229. Incomplete. Neck of medium length and long body curving directly from base of neck. Horizon 7.
230. Fragmentary. Large bowl. Thin, very worn sherds. Horizon 7.
231. Incomplete. Fairly short neck and curved body. Friable fabric. Horizon 7.
232. Fragmentary. Small rim sherd. Horizon 7.
233. Fragmentary. Rim sherd from shouldered bowl with long neck. Horizon 7.
234. Incomplete. Fairly long neck and rounded shoulder. Irregular line at base of neck. Horizon 7.
235. Incomplete. Fairly short neck and small, ledge-like shoulder. Horizon 7.

Cordoned bowls
236. Almost complete. Large, shallow, thick-walled dish with slashed cordons at intervals close to rim, and slashed base. Smoothed walls. Horizon 7.
237. Almost complete profile. Shallow bowl with slashed cordon close to rim. Horizon 7.

Jugs
238. Complete profile. Jug of coarse fabric with slightly curved shoulder. Zigzag line below rim, otherwise plain neck, alternating panels of verticals and zigzag on shoulder and handle. Late Horizon 2. Earliest Horizon on site, also represented by Nos 113, 126, 239, 240.
239. Incomplete. Fairly coarse jug. Zigzag line below rim. Horizon 2.
240. Incomplete. Undecorated neck of coarse jug. Horizon 2.
241. Coarsely made handle of an otherwise unrecognised jug. Possibly Horizon 2.

Tureens
242. Fragmentary. Shoulder of tureen. Lozenge. Horizon 3/4.
243. Fragmentary. 2 shoulder sherds of a tureen with a sharply angled profile and shoulder decorated with panels of hurdle motif, zigzag and vertical lines. Horizon 3.
244. Incomplete. Zigzag line below rim with groups of vertical

lines on neck, possibly in vicinity of handle. Triangles pendant from broad straight line on shoulder. Horizon 3.

245. Incomplete. 3 lines of zigzag below rim, groups of verticals on either side of handle, filled triangles pendant from broad straight line on the shoulder. Handle is decorated with zigzag. Horizon 3.

246. Complete profile. Double zigzag line below rim, *tvaerstik* line at junction of shoulder and neck. Fine *Tiefstich* lines in vicinity of handle, pseudo-triangles on remainder of shoulder. Decoration on handle comparable to Issendorf (Tempel, 1972: fig. 4). Horizon 3.

247. Complete profile. Spouted tureen. Two lines of large zigzag below rim and third on shoulder. Groups of short verticals descend from shoulder onto body. Inverted arc motif below position of handle. Horizon 3?

248. Incomplete. Small, finely made tureen. Skating motif below rim, spaced panels of three vertical lines on neck, possibly only in vicinity of handle. Broad straight line at base of neck and hurdle pattern on shoulder. Horizon 3.

249. Incomplete. Conical neck, long, slightly curved shoulder and curved body. Rim tip is slightly everted. Zigzag line below rim, scattered groups of vertical zigzags on neck, filled triangles pendant from straight line on shoulder with groups of vertical lines running over shoulder onto body in vicinity of (absent) handle. Decorated panel below handle. Service set with funnel beaker No. 57. Horizon 3.

250. Incomplete. Conical neck and rounded shoulder. Everted rim tip. Position of handle shown by negative. Decoration is carefully executed throughout. Band of herringbone below rim, small rusticated triangles pendant from a straight line on shoulder except in immediate vicinity of handle. At regular intervals along shoulder and descending onto body are panels of vertical chevrons and vertical lines pendant from zigzag, the number of verticals incorporated in the design increasing from one to three with distance from handle. Below handle the panel is flared. Horizon 3.

251. Incomplete. Conical neck and short, rounded shoulder. Line of *hartstempel* beneath rim and above groups of finely incised inverted pointed multiple arcs. Filled triangles pendant from line on shoulder and vertical lines on either side of handle. Handle is also decorated with vertical lines. Horizon 3.

252. Incomplete. Slightly flaring neck and short rounded shoulder. *Tvaerstik*. Two lines below rim. Inverted arcs on neck, line of *tvaerstik* at base of neck and stab marks on shoulder. Vertical *tvaerstik* below handle. Some paste inlay visible. Horizon 3.

253. Incomplete. Sharp shoulder. *Tvaerstik*. At least 3 lines beneath rim and some vertical lines on neck. Triangles on shoulder and vertical lines of double stamp on body. Horizon 3.

254. Incomplete. Conical neck and short, rounded shoulder. *Tiefstich* and *tvaerstik*. 3 lines below rim, long panels of sloping *Tiefstich* on neck probably on either side of (absent) handle, stab marks on shoulder. Some paste inlay visible. Service set with Nos 145, 146. Horizon 3.

255. Incomplete. Conical neck and short sharp shoulder. *Tiefstich* and *tvaerstik*. Two lines of *hartstempel* below rim. Pairs of *Tiefstich* and *tvaerstik* on neck, pseudo-triangles pendant from straight line on shoulder. Negative of handle. Horizon 3.

256. Complete profile. Short, sharp shoulder. Line of pseudo-*tvaerstik* (*Tiefstich* with short vertical strokes across it) below rim. Groups of vertical lines on neck, small pseudo-triangles pendant from straight line on shoulder and vertical pseudo-*tvaerstik* beneath handle descending onto body. Handle bears stab-like impressions. Horizon 3.

257. Almost complete profile. Slightly conical neck and short, rounded shoulder decorated with *tvaerstik* and quatrefoil impressions. Blocks of three horizontal *tvaerstik* lines below rim, small filled triangles pendant from straight line on shoulder and panels of vertical *tvaerstik* pendant from zigzag and quatrefoil impressions. Horizon 3/4.

258. Almost complete profile. Conical neck and short sloping shoulder. *Tiefstich* and *tvaerstik*. 3 lines of *tvaerstik* below rim, narrow blocks of 2 horizontal lines of *tvaerstik* on neck, filled triangles pendant from straight line on shoulder and panels of *Tiefstich* vertical lines and vertical chevrons descending from shoulder, similar to No. 250. Probable handle bears single line of *Tiefstich*. Horizon 3.

259. Almost complete profile. Conical neck short, sharp, slightly sloping shoulder and deep body. *Tvaerstik*. Narrow blocks of horizontal lines on neck, filled triangles pendant from line on shoulder. Service set with pail No. 130. Horizon 3.

260. Incomplete. Conical neck and very short, rounded shoulder. *Tvaerstik*. 3 lines below rim covering most of neck, line of *Tiefstich* at base of neck with pendant stab impressions on shoulder. Horizon 4.

261. Incomplete. Short, rounded shoulder. *Tvaerstik*. 3 lines below rim, lozenge motif on neck, stab marks at junction of neck and shoulder, lozenge motif on upper body. Some paste inlay visible. Horizon 4.

262. Incomplete. Conical neck and short, sharp shoulder. *Tvaerstik*. 3 lines below rim, groups of stab marks on neck, 'X' motif on shoulder. Horizon 3/4.

263. Almost complete profile. Large, thick-walled tureen. Slightly conical neck, short sloping shoulder. Alternating blocks of vertical chevron and horizontal lines on neck, short verticals on shoulder, and in immediate vicinity of base (which is missing). Below handle is long panel of decoration similar to the vertical neck panels. Horizon 4?

264a. Incomplete. Long vertical neck, short, sloping shoulder. *Tvaerstik*. 3 lines below rim, large double zigzag on neck, short verticals on shoulder, some verticals on body. Horizon 4.

264b. Fragment. Body sherd with decoration similar to No. 264a but different fabric.

265. Incomplete. Two handled tureen. Slightly conical neck and short, rounded shoulder. *Tvaerstik*. 3 lines below rim. Lozenge and multiple arcs on neck. Shoulder is decorated with various *Tiefstich* marks. On either side of handles groups of lines descend onto body. Handles decorated with vertical lines and chevron. Horizon 4.

266. Almost complete profile. Fairly long neck and very short shoulder. Very deep *Tiefstich*. 4 lines below rim, multiple arcs on neck, very deep line along shoulder and blocks of verticals on body. Multiple arc below handle. Traces of paste inlay visible. Horizon 4.

267. Incomplete. High, cylindrical neck and short, sharp, ridge-like shoulder and small lug handle. Very deeply impressed *Tiefstich*. 4 lines below rim, multiple arcs on neck. At base of neck deeply impressed 'pseudo-skating' line. Alternating multiple arcs and blocks of verticals on body. Horizontal lines on handle. Horizon 4.

268. Incomplete. Conical neck and short, rounded shoulder. *Tvaerstik*. 3 lines below rim, multiple arcs and short lines on neck, line at base of neck, multiple arcs and vertical lines (in vicinity of the handle) from shoulder onto body. Horizon 4.

269. Incomplete. Horizontally perforated, undecorated lug on shoulder. *Tiefstich*. 3 horizontal blocks below rim, arcs on neck, 'C' motif on shoulder and short blocks on body. Horizon 4.

270, 271. Two near identical tureens, one complete profile (No. 271). Fairly high cylindrical necks, very small shoulder, and footring base. Horizontal blocks below rim, arcs on neck, line at base of shoulder and short vertical lines running over shoulder on either side of handle. Multiple arc motif on handle and largish multiple arc below. Service set. Horizon 4.

272. Incomplete. Very short, rounded shoulder. *Tvaerstik*. 3 lines below rim, multiple arcs on neck, line at base of neck and vertical *Tiefstich* lines running over shoulder onto body. Vertical lines on lower half of handle. Horizon 4.

273. Incomplete. High cylindrical neck and extremely short, ledge-like shoulder. *Tvaerstik*. 3 lines below rim, multiple arcs and horizontal lines on neck and line of *tvaerstik* at base of neck. Horizon 4.

274. Incomplete. Neck only preserved. *Tvaerstik*. 4 lines below rim, double 'N' type motif on neck, horizontal line at base of neck. Traces of paste inlay visible. Horizon 4.

275. Incomplete. Very short, ledge shoulder. Line of *tvaerstik* at junction of neck and shoulder, stab marks on shoulder. Horizon 4.

276, 277. Incomplete. 2 almost identical tureens with high, slightly conical necks and almost unmarked shoulders. 3 lines below rim, groups of 3 (No. 276) or 4 (No. 277) arcs on neck. Shoulder is marked by horizontal line below which are scattered groups of small figure-of-eight motif. Arcs below lug-like handles. No. 276 is slightly larger than No. 277. Both vessels were found together. Horizon 4.

278. Almost complete. Cylindrical neck, sloping shoulder and fairly angular profile. *Tvaerstik*. Horizontal blocks on neck and shoulder. Horizon 4.

279. Almost complete profile. Slightly flaring, long neck and small shoulder. *Tvaerstik*. 3 lines below rim, vertical lines on neck adjacent to handle, line at base of neck, vertical lines running onto body in vicinity of handle. Inverted 'V' motif on small, lug-like handle. Horizon 3/4?

280. Almost complete profile. Slightly flaring neck and rounded shoulder. *Tvaerstik*. 4 lines below rim covering most of neck, discontinuous horizontal blocks on shoulder, inverted 'V' motif on handle. Horizon 4.

281. Incomplete. Cylindrical neck and fairly long, sloping rounded shoulder. Crude *tvaerstik*. Horizontal lines on neck, alternating panels of horizontal lines and arcs on shoulder. Horizon 4.

Tureen-amphora (see also Nos 320-325)

282. Complete profile. Unperforated lugs along shoulder. *Tvaerstik*. 3 lines below rim, line of double arcs on neck, *Tiefstich* line at base of neck, alternating long blocks and 2 lines of stabs. Vertical stabs at base. Traces of paste inlay visible. Lugs bear vertical *Tiefstich* lines. Well finished. Horizon 4.

Amphorae, miscellaneous

283. Incomplete. Two handled tureen with slightly conical neck and long, sloping, rounded shoulder. *Tvaerstik*. Horizontal blocks on neck with multiple arcs above handles. Horizontal lines on shoulder except below handles where panels of vertical lines are visible. Horizontal lines on handles. Horizon 4.

284. Incomplete. Flaring neck and long, sloping shoulder, 4 small lugs at base of neck. *Tiefstich*. Undecorated neck and lugs. Alternating panels of arcs and horizontal and vertical lines on shoulder. Horizon 4.

285. Almost complete profile. 4 lugs straddling shoulder. *Tiefstich*. 2 lines below rim, multiple arcs on neck, line along shoulder, vertical blocks on body. Small arcs on lugs. Horizon 4.

286. Incomplete. Small lug. Horizontal lines on neck.

287a-c. Fragmentary. *Tiefstich* and *tvaerstik*. 3 lines below rim; remainder of decoration unreconstructable, *Tiefstich* at base. Traces of paste inlay visible. Horizon 4.

Amphorae type I

288. Almost complete profile. Long cylindrical neck and long sloping shoulder with very small lugs at base of neck. *Tvaerstik*. 3 lines below rim and 2 at base of neck, with short fringe; alternating blocks of horizontal lines and lozenge motif on shoulder. Traces of paste inlay visible. Thin-walled well made vessel. Horizon 4.

289. Almost complete profile. No sharp distinction between neck and shoulder. Lugs high on neck. *Tiefstich* and *tvaerstik*. *Tvaerstik* lines on neck, blocks of horizontal *tvaerstik* alternating with multiple arcs of *Tiefstich* on shoulder and handle. Horizon 4.

290. Incomplete. Slightly flared neck and sloping shoulder. No sherds of body. *Tvaerstik*. Horizontal lines below rim and on most of neck. Traces of paste inlay visible. Horizon 4.

291. Incomplete. Small, undecorated lugs at junction of neck and shoulder. *Tiefstich* and *tvaerstik*. 4 lines of *tvaerstik* on neck, blocks of vertical *Tiefstich* on body. Horizon 4.

292. Incomplete. Cylindrical neck and sloping shoulder. 4 small lugs at base of neck. *Tvaerstik*. Horizontal lines on neck, vertical blocks on shoulder and body. Horizon 4.

293. Incomplete. Cylindrical neck and rounded body. *Tvaerstik*. Horizontal lines on neck; horizontal blocks on body; arcs on lugs. Horizon 4.

294. Complete profile. Vertical neck, round body, 4 small perforated lugs. Undecorated. Type 1/2.

295. Incomplete. Funnel beaker-like amphora. *Tiefstich*. 3 lines below rim, horizontal blocks on neck, blocks of vertical lines from shoulder. Low, pierced lugs on shoulder. Horizon 4.

296. Incomplete, similar to No. 295, but smaller. *Tiefstich*. Horizontal blocks on neck, zigzag on shoulder and vertical lines on body. Vertical lines on lug. Horizon 4.

Amphorae type 2

297. Incomplete. Slightly flaring neck and round body. *Tiefstich*. 3 lines below rim, lozenge on shoulder with horizontal lines on shoulder and vertical blocks on body. Arc below lug. Horizon 4.

298. Incomplete. Funnel beaker-like amphora. *Tiefstich*. Multiple zigzag on neck. Hatched line at base of neck, vertical lines on body. At least one lug. Horizon 5.

299. Incomplete. 4 plain lugs, in 2 pairs. Horizontal lines on neck, fringe on shoulder, vertical panels on body. Fine *Tiefstich*. Horizon 4/5.

300. Complete profile of extremely fine, well finished amphora with cylindrical neck and rounded body. Lugs at base of neck. Very fine, black, burnished fabric. Very regular round tipped *Tiefstich*. Horizon 5.

301. Incomplete. Flaring neck and rounded shoulder. *Tiefstich*. 3 lines below rim, undecorated neck, stab marks along base of neck, panel of zigzag on shoulder and vertical blocks on body. Traces of paste inlay visible. Horizon 5.

302. Incomplete. Fringe at base of neck. Triple line of zigzag on shoulder, vertical blocks on body. Horizon 5.

303. Incomplete. Fine *Tiefstich*. 2 lines below rim with pendant fringe, line at base of neck with pendant fringe, double zigzag along shoulder, vertical blocks on body. Horizon 5.

304. Incomplete. Cylindrical neck and long, rounded shoulder. *Tiefstich* and *tvaerstik*. 3 lines of *tvaerstik* below rim, lozenge motif on neck, non-reconstructable pattern on shoulder and line of *tvaerstik* above lozenge on upper body. Arcs on handle. Might in fact be a tureen instead of amphora. Horizon 4.

305. Complete profile. Cylindrical neck, rounded body, 2 lugs at base of neck. Line at base of neck with pendant fringe. Grooves. Horizontal lines below rim with double line of stab marks. Line at base of neck with stab marks. Alternating panels of verticals and zigzags. Double line of stab marks at base. Traces of paste inlay visible. Horizon 5.

306. Incomplete. Slightly flaring neck, gently sloping shoulder. *Tiefstich*. 4 horizontal lines below rim, horizontal line on shoulder, vertical panels of short and long lines. Horizon 5.

307. Incomplete. Slack profiled funnel beaker-like amphora. Small undecorated lug of shoulder. Double zigzag on neck, vertical panels on body. Horizon 4?

308. Incomplete. Small, thin-walled amphora with at least two lugs spanning shoulder. Multiple zigzag on neck, fringe on shoulder, alternating blocks of multiple zigzag and verticals on body. Undecorated lug. Horizon 5.

309. Incomplete. Small, thin-walled amphora with at least 2 lugs. 3 lines below rim, zigzag on neck, alternating blocks of zigzag and verticals on body. Vertical lines on lugs. Horizon 5.

310. Incomplete. 2 lugged amphora. *Tiefstich*. 3 lines below rim, line of zigzag on neck, fringe from base of neck, blocks of verticals on body. Undecorated lugs. Horizon 5.

311. Incomplete. Small, thin-walled amphora body. Fringe along shoulder, alternating blocks of verticals and multiple zigzag on body. Horizon 5.

312. Incomplete. Small thin-walled amphora of exceptionally fine fabric. Vertical blocks on body. Horizon 5?

313. Fragmentary. Amphora body. *Tiefstich*. Alternating long and short blocks of verticals. Horizon 5?

314. Incomplete. Small, thin-walled neck of amphora. 3 lines

below rim, double zigzag on neck. Horizon 5.

315. Fragment. Lug from amphora not otherwise recognized.

316. Incomplete. Flaring neck and rounded shoulder. *Tiefstich*. At least 2 lines below rim, horizontal panels alternating with single arcs on neck, no lower body sherds. Fairly similar to No. 317, but different fabric. Horizon 4?

317. Almost complete profile. Small amphora with probably only 2 lugs. *Tiefstich*. Horizontal lines on neck with arcs, vertical blocks on body. Horizon 4.

318. Almost complete profile of vessel with short, cylindrical neck and round body. 2 horizontally pierced lugs at junction of neck and body, line of pointed *Tiefstich* just below rim.

319. Incomplete vessel with sloping shoulder and rounded body. Horizontally pierced large lug(s) at junction of neck and shoulder. Undecorated.

Tureen-amphorae (see also No. 282)

320. Incomplete. *Tiefstich*. Single line below rim, horizontal blocks on neck, vertical blocks on body. Horizon 5.

321. Incomplete. *Tiefstich*. 4 lines below rim, blocks of close-set, multiple zigzags on neck, line at base of neck with pendant fringe. Very slight traces of paste inlay visible. Horizon 5.

322. Incomplete. No neck sherds. Hollow stamp impressions. Triple line at base of neck, broad panels on body. Horizon 5.

323. Incomplete. Bosses in place of lugs. Hollow stamp impressions. Single line below rim, blocks on neck, straight line at base. Horizon 5.

324. Incomplete. Bosses in place of lugs. Hollow stamp impressions. 2 lines below rim with blocks on neck. Line at base of neck with close-set panels below. Horizon 5.

325. Incomplete. Similar in shape to Nos 323 and 324. Low unpierced bosses in place of lugs. Undecorated. Horizon 5?

Shouldered lugged and non-lugged vases

326. Incomplete small bowl. Slight, conical neck and single, horizontally perforated lug.

327. Incomplete bowl with lugs. Similar to No. 326.

328. Incomplete undecorated funnel beaker-like vessel. 2 pairs of horizontally perforated lugs on neck.

329. Incomplete small vessel. Flaring rim and at least one small lug at base of neck.

330. Incomplete small vessel. Flaring neck and horizontally pierced lug on neck.

331a-e. Sherds with small, horizontally pierced lugs, from vessels not otherwise identified.

332. Fragments of vessel. Similar to Nos 342-346.

333. Incomplete small funnel beaker-like vessel.

334. Incomplete small, thin-walled vessel.

335. Incomplete small vessel.

336. Fragment of small funnel beaker-like vessel.

337. Incomplete vessel. Slightly flaring neck, long shoulder. No body sherds recognized but break at end of neck suggests sharply angled body.

338. Incomplete small, funnel beaker-like vessel.

339. Incomplete small, funnel beaker-like vessel.

340. Incomplete small vessel.

341. Incomplete funnel beaker with flaring rim and ledge-like shoulder.

342. Incomplete vessel with short, cylindrical neck, sloping shoulder and high body. Two lugs at base of neck.

343. Incomplete vessel with short, slightly flaring neck and long sloping shoulder. Probably lugged although none recognized.

344. Incomplete undecorated tureen-like vessel.

345. Incomplete vessel. Slightly flaring neck with lug at junction of neck and shoulder.

346. Incomplete vessel with short neck and round body. 2 lugs spanning junction of neck and shoulder. Heavy fabric.

347. Lower portion of vessel.

348. Incomplete, possibly necked, vase. Black fabric with medium-sized grits clearly visible in the surface.

Collared flasks

349. Incomplete. Fairly short neck with wide throat. Collar close to base of neck. Spherical body. Grouped lines on shoulder.

350. Incomplete. Wide throat with broad collar less than half way down neck and flat shoulder.

351. Incomplete. Decorated collared flask with biconical body. 4 horizontal lines at base of neck, alternating blocks of short and long verticals on shoulder.

352. Fragmentary. Undecorated.

353. Incomplete. Shoulder of undecorated, biconical body.

354. Incomplete. Complete short neck with collar.

355. 8 assorted sherds of collared flasks apparently not represented in the above list, 1 collar only illustrated.

356. Almost complete profile. Short neck with throat and long biconical shoulder and body.

357. Incomplete. Spherical body.

358. Incomplete. Spherical body. Fairly wide throat with collar midway between rim and base of neck.

359. Incomplete. Narrow throat, biconical body.

Biberons

360. Incomplete. Biberon.

361. Incomplete. Biberon.

Miscellaneous

362. Incomplete. Flat shoulder. Highly burnished, black, leathery finish.

363. Incomplete jug with spout formed by pressing out rim. The profile is not unlike some of the more uncouth funnel beakers.

Featured sherds

30 complete bases, including 7 decorated, 8 low foot rings, 1 protruding foot, and 1 perforated base. In addition to these there are 45 fairly complete bases and 41 sherds of bases. 30 decorated rims apparently representing vessels other than those described in the foregoing catalogue, and 20 decorated body sherds are also illustrated. Approximately 3000 undecorated and, for the most part, featureless sherds could not be assigned to specific pots and are not illustrated. Although as a percentage of the total number of sherds this appears quite high, many of these sherds are extremely small fragments.

3.6. Post TRB finds

The sherds of an EGK amphora of *Strichbündel* type (No. 364) found in the chamber area probably represent the end of continuous activity, dating to about 2900-2800 cal BC. Amphorae of this type occur sporadically in *hunebeds*; a similar find was made in the nearby *hunebed* G3 (Brindley, 1983: fig. 7). Two smaller EGK amphorae may have been present (Nos 365-366). The number of fragments is too small, however, to allow for reconstruction of shape and decoration.

Fragments of three Bell Beakers were found. Two of these (Nos 367-368) are well-made classical examples of Maritime Bell Beakers, which can be dated to about 2500 cal BC. The third one (No. 369) with simple linear comb ornament is a local type, and probably slightly younger.

The sherd of a *kogelpot* (No. 370) which was found in the extraction pit of a side stone is of the type current during the late 10th and 11th centuries

AD. This fragment almost certainly dates the destruction of the *hunebed.*

4. ANALYSIS OF WHITE FILLING

Traces of white filling remained in the *Tiefstich* and *tvaerstik* lines on many sherds. Originally this must have given the pottery a highly ornamented and dramatic appearance, contrasting with the usually dark colour of the fabric. This filling must have disguised, in many cases, the precise nature of the *Tiefstich/tvaerstik* technique. The definition of *tvaerstik* has been given as decoration executed along a guide line (Bakker, 1979: p. 179). This should perhaps be amended to decoration along a guideline which itself is a dominant visual aspect of the filled decoration. Some of the variations which occur in *Tiefstich* lines during Horizon 3 are executed along a guideline, and in some cases the end result must have resembled *tvaerstik* (compare Nos 251, 255). Similarly some of the close-set *tvaerstik* lines when filled resembled *Tiefstich* lines (Nos 270, 282). The depth of the impressions, while creating a particular effect when empty, is not noticeable when filled. Contrasting width and outline of stylus heads are more important.

Analysis of the white filling of *Tiefstich/tvaerstik* impressions from six pots (5 from G2 and 1 from Bronneger D21 for comparative purposes) was carried out by Peter Zijlema, a student at the Department of Inorganic Chemistry, University of Groningen. The material was examined by means of X-ray diffraction, using a Guinier-Hägg camera. The samples were taken from pots G2:127, 128, 136, 142 and 154, and from Bronneger 1918/X.31** (collection B.A.I.). All six samples yielded the same result: hydroxy-apatite, or $Ca_5(PO_4)_3OH$. This is the main mineral in bone. A recent study by Shipman, Foster and Schoeniger (1984) has shown that after heating to 600° C or more, the crystal size of hydroxy-apatite increases, and that the lines of the X-ray diffraction pattern become sharp and well-defined. All six samples showed these sharp lines. Therefore the filling consisted of a paste of bone, burnt to 600°C or more. It is very likely that burnt bone was used, because this is more easily crushed and because fresh bone would have shrunk during the firing process and fallen out.

5. DISCUSSION OF POTTERY BY TYPE

5.1. Summary of typochronology used in this paper.

A series of horizons within the pottery sequence are proposed on the basis of the material from Glimmen G2 and a general survey of other material. These are discussed in detail elsewhere (Brindley, 1986). A short summary only is given.

Horizon 1: lugged or perforated bowls, round bodied jugs, round bodied funnel beakers, round bodied lugged beakers, collared flasks. Some, chiefly horizontal, elements immediately below the rim, vertical strip decoration immediately below this.

Horizon 2: lugged bowls/pails, slightly more angular jugs and possibly lugged beakers, funnel beakers, collared flasks, biberons. Horizontal, narrow band immediately below rim. Vertical strips broaden throughout horizon and are interspersed with vertical lines. Strips on shoulders broaden into panels, and are occasionally replaced by triangles.

Horizon 3: lugged bowls/pails, tureens, lugged beakers, funnel beakers, collared flasks, biberons. Broad, well-defined horizontal zone below rim. Broad strips and well-defined panels on body. Shoulders frequently include triangles in place of panels.

Horizon 4: bowls/pails, tureens, amphorae type 1, funnel beakers, lugged beakers, collared flasks, biberons. Upper zone loses its vertical element, which may be replaced by multiple arcs, large zigzag, or horizontal blocks. Alternating blocks replace panels of the lower body. Shoulders disappear as an area of specific decoration.

Horizon 5: bowls, tureen-amphorae, amphorae type 2, a small number of funnel beakers, funnel beaker cups, collared flasks. Two regional styles identified. Heek-Emmeln style, chiefly in the Emsland-Westphalia area, is characterized by bands of large zigzag and horizontal blocks incorporating up to 13 lines; Uddelermeer-Anlo style, predominating in Northern and Central Netherlands and characterized by small closeset zigzag in bands or blocks, and blocks of *Tiefstich* and of hollow stamp.

Horizon 6: bowls, necked bowls, tureen-amphorae, funnel beaker cups, collared flasks. Neck is usually contracted, and decoration consists of continuous bands or oblong blocks of zigzag or related motifs, or horizontal lines. Blocks or bands on shoulder become very narrow and restricted to uppermost part only.

Horizon 7: bowls, necked bowls, lugged and non-lugged shallow bowls, funnel beaker cups, collared flasks. Neck decoration disappears entirely. Very narrow horizontal blocks of decoration on the top of the shoulder only. Many undecorated vessels. Applied cordons.

5.2. Funnel beakers and funnel beaker cups

5.2.1. *Funnel beakers* (Nos 1-102)

About 102 funnel beakers were identified. All of these have necks which flare to a greater (e.g. Nos 12, 57) or lesser extent (e.g. No. 45) and bodies with either sharp (e.g. Nos 16, 58) or rounded, high shoulders (e.g. Nos 2, 4) or, occasionally, more spherical bodies (e.g. No. 1). About seventy-two have decoration of some form, most frequently on the body but extending onto the neck (although never on the neck alone). Apart from four examples of *tvaerstik* impressions (Nos 20, 23, 31a, 51, but possibly also Nos 24, 63, 68), the decoration is in *Tiefstich*.

Neck decoration consists chiefly of zigzags, single or multiple, usually below the rim (Nos 49-52) or at the base of the neck (Nos 53-63), or as large, multiple zigzags covering most of the neck (Nos 46-48). There are also three instances of multiple arcs (Nos 65-67). Decorated necks are always associated with short, high shouldered bodies. A horizontal line (or lines) at the base of the neck is present on about 50% of the decorated beakers; occasionally this is the only form of decoration on the vessel (e.g. No. 69, a good parallel for this vessel is Emmeln No. 114 (Schlicht, 1968)). Body decoration almost always consists of multiple vertical lines (exception, No. 68 has horizontally disposed stabs on an unusually long shoulder), as a short fringe (Nos 43, 49), extending halfway down the body (Nos 56, 61), or to the base (No. 15). These lines may be grouped in alternating blocks of long and short lines (Nos 37-39) or with blank spaces (Nos 40, 44, 67).

The small, thin-walled funnel beakers (Nos 52, 54, and especially 64) bear a resemblance to some of the smaller amphorae of type 2 (Nos 298, 301-303) in terms of their size, thin fabric and general decoration. The amphorae can usually be distinguished by their more pronounced shoulders and more sophisticated decoration, but the presence or absence of lugs is taken as the deciding factor in this assemblage.

Nos 70 and 71 are apparently unfinished decorated funnel beakers. The surfaces have been only partly smoothed and bear only very slightly sketched lines which seem to be guidelines for more formal decoration never carried out.

The undecorated funnel beakers show a much greater degree of variation in terms of shape, size and finish. Some (e.g. Nos 97, 99) are very large and of good, hard fabric, and in a domestic context would be termed 'storage vessels'. Nos 86 and 102 provide good parallels for what were formerly called 'pre-Drouwen' funnel beakers (Bakker, 1979: p. 117). The sherds of No. 86 were found amongst the chamber debris and there is no reason to think that the vessel was not deposited there. No. 102 was found in a pit outside the chamber which had probably been inserted through the covering mound (see 2.5). No. 72 is remarkable for the line of small pits immediately below the rim, parallels for which were found at Beekhuizerzand (Modderman et al., 1976: fig. 9) at Emmen D40 and Eexterhalte D14 (collection B.A.I.). The variations which occur in the quality and size of these funnel beakers suggest that there was no particular choice made in terms of the pottery which might be deposited in the burial chamber.

Early funnel beakers are not well-known in the West Group in clearly documented, small assemblages. Their precise characteristics are therefore unknown, but it is likely that they resembled funnel beakers of the Fuchsberg Group (Schwabedissen, 1979) and the funnel beakers from Bistoft (Johansson, 1981), which are slightly older than or contemporary with the earliest part of the West Group. The funnel beaker No. 38 which was found in the flat grave (see 2.5) is possibly of this type, but as pointed out before, its stratigraphic relationship to the *hunebed* which (on the basis of the earliest pottery found in the chamber area) was constructed sometime after the beginning of Horizon 1, is unclear. The flat grave may also have been inserted after the construction of the main tomb.

The range of funnel beakers in use by Horizon 2 is well-documented in the assemblages of Zeijen, Diever and Hooghalen (Bakker, 1979: figs B3-5, B12-15). The funnel beakers from these sites show vertical and flaring necks, angular and rounded bodies, high, short and long shoulders and varied, although chiefly vertical, linear decoration on the bodies, all occurring side by side. Funnel beakers of these types remained in use until Horizon 5. The G2 funnel beakers with only these characteristics cannot therefore be precisely dated. However, necks which are less frequently decorated, are sometimes a more reliable indicator of date. Multiple arcs occur on the necks of Nos 65-67. This motif belongs primarily to Horizon 4, to which these funnel beakers probably belong. A more pointed form of multiple arc occurs during Horizon 3. The larger, usually multiple, zigzag which covers the main part of the neck (Nos 45-48) is a decorative motif of Horizons 4 and 5; it appears on late tureens, tureen-amphorae and bowls of Horizon 4 (Zuidwolde-Ekelberg, Mesum: Bakker, 1979: figs B16-18; Sleen-Kruidhaarsveld: Van Giffen, 1937: fig. 5) and of Horizon 5 (Heek-Emmeln Style, e.g. Wechte Nos 193-196). Another form of zigzag which occurs on the necks of funnel beakers (Nos 52, 61) also occurs on amphorae type 2 of Horizon 5 (Uddelermeer-Anlo Style). These funnel beakers are therefore the very last of the series of true funnel beakers. Zigzags immediately below the rim are not closely datable unless associated with other distinguishing motifs e.g. groups of vertical lines (Bronneger: Knöll, 1959: 28:3), which are a feature of Horizon 3 decoration. These zigzags by themselves are also known to occur during Horizon 4. Examples are known from Horizon 4 contexts (Landersum: Bakker, 1979: fig. 15:3a).

5.2.2. *Funnel beaker cups* (Nos 103-110)

In his discussion of funnel beakers, Bakker (1979: p. 54) includes small, single lugged, 'cup-like' vessels similar Nos 103-109 of this text at the end of the funnel beaker series, dating them to his phase E2 and describing them as "seems to be only a vague shadow of the earlier funnel beaker". As no suitable term has been given to these vessels, I have adopted the term 'funnel beaker cups'. There are at least seven of these cups at G2 while three small, horizontal lugs (Nos 110a-c) suggest that there were originally about ten. In all cases, the constriction between neck and the practically shoulderless body is very slight. A single small thumbshaped horizontal lug or, in one case, a small boss (No. 107), is usually present above the shoulder and the bases normally have a low standring.

Funnel beaker cups are fairly well associated with material of the later TRB phases. At G2 itself, one of these cups (No. 103) was found together with a Horizon 6 pail (No. 133) in a pit which also produced datable charcoal (4380±40 BP, GrN-6156). Funnel beaker cups form an integral part of several important late TRB assemblages (Beekhuizerzand: Modderman et al., 1976: fig. 6, pp. 3-5; Angelslo, Spier and Allardsoog: Bakker & Van der Waals, 1973: figs 5, 10). The first of these indicates that funnel beaker cups date from early in Horizon 5, the other shows a broader, more bowl-like variant of the same type in use during the latest stage of the TRB, Horizon 7; the Glimmen cup with its Horizon 6 association falls neatly between the two. The rather crude and shapeless funnel beaker (No. 102) may possibly point in the typological direction from which these funnel beaker cups originated.

5.3. Lugged beakers (Nos 111-112)

These vessels never occur in large numbers. The earlier of the two, No. 111, is represented only by some finely decorated neck sherds. It provides an excellent parallel for the small lugged beakers from Exloo D30 (Brindley, 1986: fig. 5). These lugged beakers are absolutely typical for Horizon 3. The second lugged beaker (No. 112) is well-preserved and almost complete, closely resembling examples from Emmeln (Schlicht, 1968: No. 51) and Gross Berssen (Schlicht, 1972: Nos 88, 131, 160). It belongs to Horizon 4.

See also under Amphorae for remarks on the similarity of early amphorae to lugged beakers.

5.4. Lugged bowls (dishes and shallow bowls Nos 113-128, pails Nos 129-133)

No. 113 is unusual in several ways – the flaring upper portion of the profile, the absence of vertical lines between the two lines of zigzag on the outer rim surface and the zigzag line on the inner rim surface. Parallels for this last occur at Noordlaren (Bakker, 1982/83: fig. 23:52), Wulften, Kr. Osnabrück (Knöll, 1959: 16:13), Gross Berssen (Schlicht, 1972: No. 21), Emmeln (Schlicht, 1968: No. 176), Drouwen D26 (unpublished, excavation Bakker), and an unprovenanced example in the collections of the B.A.I. which is possibly from Bronneger D21/-D22. It has no chronological significance. This bowl is the earliest bowl in the assemblage. Its narrow horizontal band below the rim and the relatively narrow panels beneath may be compared with the decorative style of Horizon 2 material. The bowls Nos 114-126 and pails Nos 129-131 form a very homogenous group in terms of decoration, shape and style and all fall within the general definition of Horizon 3. As a group they indicate some of the difficulties of the Bakker system of phases. Few of the individual bowls are assignable to specific phases which are based on a typology of two pot forms – shoulderless lugged and non-lugged bowls on the one hand and jugs/tureens/-amphorae on the other (Bakker, 1979: pp. 64-66). In terms of bowls, the phases are rather scantily outlined, with the use of true zigzag as a main defining characteristic of Phase B leading to the exclusion of many bowls on which the zigzag lines have been replaced by similar but alternative motifs or techniques (e.g. Nos 114-121) but which are otherwise identical. Similarly, the definition of Phase C bowls is unsatisfactory, being based on a change in a single decorative element – the replacement of the zigzag line(s) at the base of the upper zone by a broad, *Tiefstich* line and by two trends, the breaking of the lines immediately below the rim into short lengths and the grouping of verticals within the upper zone. Of these two trends, in fact the former is very rarely found without *tvaerstik* (i.e. Bakker's D1) ornamentation while the latter occurs only sporadically. Continuous lines below the rim and bands of continuous vertical lines occur commonly on Phase D1 bowls. The replacement of the lower zigzag line is most striking when it is replaced by a broad grooved or *Tiefstich* line and especially when this occurs in conjunction with a similar partitioning of the vertical patterns of the lower zone and the spacing of motifs within the resulting panels. Occasionally, the partition lines may even penetrate the upper zone as well, as the pail No. 130 shows, thus constituting a fine and well-organized decoration, but this is comparatively rare and cannot be regarded as chronologically important. A second difficulty with the application of Phases B and C as defined, is the number of bowls which lack some small decorative element and therefore do not fit into any of the given phases. According to Bakker's definitions, Nos 119, 120 and 125 are incompletely decorated. But such decorative schemes are not uncommon and are best treated as variants within the style.

Examination of material in the collections of the I.P.P. and the B.A.I. and the published inventories of Wechte (Knöll, 1983), Emmeln and Gross Berssen (Schlicht, 1968; 1972) show that many bowls and pails are not assignable directly to Bakker's phases and that a mixing of phase-defining elements occurs quite often, e.g. a combination of Phase B zigzag and Phase D *tvaerstik* has been noted at Mander O2 (collection O.K.T.), Bronneger D21/22 (collection B.A.I.), and at D43a (information J. Molema). In brief, Nos 114-126 and 128-131 form a homogenous group which should be considered as such.

Pails are the most poorly represented class at Glimmen G2. No. 130 is a particularly well-made and decorated example, belonging to the last stage of Horizon 3. It forms a service set together with tureen No. 259. The distribution of sherds of both vessels indicate that they were deposited together (fig. 9).

Nos 127-128 in both profile and decoration are clearly hybrid lugged/non-lugged bowls. The layout of decoration is dictated by the presence of lugs and, like Nos 114-126, consists of a horizontal zone below the rim including a band of (now contracted) verticals separated from a lower zone by a strongly marked horizontal line, with well-defined but narrow vertical panels of chevron related ornament below the lugs. Bowls of this type are well-known, good parallels occur at Landersum (Bakker, 1979: fig. B15) and Heek-Ammerter Mark, Kr. Borken (information W. Finke) in well-associated contexts and at Emmeln and Gross Berssen (Schlicht, 1968; 1972). Typologically, such bowls are important as they show the breakdown of the panel decorated bowls of Horizon 3 moving into the block decorated bowls of Horizons 4 and 5. These bowls belong to Horizon 4.

No. 132 is part of a Horizon 4 pail, with horizontal zone below the rim and narrow panels of vertical chevron below the lugs. Pail No. 133 was found in a pit together with a funnel beaker cup (No. 103) and charcoal (see 2.5). It belongs to Horizon 6. No other material from this period has been identified at G2.

5.5. Non-lugged decorated bowls (Nos 134-175)

Approximately 40 bowls fall within this category; in addition, sherds of several other very fragmentary examples can be recognized. Large portions of many of the bowls can be reconstructed but only eleven complete profiles were obtained. Four of these have simple bases, and seven have footring bases.

No. 134 is a well-preserved small bowl probably of Horizon 3. Nos 135-138 appear to be late and poorly executed examples of Horizon 3 bowls. The fragmentary condition of all four make further comment unnecessary. The decoration on Nos 139-141 tends to be zonal, consisting of horizontal lines below the rim and above a band of short vertical elements (often lines). The bottom part of this zone still consists sometimes of horizontal line(s). The lower zone consists of vertical elements, chiefly lines, usually in simplified panels or blocks. This decorative scheme is essentially a simplified version of that occurring on the lugged bowls.

The decoration on Nos 142-152, 154, 155, 160-162 consists basicaly of an upper zone of lines which are usually horizontal and continuous, and one, two, or three less well-defined zones on the lower body, most frequently made up of horizontal or vertical blocks but occasionally including other motifs. Bowls with decoration of this type occur in Horizon 4 in the assemblages of Landersum (early Horizon 4; Bakker, 1979: fig. B15) and Mesum (late Horizon 4; Bakker, 1979: B18). The lozenge element on bowls Nos 160-161 may indicate the first half of this horizon. Block pattern also occurs in both Horizon 5 styles.

At first sight the bowl No. 166 seems to combine the decorative layout of Horizon 3 bowls (a well-defined upper zone, terminating in a horizontal line, and a lower zone consisting of long panel-like blocks of lines) with small, closely set zigzag in horizontal blocks suggestive of the Uddelermeer-Anlo Style of Horizon 5. On closer inspection, however, it is clear that the lower vertical blocks are pendant from short lines of zigzag, and that there is no question of an upper zone terminating in a single zigzag line. This bowl may belong to late Horizon 4 or early Horizon 5. The bowls Nos 167-168 include large zigzag in the upper zone and some *tvaerstik* lines in the long vertical blocks, suggesting late Horizon 4. Bowl No. 171 with multiple smaller zigzag lines belongs to the Heek-Emmeln Style of Horizon 5. Bowl No. 172 with double small zigzag line represents the Uddelermeer-Anlo Style of this horizon. The poorly preserved bowls Nos 169 and 170 also belong to this style. Bowls occur in two service sets with identical fabric, decoration style and in the case of the bowls, shape (Nos 139 and 140; Nos 145 and 146, together with tureen No. 260).

The shoulderless bowls Nos 174-175 should be considered with the necked and cordoned bowls of Horizon 7.

5.6. Non-lugged undecorated bowls (Nos 176-212)

There is a minimum of thirty-seven undecorated bowls in this assemblage. With two exceptions (Nos 176, 201), only bowls which are reconstructable from the rim downwards are included in this group. Some of the reconstructed bases may also belong to undecorated bowls, but the custom of decorating only the upper portion of funnel beakers, some bowls, tureens, jugs and amphorae makes it impossible to distinguish undecorated bowls from undecorated portions of otherwise decorated vessels.

The undecorated bowls show considerable variety in terms of fabric, finish, shape and size. No. 180 has a single, pre-firing perforation in its base. A similar but not quite so extensive collection of undecorated bowls was found at Emmeln (Schlicht, 1968: Nos 877-883, 901-902, 908-914, 916-921, 931-938, total 30). While the proportion of undecorated but recognizable diagnostic types increases during Horizons 6 and 7, undecorated bowls and other forms do occur earlier as the associated finds from several small assemblages and flat graves demonstrate (Diever 1, Hooghalen, Zeijen E: Bakker, 1979: figs B3, B4, B14; Schipborg: Waterbolk, 1958: fig. 2:7-9).

5.7. Necked and cordoned bowls

Horizon 7 is represented by c. 30 necked bowls, cordoned bowls and simple bowls (Nos 174-175). These are characterized by limited and restrained decoration.

5.7.1. *Necked bowls* (Nos 213-235)

There are about twenty-six bowls of this form in this group, eight of which are represented by single sherds only, and fourteen of which are decorated. The decoration is very sparse and always occurs at the top of the shoulder in three main forms – vertical slashes on a narrow, applied, sometimes discontinuous cordon, as panels of neat, discrete or joined and opposed slashes (Nos 215, 216), or as a horizontal line with pendant short strokes, sometimes in discontinuous portions. All these bowls belong to Horizon 7.

5.7.2. *Cordoned bowls* (Nos 236-237)

With the above, and also belonging to Horizon 7 should be considered the cordoned bowls Nos 236 and 237, and the shoulderless bowls Nos 174 and 175. The cordoned bowls seem to have developed out of the bowls with narrow, horizontal lugs with vertical lines of Horizon 6, which are not represented in G2.

5.8. Jugs and tureens

In general typological terms, tureens develop out of jugs. Both rounded and angular profiles occur during Horizon 2. Although the original rounded form disappears during Horizon 3, the range of profiles extends to include wide and angular shoulders. Horizon 4 tureens once more include both relatively wide forms but with small shoulders. In general, shape by itself is not a reliable indicator of horizon; in fact, it seems to have been dictated by the individual's choice of decorative scheme as much as by any other factor. However, ultimately, the tureen becomes indistinguishable from the tureen-amphora (see Amphorae).

5.8.1. *Jugs* (Nos 238-241)

The assemblage includes several vessels which are considered to be jugs. Nos 238 and 240 can be distinguished from the tureen series as a whole because of their rather coarse fabric and sparse decoration. No. 239 is probably a jug neck because of its general similarity to these two. The zigzag immediately below the rim is also occasionally found on early funnel beakers but in this case, the flat top of the rim and the lack of neck curvature suggest that it is more likely to be a jug. These jugs, together with dish No. 113 probably represent the earliest material on the site, Horizon 2.

5.8.2. *Tureens* (Nos 242-281)

Bakker (1979: p. 176) defines a tureen as having 'one lug'. At least one vessel included here as a tureen has two handles (No. 265). Two-handled tureens, although not common, are known and several were found at Emmeln (Schlicht, 1968: Nos 275, 370, 374, 392, 416).

A minimum of forty-one tureens was identified. In broad terms, they show a general trend from wide and angular to less prominently shouldered, deeper forms, but this trend is by no means absolute. In terms of decoration, on the neck there is a general movement away from a combination of horizontal and vertical elements, to chiefly horizontal elements, or to horizontal elements associated with multiple arcs. On the shoulders, triangles, pseudo-triangles and related shoulder decoration (e.g. broad, deep stabs), are replaced by a single, deeply intended horizontal line, frequently used to mark and sometimes almost to create the often practically non-existent shoulder (e.g. Nos 276, 277). Decoration below the shoulder and especially below the handle is common.

All but one of the recognized 'service sets' (see 2.6) include at least one tureen. These are listed in table 1 (including the non-tureen service set).

Tureens Nos 245 and 246 with relatively deep bodies belong to the earliest part of Horizon 3. The few lines of *tvaerstik* on each are at home in both Horizon 2 and 3 contexts. In addition, tureen No. 246 has a horizontal ladder on its handle, the best parallel for which occurs on the Horizon 2 Weerdinge jug, which also includes *tvaerstik* lines (Brindley, 1986: fig. 4:10). A similar feature occurs on the handle of one of the tureens from Issendorf (Tempel, 1972: fig. 4). The tureen series as a whole is dominated by two motifs, shoulder triangles, usually associated with broader shoulders and strap handles, and multiple arcs on the necks, usually associated with small to virtually non-existent shoulders and small lugs. The two motifs are largely mutually exclusive (exceptions Nos 251, 252). In general, the shoulder triangle is the earlier motif, with a well-documented ancestry in the later stages of Horizon 2 (Brindley, 1986: fig. 4). It has a widespread distribution, occurring in a number of forms during Horizon 3. True and pseudo-triangles and shoulder stamps occur simultaneously as is demonstrated by the pseudo-triangles of No. 245 and the shoulder stamps of No. 246 which are arguably early within Horizon 3, and the true triangles of Nos 257-259 which belong to the end of the same horizon. Triangles are predominantly a Horizon 3 motif and do not seem to occur subsequently. The multiple arc motif is more difficult to assess. Tureens with this motif have hardly ever been found in flat graves or small assemblages, with the exception of Oldenburg-Donnerschwee Grambergs Gründe (Knöll, 1952: fig. 15:12, 15, 16). The motif rarely occurs on forms other than tureens. In some assemblages, a zigzag (sometimes multiple) is used as an alternative (both occur at Wechte: Knöll, 1983: Nos 144, 148, 161 resp. 100, 145; and Emmeln: Schlicht, 1968: Nos 227, 456, 457 resp. 371, 360, 859, and Emmen D43, collection R.M.O.L.). The multiple arc is by far the more popular at Glimmen G2; only one possible example of a very slack zigzag was noted (No. 264a). It has its ancestry on Horizon 3 tureens where it occasionally occurs in a pointed form (No. 251 has both pointed multiple arcs and shoulder triangles). It developed from single triangles at the base of the neck, examples of which are also known from early Horizon 3 (Seeste tureen: Bakker, 1979: fig. 31:1; D43 (72S); Emmeln: Schlicht, 1968: amphora No. 16). The origin of the zigzag is not clear. Tureen No. 10 from Gross Berssen (Schlicht, 1972) shows a zigzag together with pointed arcs on the neck, shoulder triangles and a broad handle. This is an early example of the larger zigzag motif, late Horizon 3/early Horizon 4. At Emmeln and Emmen D43 where both motifs occur, intermediary 'W'and 'M' motifs which look very like short lengths of zigzag are found (e.g. Emmen D43 76S, 108S; Emmeln: Schlicht, 1968: No. 217). These intermediary forms, however, are more likely to be due to mutual influence than to typological development of one from the other.

Shoulder triangles and neck multiple arcs seem in general to be mutually exclusive, probably for chronological reasons. Shoulder triangles do not occur on amphorae of any form, whereas the multiple zigzag and, although less frequently, the multiple arc are well known. This suggests that the shoulder triangle had fallen out of favour by the time these two decoration motifs emerged. Furthermore, there are indications that while tureens with multiple arcs/zigzag represent early Horizon 4 material, only multiple zigzags occur during the later stages of the horizon, and continue into Horizon 5.

Apart from multiple arcs, Horizon 4 tureens show two other main motifs – lozenge lines (Nos 261, 262, 265) and block patterns. Lozenges also occur frequently on bowls of the early part of this horizon (e.g. Heek-Ammerter Mark, Kr. Borken: information W. Finke; Emmeln: Schlicht, 1968: e.g. Nos 299, 307; Gross Berssen: Schlicht, 1972: No. 136). Very narrow horizontal blocks sometimes occur with shoulder triangles during Horizon 3 (Nos 258-259). These narrow blocks are probably the prototype for the block patterns which occur during Horizons 4 and 5, and which consist of more than two horizontal lines (i.e. 3-8 lines and up to 13 lines in some examples of the Heek-Emmeln Style), on shoulder and neck, on amphorae type 1 and tureen-amphorae, and on bowls. Block patterns may alternate with empty spaces, multiple arcs, or blocks of alternating vertical and horizontal lines.

5.9. Amphorae (Nos 282-325)

This group was defined by Bakker as 'amphorae (no lugs, two or more lugs, but shape is more important)' (1979: table 4). The shape is shown by two outlines, one of a large Phase F/Horizon 6 amphora and the second, a large vertical necked vessel with four small lugs at the base of the neck and an angular profile, with long shoulder equal in length to the depth of the body. Elsewhere in the text, amphorae are given as the pilot type of Phase E (Bakker, 1979: p. 70). This definition is meaningless. Amphorae occur in two typologically distinct forms (both of which normally have lugs or bosses) here renamed Amphora (types 1 and 2) and Tureen-amphora.

Amphora (types 1 and 2) consist of a cylindrical neck with angular (type 1, e.g. No. 288) or rounded (type 2, e.g. No. 300) body. The lugs usually occur at the junction of the neck and shoulder, but may occur on the shoulder itself (No. 295). The precise origin of type 1 amphorae is unclear. An undecorated example occurs in the late Horizon 2 assemblage of Hooghalen. Decorated examples occur sporadically during Horizon 3 (Angelslo No. 1966:411, unpublished, coll. P.M.D.), Emmeln No. 166 (Schlicht, 1968), Bottrop-Kirchhellen (*Ausgrabungen und Funde in Westfalen-Lippe* 3, 1985: Abb. 95), Werlte Kr. Aschendorf-Hümmling (Schlicht, 1968: Fig. 7). They can be distinguished from the rather similar lugged beakers chiefly by their cylindrical necks with tureen style decoration (i.e.

single neck zone), whereas lugged beakers have slightly flaring necks with two horizontal zones (see under Lugged Beaker for references). The lugged vessel from Emmeln (No. 17) combines the amphora shape with the decoration of the lugged beaker. In general, the later examples are easier to distinguish from each other.

Amphorae increase in numbers from the beginning of Horizon 4 and because of their frequent occurrence in flat graves, particularly in the Westfalia region, form an important constituent of the ceramic range. The earliest amphorae at G2 belong to type 1 Horizon 4, with horizontal *tvaerstik* and occasional multiple arcs, and empty zones on the necks (Nos 288-290). Possibly slightly later are amphorae with block decoration on the neck and shoulder, with the vertical blocks continuing over the shoulder onto the upper body (Nos 291-293, 295, 296). Also belonging to the same horizon are amphorae with narrow panels of chevron below the lugs (Nos 289, 293) or lozenge (No. 288).

Type 2 amphorae (Nos 297-319) have cylindrical necks, sometimes high shoulders, rounded bodies and predominantly *Tiefstich* or grooved decoration. The motifs include blocks of tightly packed small zigzag and horizontal lines (Nos 301-303, 305, 308, 309, 311, 314) which belong to the Uddelermeer-Anlo Style of Horizon 5 when similar patterns are found on bowls and tureen-amphorae. These examples may belong to the earlier part of the horizon when lines below the rim, a gap immediately above the base of the neck and decoration covering a larger proportion of the body (cf. amphorae type 1 Nos 287-290) suggest the continuing influence of Horizon 4. Type 2 amphorae (e.g. Nos 298, 307, 312, 313) are sometimes difficult to distinguish from similarly decorated funnel beakers. The close resemblances between some funnel beakers and amphora of type 2 is also seen at Hardenberg (compare the apparently lugless amphora Knöll, 1959: Tafel 42:22 with funnel beaker No. 24 on the plate). Amphorae of type 2 do not seem to occur later than Horizon 5.

Tureen-amphorae (Nos 282, 320-325) evolve out of tureens of Horizon 4, probably towards the end of that horizon. The necks are usually vertical, but may also be slightly flaring or conical. The shoulder is small and may be accentuated by the small lugs or bosses which occur on it. The body starts to narrow gently immediately below the shoulder. Early tureen-amphorae occasionally bear multiple arc decoration (No. 282). The large multiple zigzag (Mesum: Bakker, 1979: fig. B18:3,6) indicates Heek-Emmeln Style, while the horizontal band and block patterns in lines, hollow stamp impressions or small closeset zigzag indicate Uddelermeer-Anlo Style.

From these tureen-amphorae develop all the shouldered pots (except the funnel beaker cups) of the following Horizon 6, none of which occur at G2. The only identifiable representatives of Horizon 6 are the shoulderless pail No. 133 and the funnel beaker cup No. 103.

5.10. Undecorated shouldered vases with/without lugs (Nos 326-348)

About twenty-two vases have been included under this heading. These are of various shapes but usually have a neck and shoulder and are about one and a half times as high as the maximum width. Many have an angular profile. About half of them have lugs, either two or four, evenly spaced or in pairs (Nos 326-331a-e, 342, 345, 346). Some of the remainder may also have had lugs originally. The absence of decoration and lack of familiarity with the shapes of these vessels probably means that some have not been recognized. Some of the vessels may be related to amphorae – undecorated amphorae of all groups are relatively well-known (Nos 294, 318, 319, 325). Undecorated shouldered and lugged vases are a frequent constituent of flat graves of all horizons. Others can be paralleled amongst the undecorated shouldered and frequently lugged wares from Beekhuizerzand (Modderman et al., 1976: figs 8-9) and Emmeln (Schlicht, 1968: Nos 874-907).

While an undecorated element certainly exists during the earlier stages of the TRB, it seems to have become increasingly popular after Horizon 5. However, it is always unwise to classify and date different pots on the basis of the absence of decoration.

5.11. Collared flasks (Nos 349-359)

The assemblage includes about eleven collared flasks, of all shapes: spherical (No. 349), biconical (No. 356), flat-shouldered (No. 350) and intermediary. Two are decorated. All the necks are short and have been constructed by drawing the clay up around a narrow stick. The collars all appear to have been applied.

As Bakker and Van der Waals (1973) have shown, collared flasks occur during all phases of the West Group TRB. Bakker (1979: p. 56) states that the shape of the flask is an unreliable indicator of date, although the truely spherical examples are probably early (his phases A-C).

5.12. Biberons (Nos 350-351)

Parts of two biberons were reconstructed.

According to Bakker, biberons occur in his Phases B+C and D, but not later (1979: p. 57). Associations in flat graves indicate at least Horizons 2 and 3 and the earlier part of Horizon 4.

5.13. Miscellaneous (Nos 362-366)

This section includes reconstructable pots which cannot be placed in any of the above categories. No.

362 is a small thick-walled vessel of indeterminate shape. Its fabric is very similar to that of some of the collared flask (very dark in colour, burnished, hard and thick). The internal faces are unsmoothed which suggests that it had a narrow mouth.

No. 363 is a small, undecorated jug, for which no precise parallels were found. Decorated tureens with a similar pressed out spout were found at Emmeln (Schlicht, 1968: Nos 170, 253, 254), Gross Berssen (Schlicht, 1972: No. 29) and Mander O2 (coll. O.K.T.).

5.14. Decorated rims, decorated body sherds, bases

The material illustrated in figures 49-50 includes sherds which are not clearly identifiable as specific vessels or types, usually because of their small size, but which also do not belong to any of the reconstructed pots.

6. CHRONOLOGY AND DATING OF ACTIVITY AT THE TOMB

Only one aspect of the chamber deposit (i.e. flint, amber, bone and ceramic offerings) is useful to the discussion of the chronology of activity on the site. The discussion below is therefore limited almost entirely to the decorated pottery. The pottery in the burial chamber belongs to Horizons 2-5 and 7 as described by Brindley (1986). The earliest pottery – one bowl (No. 113), and possibly a second (No. 126), and three jugs (Nos 238-240) – belongs to the later part of Horizon 2. The ornamentation of the bowl as a whole in particular has more in common with Horizon 3 bowls than with the simpler ornament of earlier Horizon 2, e.g. the bowl from Elspeet A (Bakker, 1979: fig. B6:3). It is probable that some of the funnel beakers also belong to this period of activity, including the funnel beaker (No. 38) from the flat grave (see 2.5). The small quantity of material probably also indicates that the chamber was in use for only a relatively short period before the beginning of Horizon 3.

To Horizon 3 belong most of the lugged bowls (Nos 114-125) and pails (Nos 129-131), some non-lugged decorated bowls (Nos 134-138), one lugged beaker (No. 111), fourteen tureens (Nos 243-256, 258, 259), and some funnel beakers including No. 57 (which is part of a service set with tureen No. 249). Some of the collared flasks and the biberons may also belong to this horizon. Although most aspects of Horizon 3 are included in this array, the number of pails is unusually small and the absence of shoulder panel decoration on the tureens is puzzling. However, neither of these two absences are likely to indicate anything other than personal choice and chance selection in the deposition of pottery.

Horizon 4 material includes a number of recognizable funnel beakers (possibly Nos 45-47 with large zigzags and Nos 65-67 with multiple arcs). Also included are a lugged beaker (No. 112), one pail (No. 132), bowls (Nos 127, 128, 139-165), tureens (Nos 260-281), one tureen-amphora (No. 282), amphorae type 1 (Nos 288-296) and type 2 (304, 307, 316, 317). Both collared flasks and biberons are also found during this horizon. The most striking absence is the large zigzag from the necks of tureens.

Both the Heek-Emmeln and the Uddelermeer-Anlo Style of Horizon 5 are represented, the former by only five examples: the funnel beakers Nos 48 and 64, the decorated bowl No. 171, the amphora type 1 No. 298 and the amphora type 2 No. 308. The two bowls Nos 167 and 168 may represent very early examples of this style, or belong to late Horizon 4. To the Uddelermeer-Anlo Style belong the amphorae type 2 Nos 299-303, 305, 306, 309-314, and possibly Nos 318 and 319, and the tureen-amphorae Nos 320-325. The funnel beakers Nos 51 and 52 and bowl No. 172 probably belong to this style also. Both funnel beaker cups and collared flasks form part of the normal Horizon 5 assemblage. Within the context of Horizon 5 material as a whole, the G2 pots appear to belong to the earlier half of the horizon, having more in common with Horizon 4 material than with the well-developed Horizon 5 material which is found for instance at Harderwijk (Manssen, 1980: fig. 2) and Anlo (Waterbolk, 1960: figs 37, 38). It seems therefore, that the chamber area was not used for sometime after the first part of Horizon 5 until Horizon 7.

The only material of Horizon 6 is that found in a pit inserted at the foot of the cover mound and consisting of a pail and a funnel beaker cup (Nos 133 and 103).

However, during Horizon 7 the chamber was once more used for offerings and deposits. To this horizon belong the bowls Nos 174, 175, necked bowls Nos 213-235 and lugged and cordoned bowls Nos 236 and 237, as well, probably, as some collared flasks and funnel beaker cups.

The available radiocarbon dates for the TRB West Group show that it lasted approximately 550 years, beginning about 3400 cal BC and ending c. 2850 cal. BC (Brindley, 1986). It is very probable that Horizons 1 and 2 were of relatively short duration, i.e. 50 years each. Horizons 3, 4 and 5 lasted probably c. 100, c. 150 and c. 100 years respectively, leaving 100 years for Horizons 6 and 7 together. Horizon 6 was probably of fairly short duration. This timescale is based on a consideration of the calibration curve of the period and the relative quantities of pottery known from each horizon.

Glimmen G2 was probably constructed c. 3300 cal BC. The chamber seems to have been in continuous use for about 300 years, before being abandonned. During the next 100 years, a single offering was placed at the foot of the mound during Horizon 6. Activity recommenced during the final stages of the culture, around c. 2900 cal BC and may have lasted another 50 years.

The chamber therefore, was used for a total of c. 350 years (end of Horizon 2, Horizons 3, 4, first half of 5, and 7), during which c. 400 pots were deposited – i.e. slightly more than one per year. However, the service sets show that more than one pot may have been deposited during a single visit and possibly as many as five or six, figures which occur quite commonly in flat graves. It is therefore very unlikely that the chamber was visited on more than on annual basis, and very possibly, less frequently. When discussing the pottery from the *hunebed*, it is important to remember that pottery of any one horizon represents only a few pots from a settlement assemblage which is otherwise unknown.

7. SUMMARY

The excavation of the destroyed *hunebed* G2 on the Glimmer Es revealed the well-preserved groundplan of a chamber with seven pairs of side stones, entrance on the south side with at least one pair of portal stones, some evidence for a kerb and the outline of the former covering mound. Small portions of the original floor of the chamber also remained in situ. The chamber filling which was fragmented but otherwise more-or-less complete, contained sherds of c. 400 pots, axes, arrowheads, *bikkels*, amber beads and a small amount of cremated bone. This represents the original contents of the chamber. Secondary activity was represented by sherds of several Single Grave amphorae, three Bell Beakers, two oval flint knives, and possibly a small hemispherical amber bead. The *hunebed* does not seem to have been further disturbed until it was robbed of all of its stone, probably for church building in the late 10th or 11th century, the date suggested by the sherd of a *kogelpot* found in the recent filling of an orthostat socket.

The finds, chiefly the pottery, indicate that the tomb was constructed sometime after the initial introduction of TRB pottery to this area. It was used continually until the end of the TRB, about five hundred years, except possibly for a short period, during which the only diagnostic pottery was deposited in a pit at the foot of the covering mound. Some sporadic activity continued after the end of the TRB for a short time, before the tomb was finally abandonned.

8. REFERENCES

BAKKER, J.A., 1979. *The TRB West Group. Studies in the chronology and geography of the makers of hunebeds and Tiefstich pottery*. Amsterdam.

BAKKER, J.A., 1982/83. Het hunebed G1 te Noordlaren. *Groningse Volksalmanak*, pp. 115-199.

BAKKER, J.A. & J.D. VAN DER WAALS, 1973. Denekamp-Angelslo. Cremations, collared flasks and a Corded Ware sherd in Dutch final TRB contexts. In: G. Daniel & P. Kjaerum (eds.), *Megalithic graves and ritual*. Papers presented at the III Atlantic Colloquium, Moesgård 1969. Moesgård, pp. 17-50.

BRINDLEY, A.L., 1983. The finds from hunebed G3 on the Glimmer Es, mun. of Haren, prov. of Groningen, the Netherlands. *Helinium* 23, pp. 209-236.

BRINDLEY, A.L., 1986. The typochronology of TRB West Group pottery. *Palaeohistoria* 28, pp. 93-132.

DECKERS, P.H. (forthcoming). Coded culture; studies in Neolithic flint. Part 3. The flint material from the Funnel Beaker Culture, Middle Neolithic of the Northern Netherlands. *Palaeohistoria*.

GIFFEN, A.E. VAN, 1937. Twee grafvelden uit de hunebedden-en bekertijd in het Kruidhaarsveld bij Sleen, gem. Sleen. *Nieuwe Drentse Volksalmanak* 55, pp. 74-77.

JOHANSSON, L., 1981. Bistoft LA 11. Siedlungs- und Wirtschaftsformen im frühen Neolithikum Norddeutschlands und Südskandinaviens. *Offa* 38, pp. 91-129.

KNÖLL, H., 1952. Neue und alte tiefstichkeramische Flachgräber aus Oldenburg. *Germania* 30, pp. 303-313.

KNÖLL, H., 1959. *Die nordwestdeutsche Tiefstichkeramik und ihre Stellung im nord- und mitteleuropäischen Neolithikum*. Münster.

KNÖLL, H., 1983. *Die Megalithgräber von Lengerich-Wechte (Kreis Steinfurt)*. Münster.

LANTING, J.N., 1973. Laat-neolithicum en Vroege Bronstijd in Nederland en N.W.-Duitsland: continue ontwikkelingen. *Palaeohistoria* 15, pp. 215-317.

LANTING, J.N., 1974-75. De hunebedden op de Glimmer es (gem. Haren). *Groningse Volksalmanak*, pp. 167-180.

MANSSEN, W.J., 1980. A flat grave cemetery of the Funnel Beaker culture near Harderwijk. *Helinium* 20, pp. 105-113.

MODDERMAN, P.J.R., J.A. BAKKER & H.A. HEIDINGA, 1976. Nederzettingssporen uit Midden-Neolithicum (TRB), Late Bronstijd en Middeleeuwen in het Beekhuizer Zand onder Harderwijk, prov. Gelderland (met bijdragen van C.C. Bakels en J. Ypey). *Analecta Praehistorica Leidensia* 9, pp. 39-74.

SCHLICHT, E., 1968. *Die Funde aus dem Megalithgrab 2 von Emmeln, Kreis Meppen* (= Göttinger Schriften zur Vor- und Frühgeschichte 9). Neumünster.

SCHLICHT, E., 1972. *Das Megalithgrab 7 von Gross Berssen, Kreis Meppen* (= Göttinger Schriften zur Vor- und Frühgeschichte 12). Neumünster.

SCHLÜTER, W., 1985. Das Grossteingrab von Hilter a.T.W., Landkreis Osnabrück. In: K. Wilhelmi (ed.), *Berichte zur Denkmalpflege in Niedersachsen. Ausgrabungen 1979-1984*. Stuttgart, pp. 122-130.

SCHWABEDISSEN, H., 1979. Zum Alter der Grossteingräber in Norddeutschland. In: H. Schirnig (ed.), *Grossteingräber in Niedersachsen*. Hildesheim, pp. 143-160.

SHIPMAN, A., G. FOSTER & M. SCHOENIGER, 1984. Burnt bones and teeth: an experimental study of color, morphology, crystal structure and shrinkage. *Journal of Archaeological Science* 11, pp. 307-325.

TAAYKE, E., 1985. Drie vernielde hunebedden in de gemeente Odoorn. *Nieuwe Drentse Volksalmanak* 102, pp. 125-144.

TEMPEL, W.-D., 1972. Erdgräber der Trichterbecherkultur in der Gemarkung Issendorf, Kreis Stade. *Neue Ausgrabungen und Forschungen in Niedersachsen* 7, pp. 46-59.

VOSS, J.A., 1982. A study of western TRB social organization. *Berichten van de Rijksdienst voor het Oudheidkundig Bodemonderzoek* 32, pp. 9-102.

WATERBOLK, H.T., 1958. Neolithische vlakgraven in Drenthe. *Nieuwe Drentse Volksalmanak* 76, Van Rendierjager tot Ontginner, pp. 3-17.

WATERBOLK, H.T., 1960. Preliminary report on the excavations at Anlo in 1957 and 1958. *Palaeohistoria* 8, pp. 59-90.

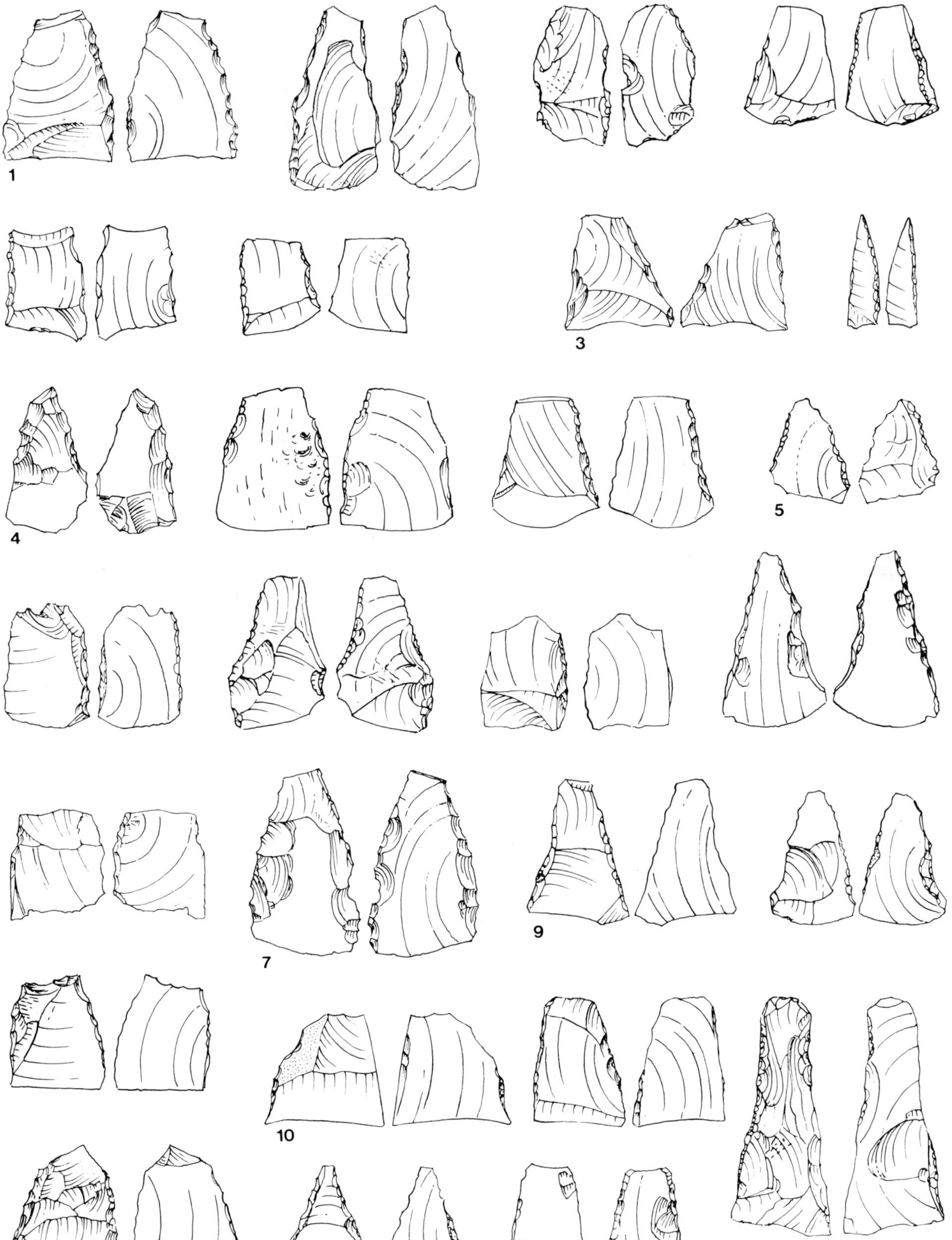

Fig. 17. Trapezoidal arrowheads.

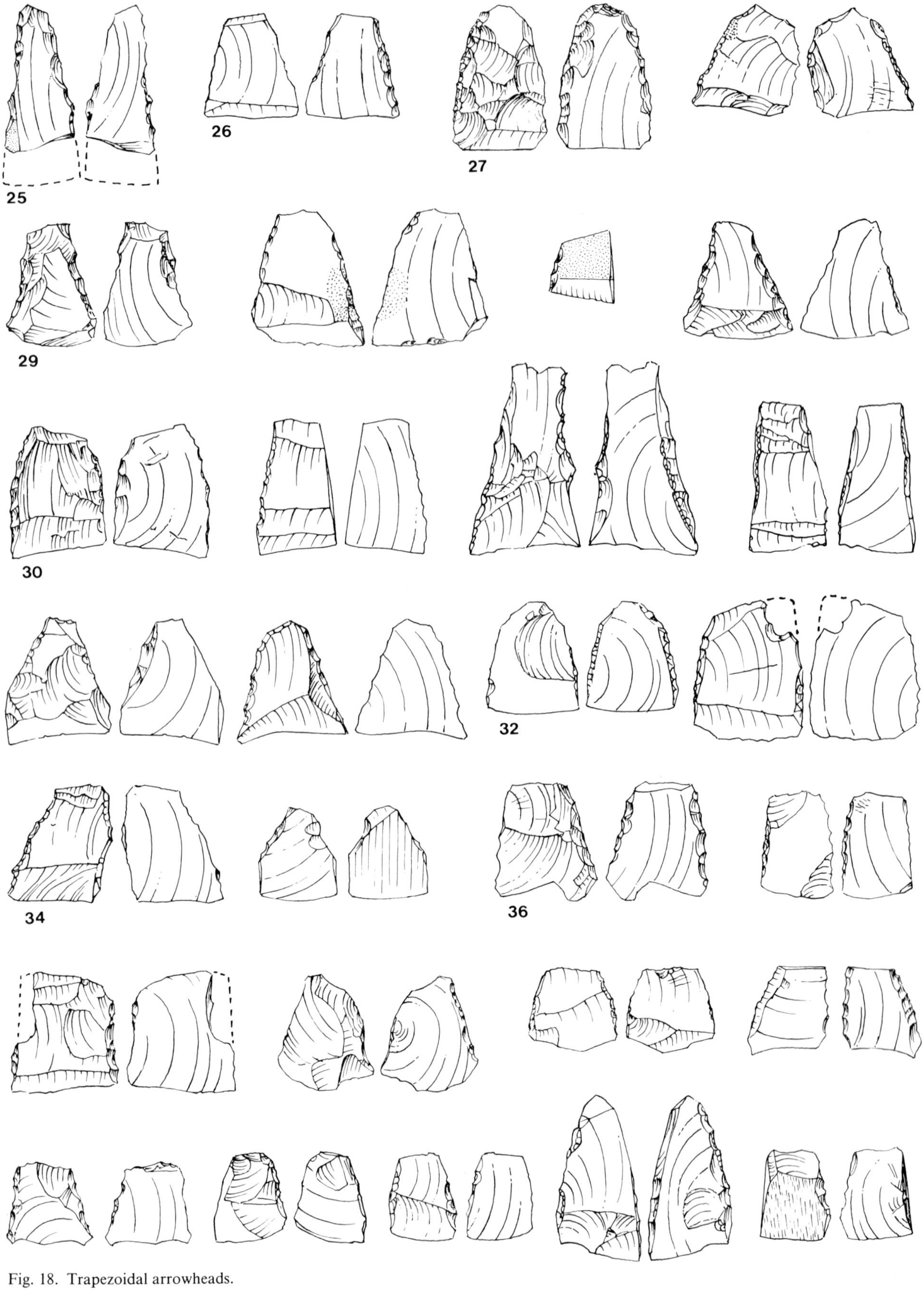

Fig. 18. Trapezoidal arrowheads.

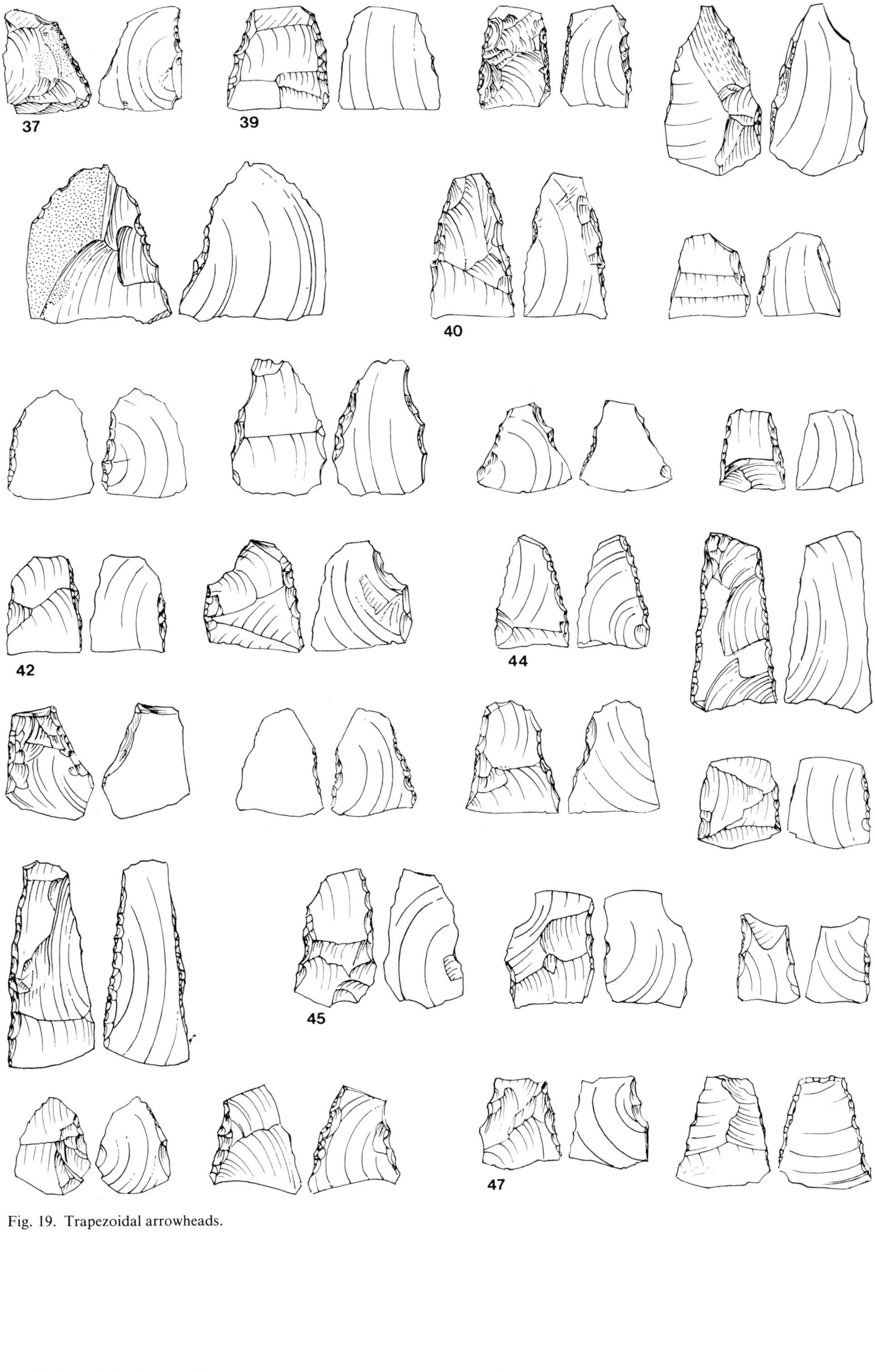

Fig. 19. Trapezoidal arrowheads.

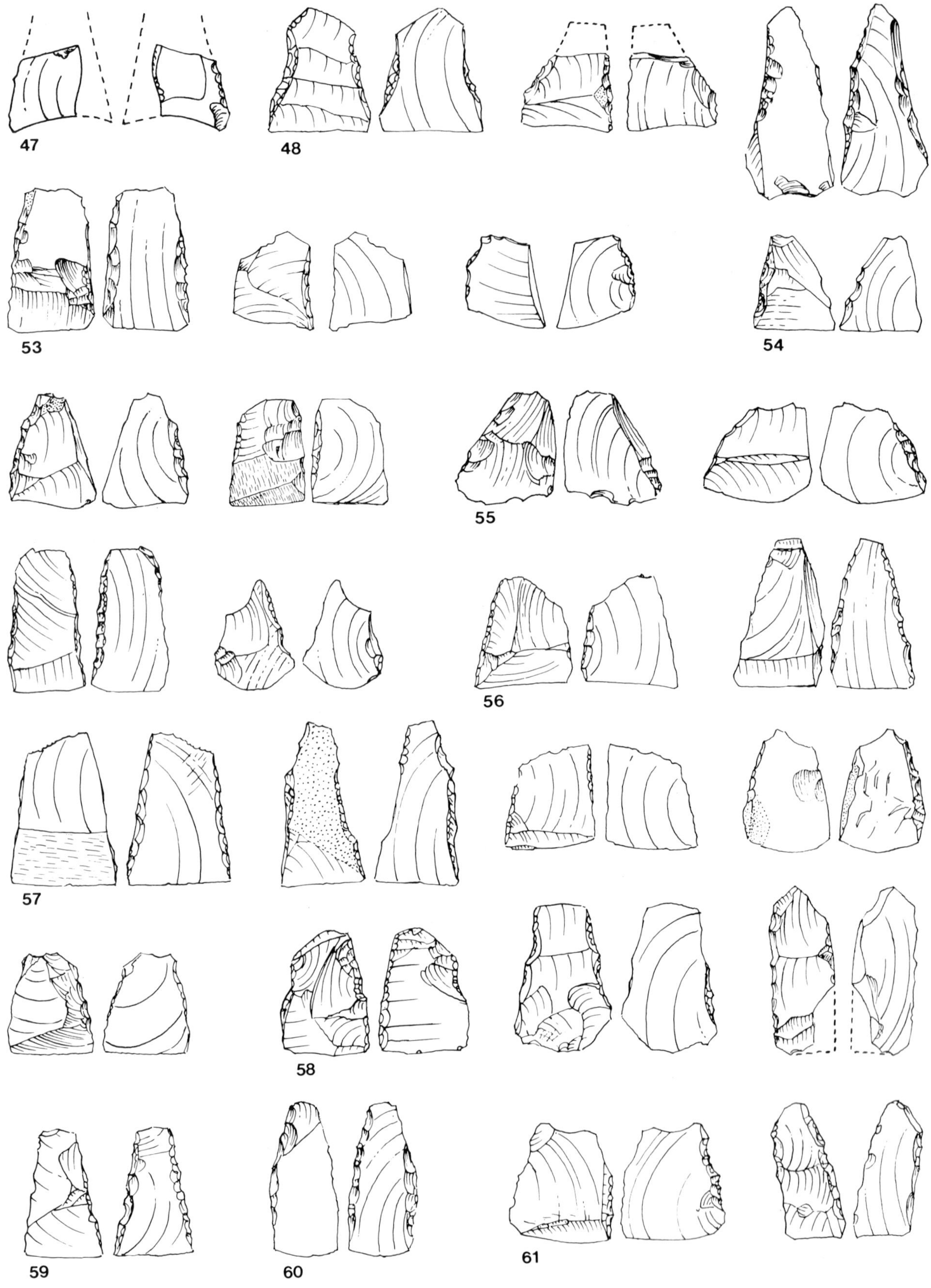

Fig. 20. Trapezoidal arrowheads.

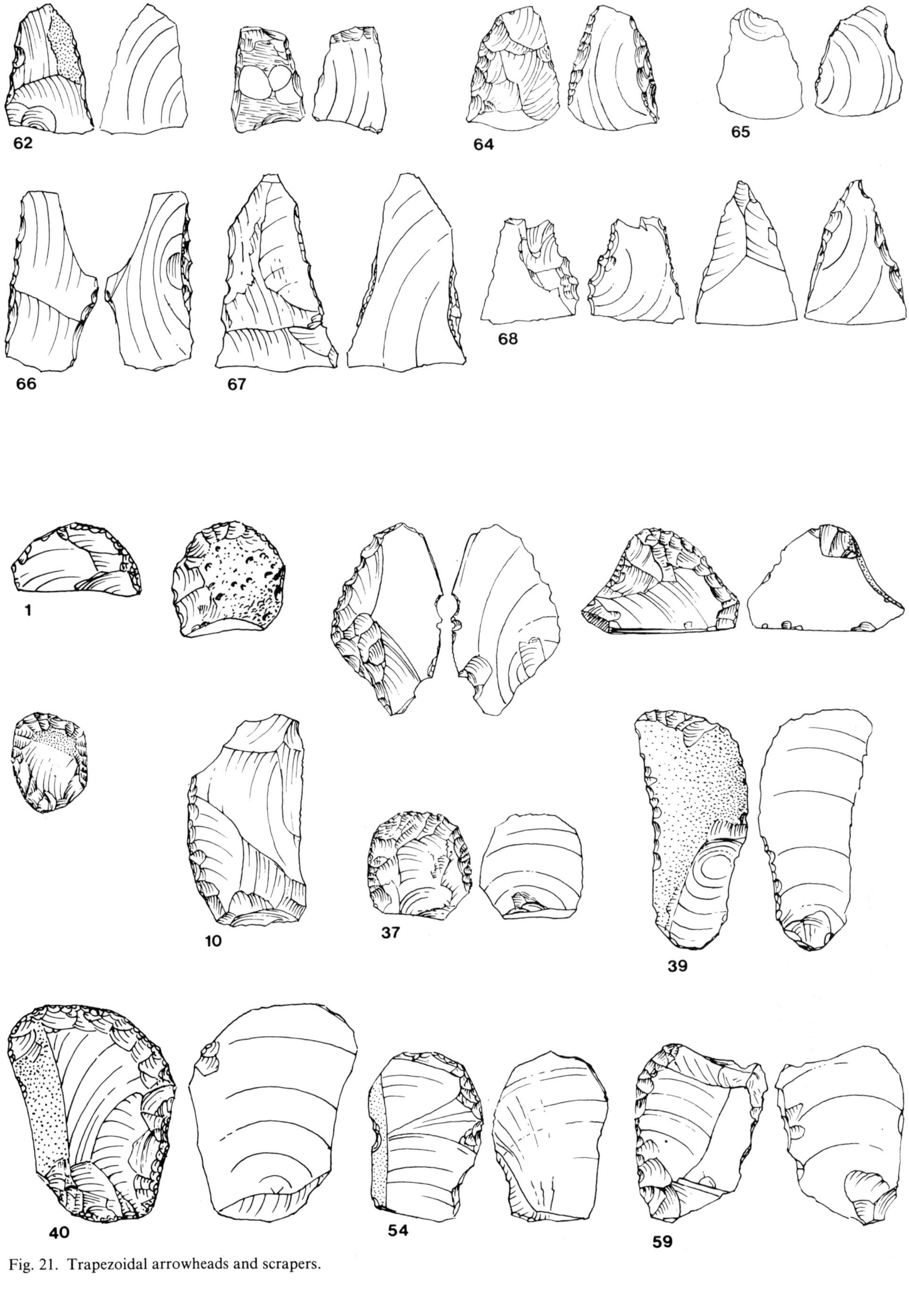

Fig. 21. Trapezoidal arrowheads and scrapers.

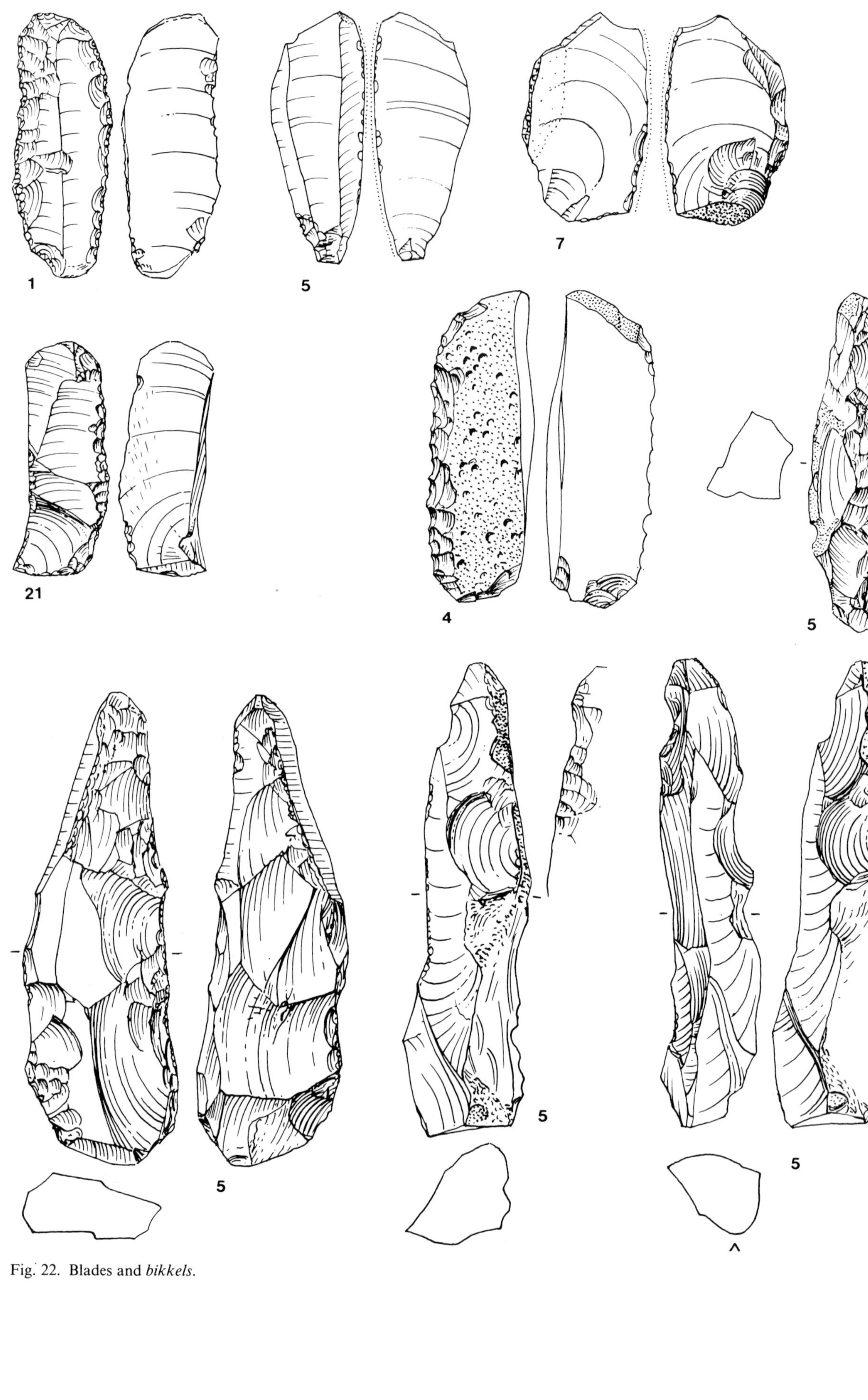

Fig. 22. Blades and *bikkels*.

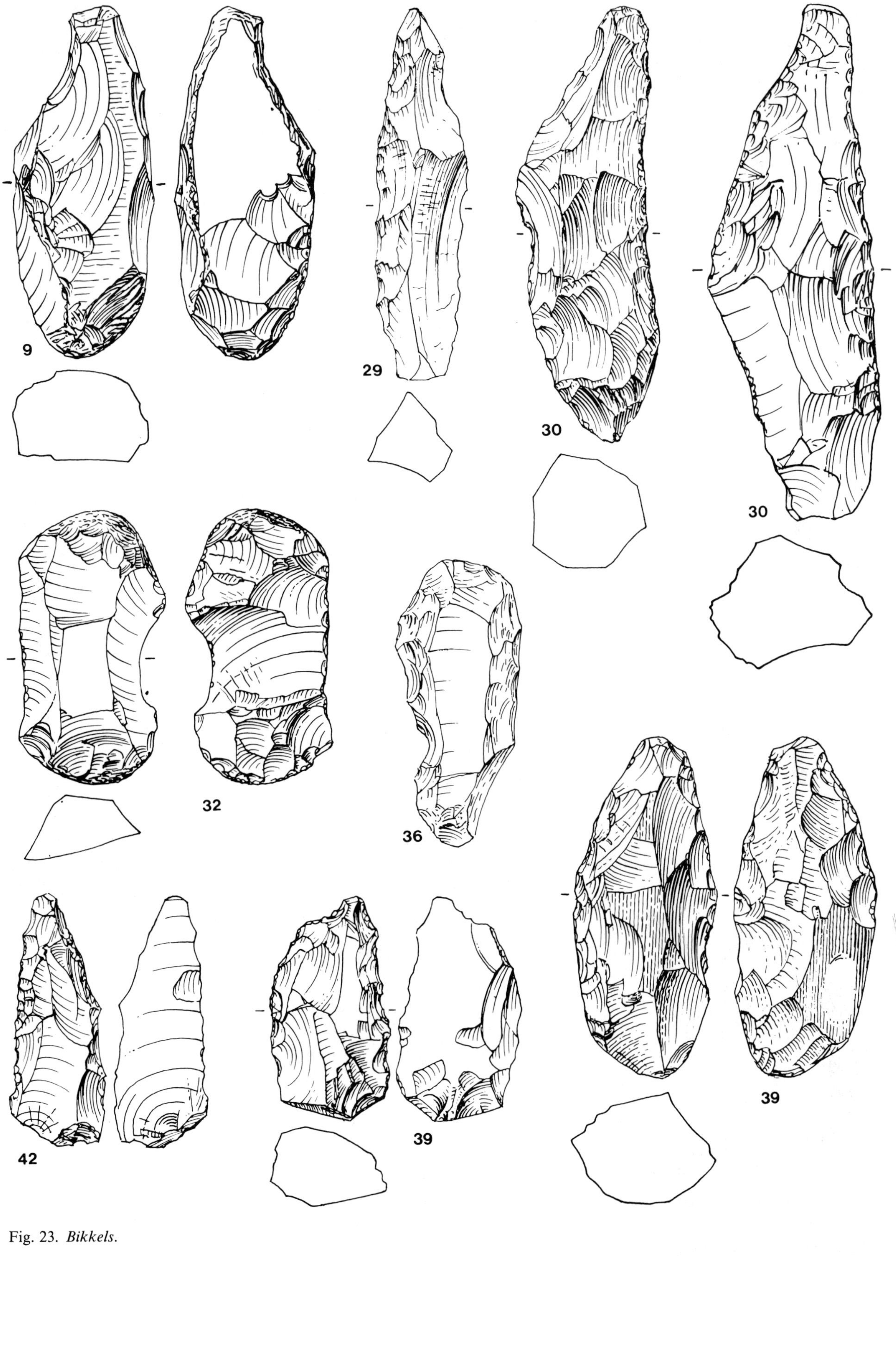

Fig. 23. *Bikkels.*

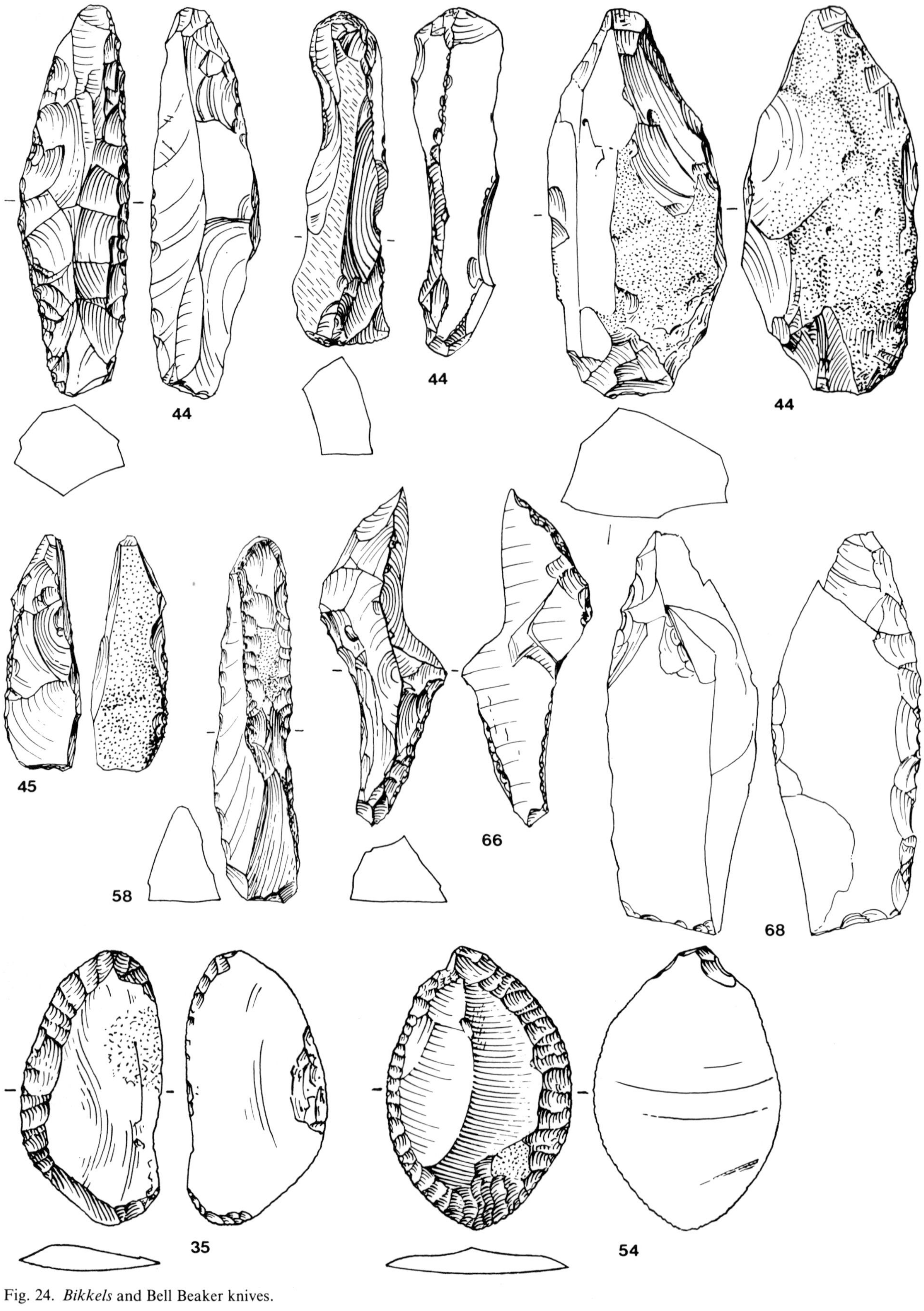

Fig. 24. *Bikkels* and Bell Beaker knives.

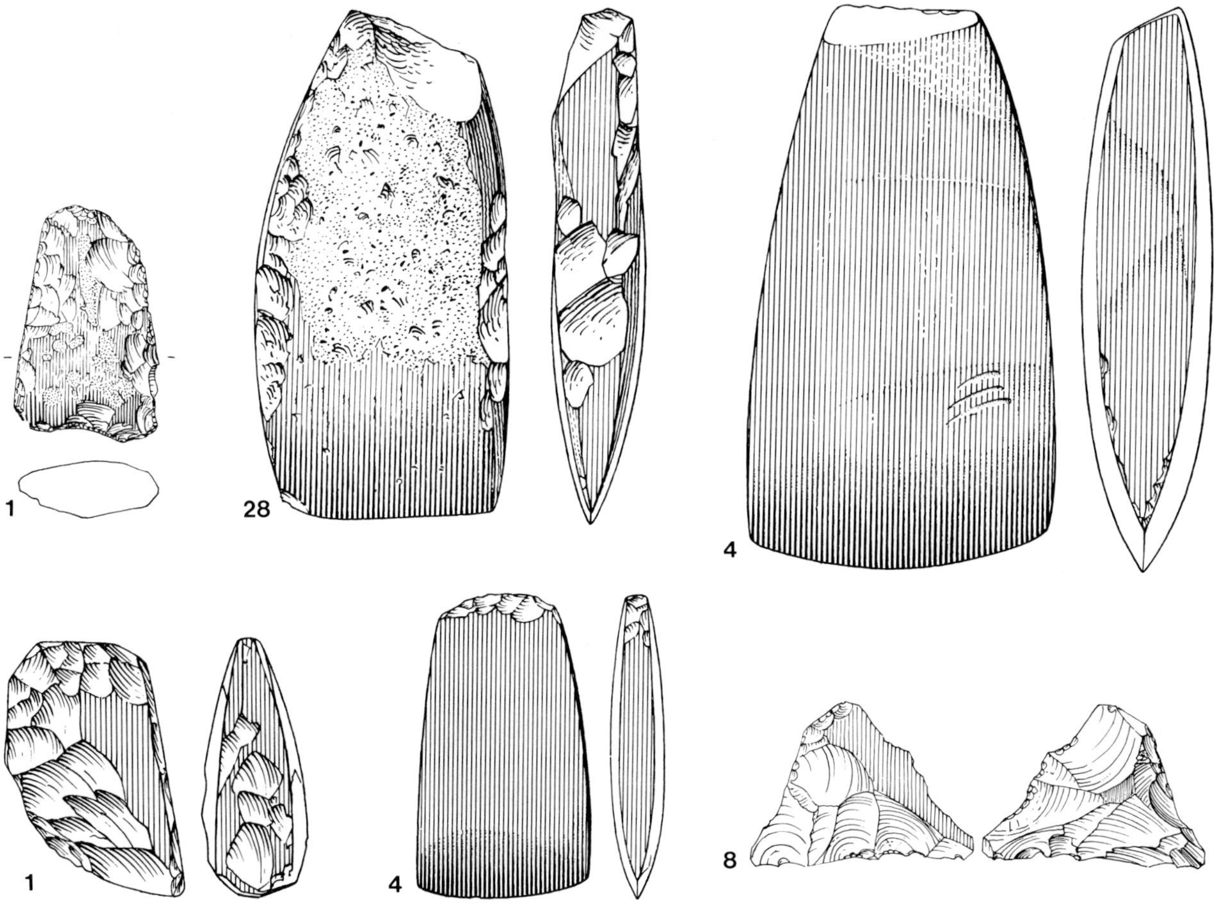

Fig. 25. Axes and axe fragments. Scale 2:3.

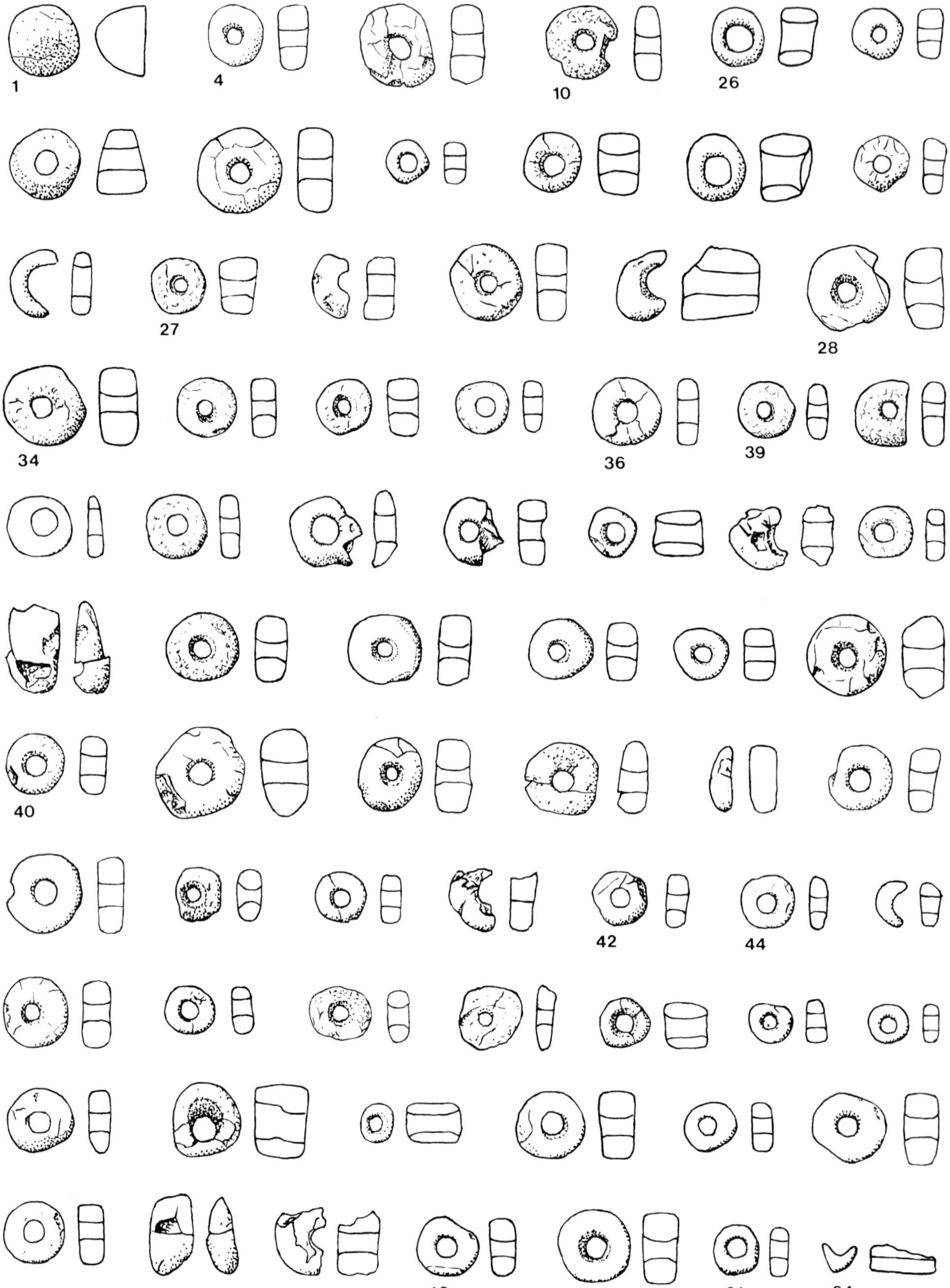

Fig. 26. Amber beads.

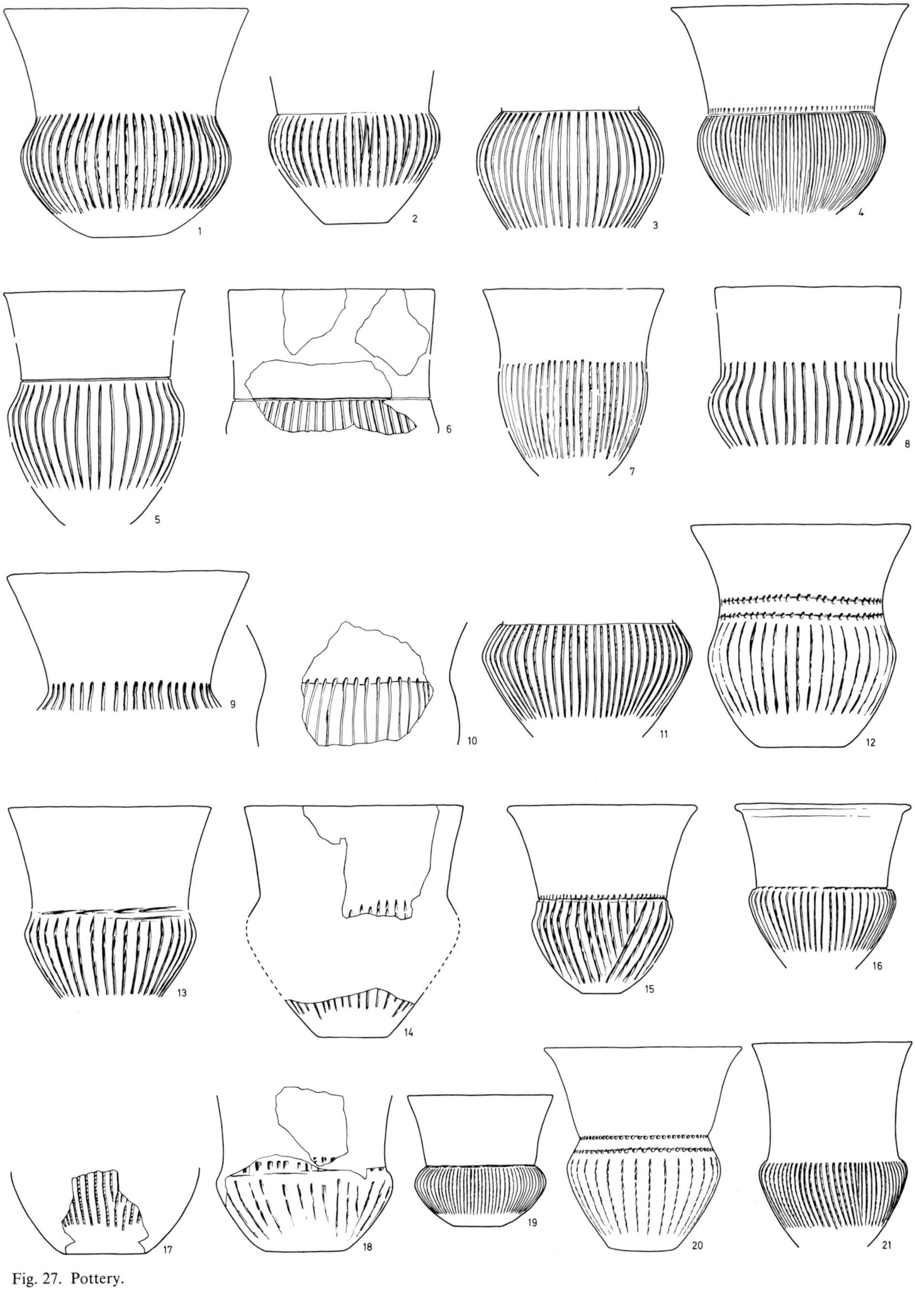

Fig. 27. Pottery.

Fig. 28. Pottery.

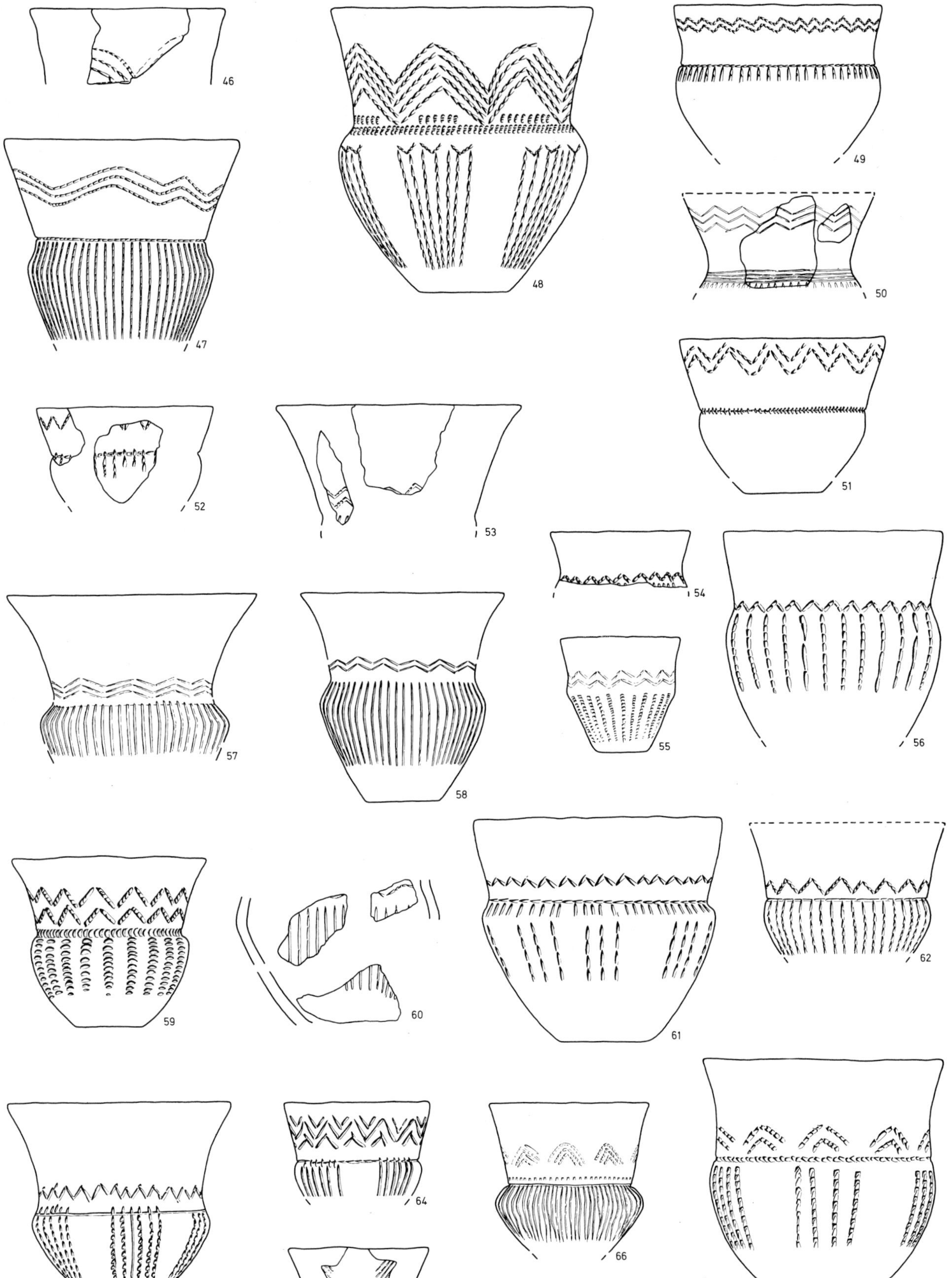

Fig. 29. Pottery.

Fig. 30. Pottery.

Fig. 31. Pottery.

Fig. 32. Pottery.

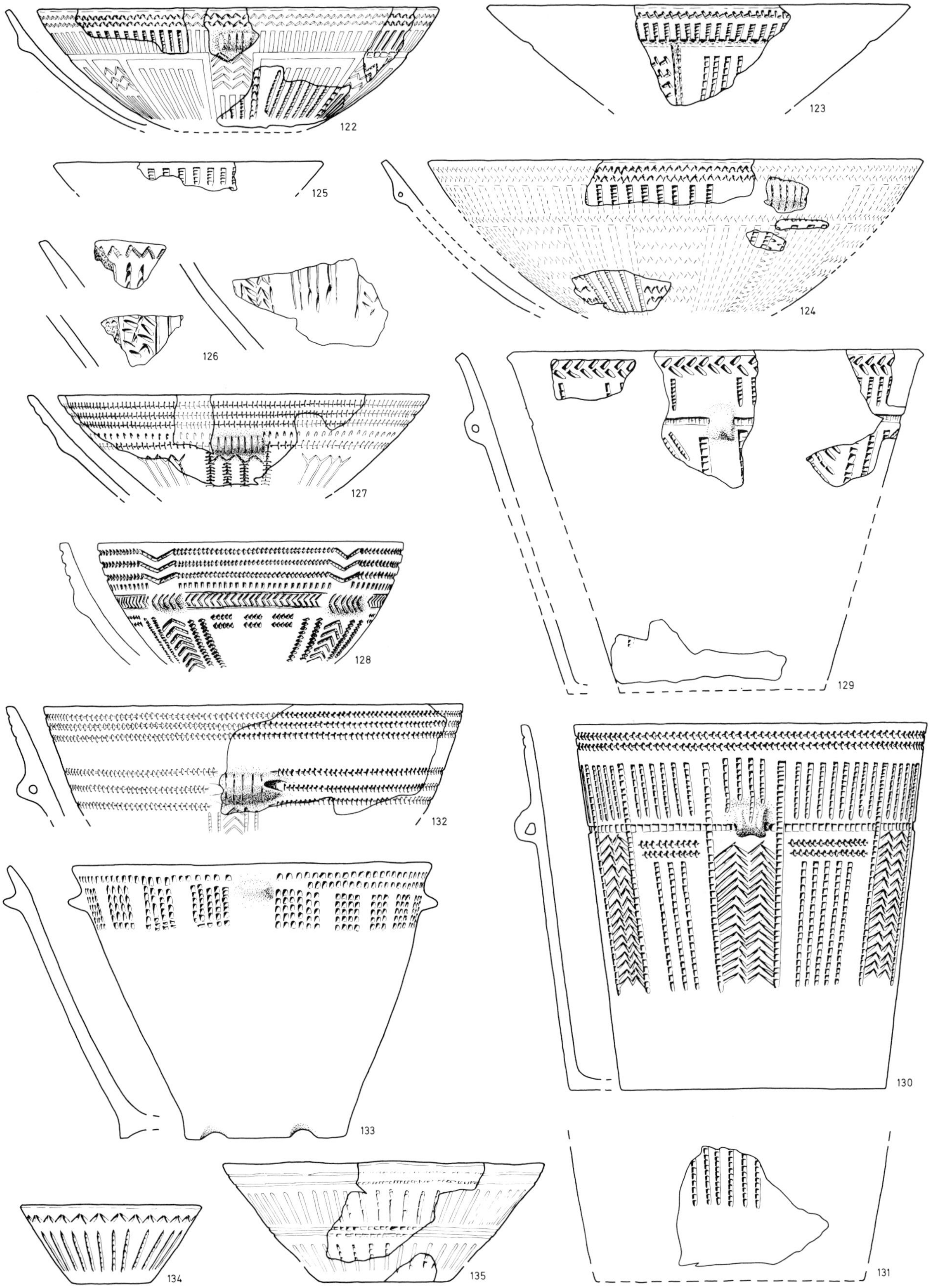

Fig. 33. Pottery.

Fig. 34. Pottery.

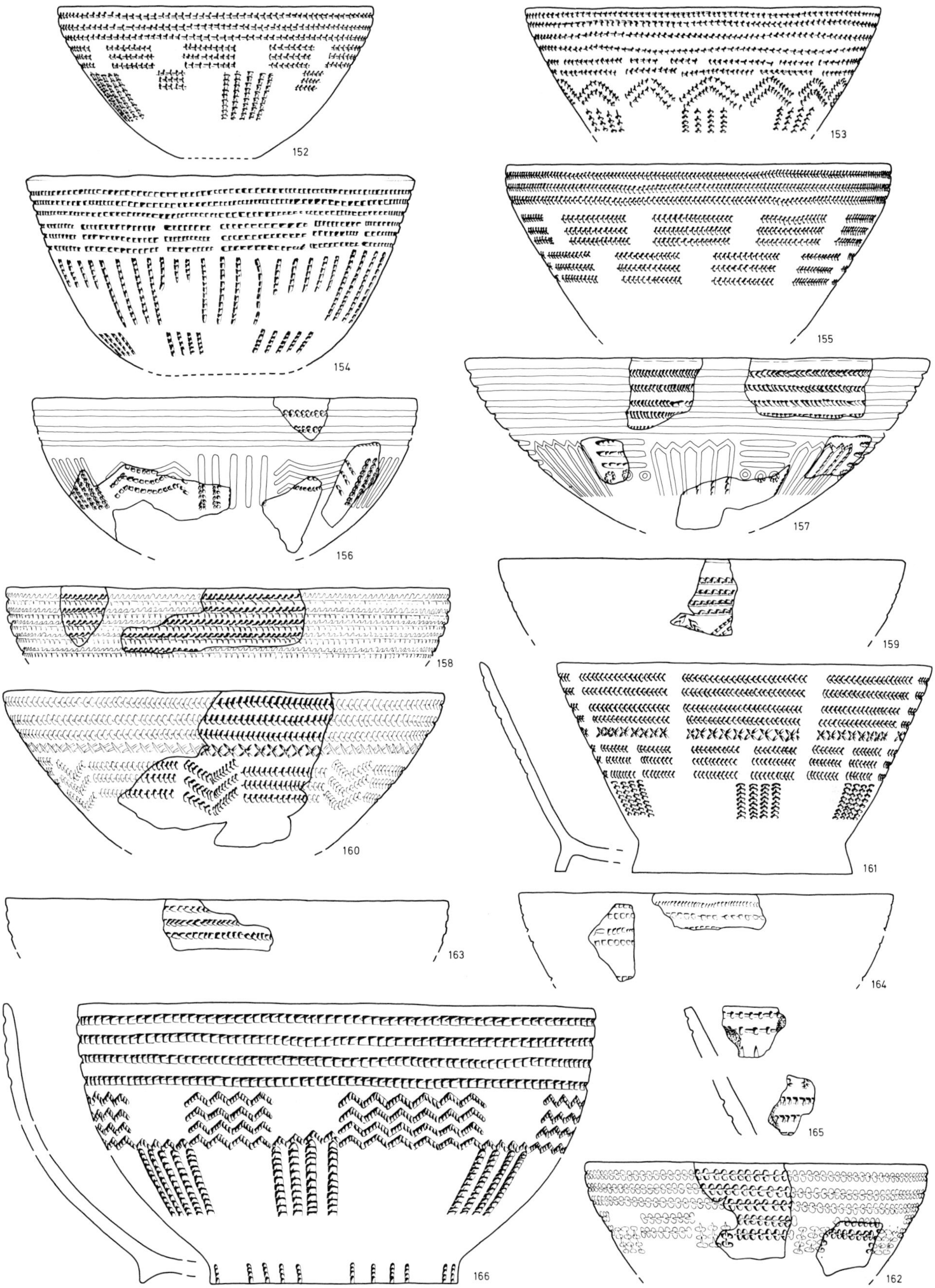

Fig. 35. Pottery.

Fig. 36. Pottery.

Fig. 37. Pottery.

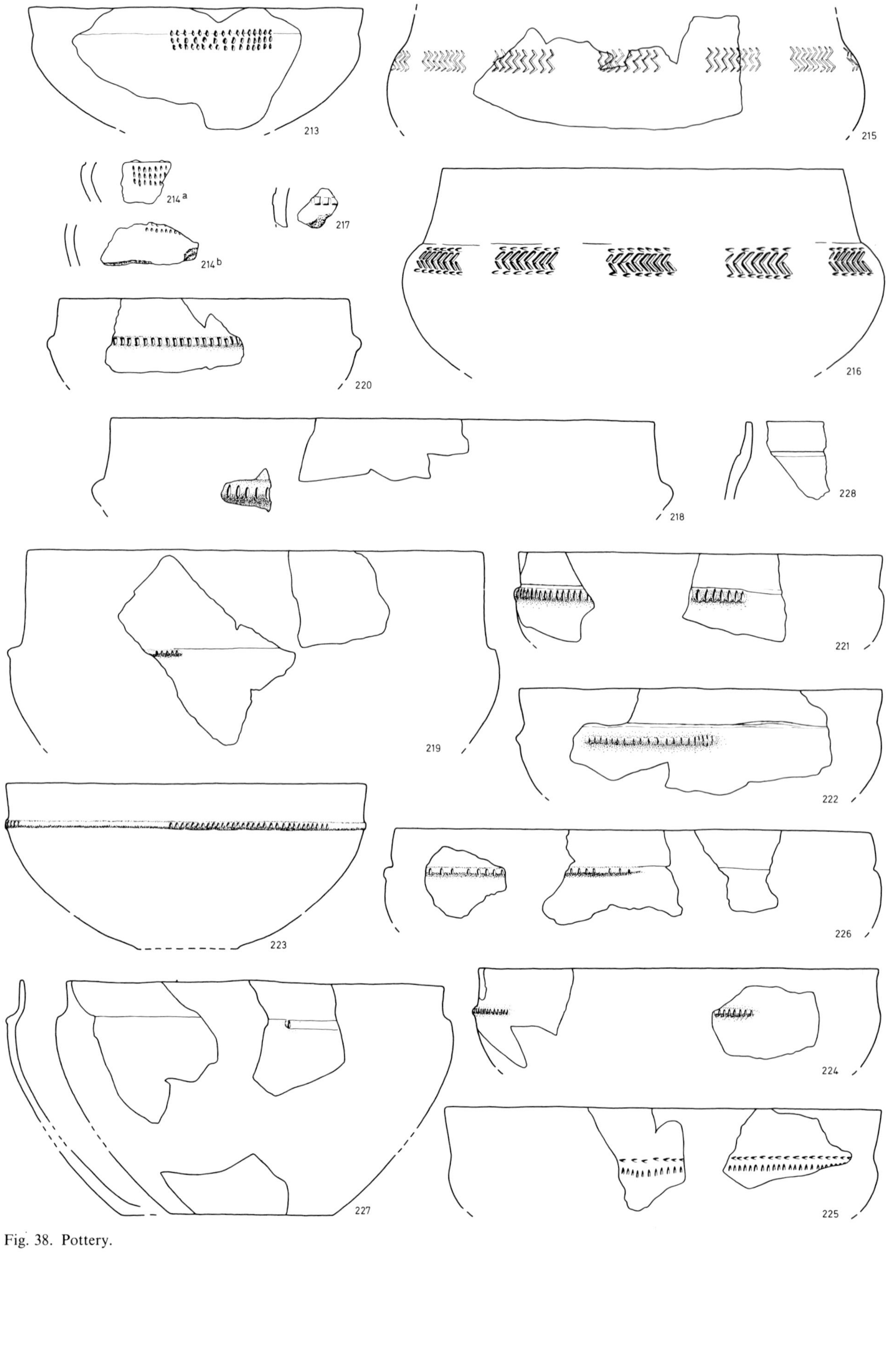

Fig. 38. Pottery.

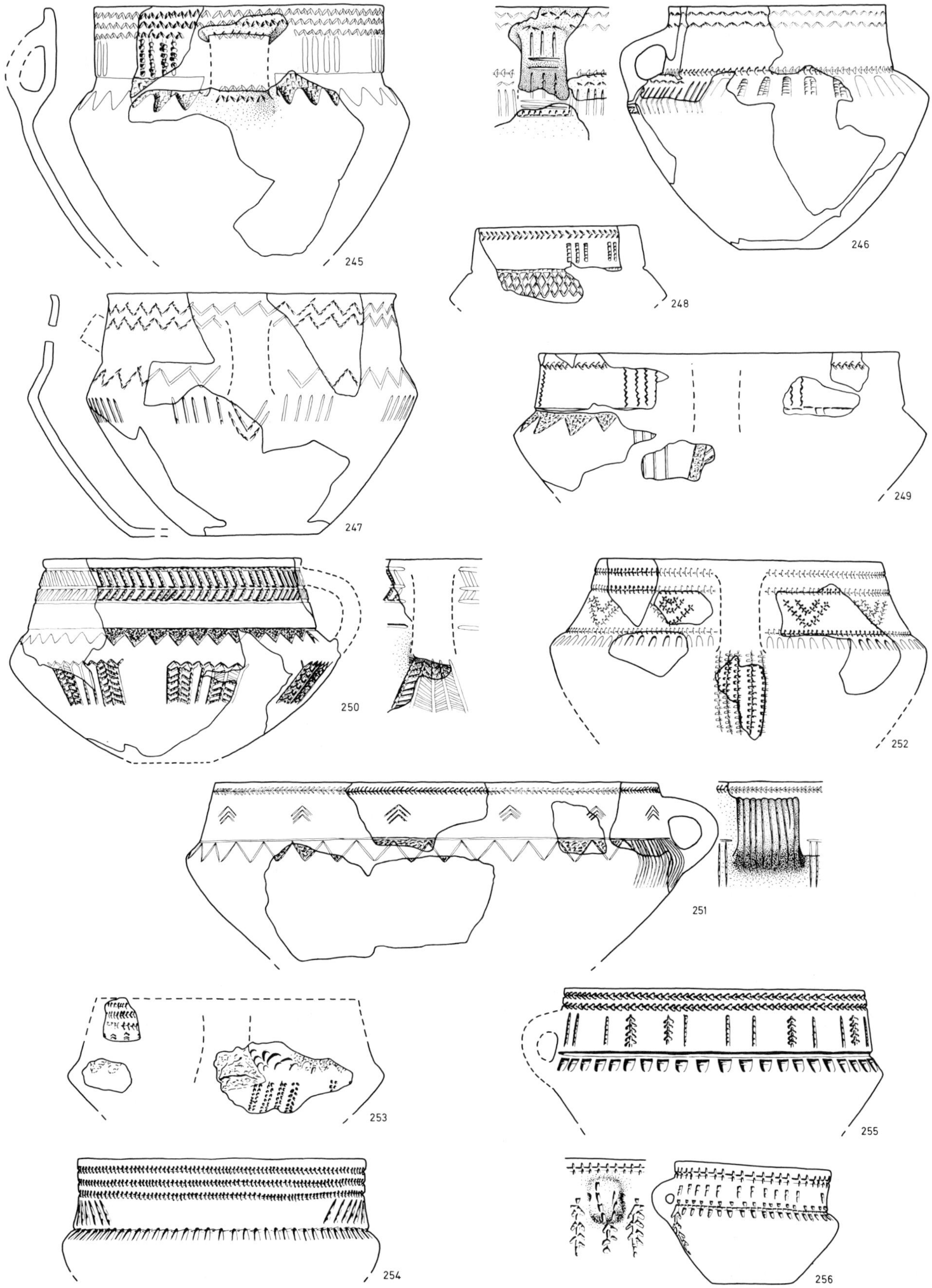

Fig. 39. Pottery.

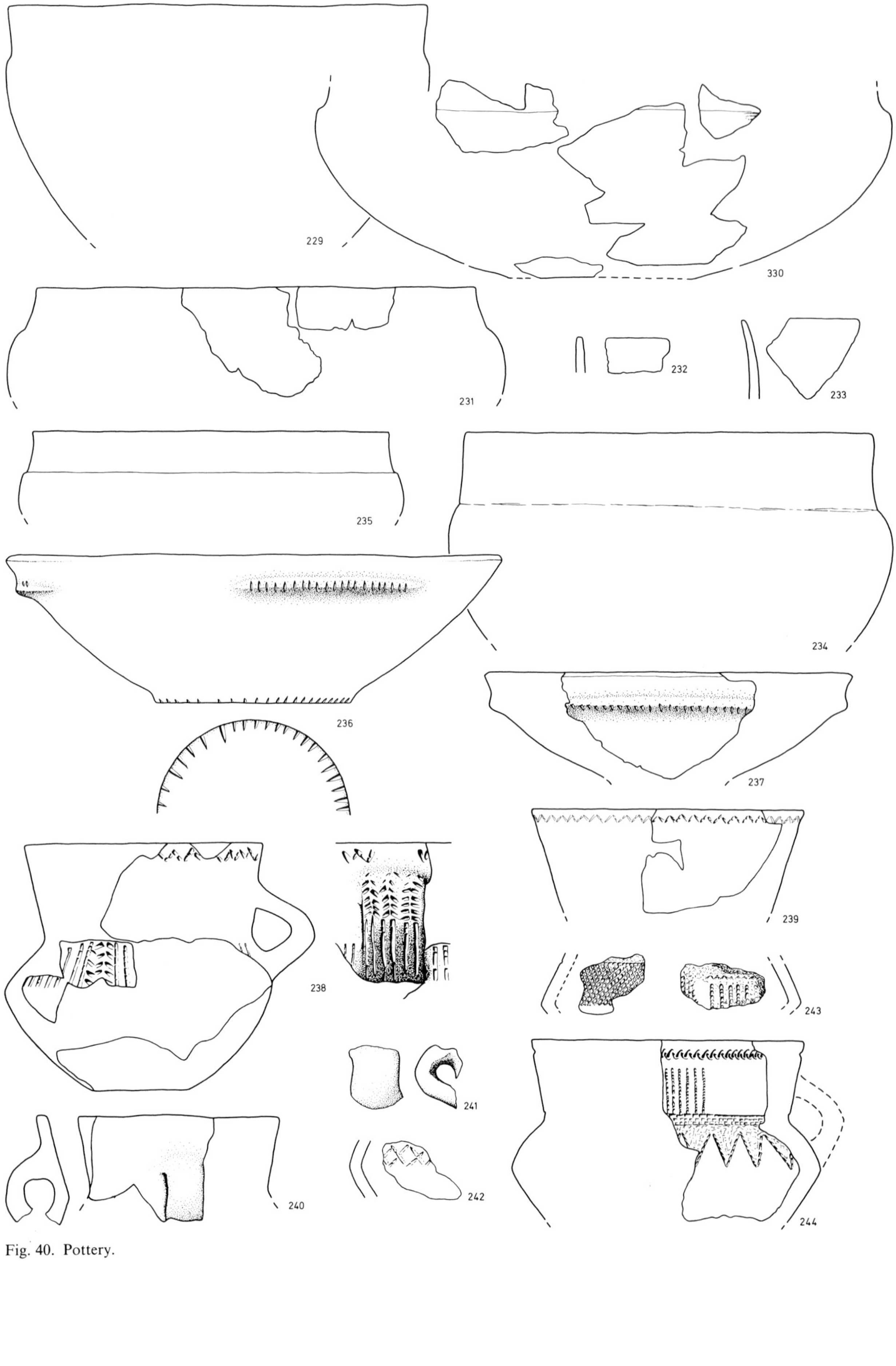

Fig. 40. Pottery.

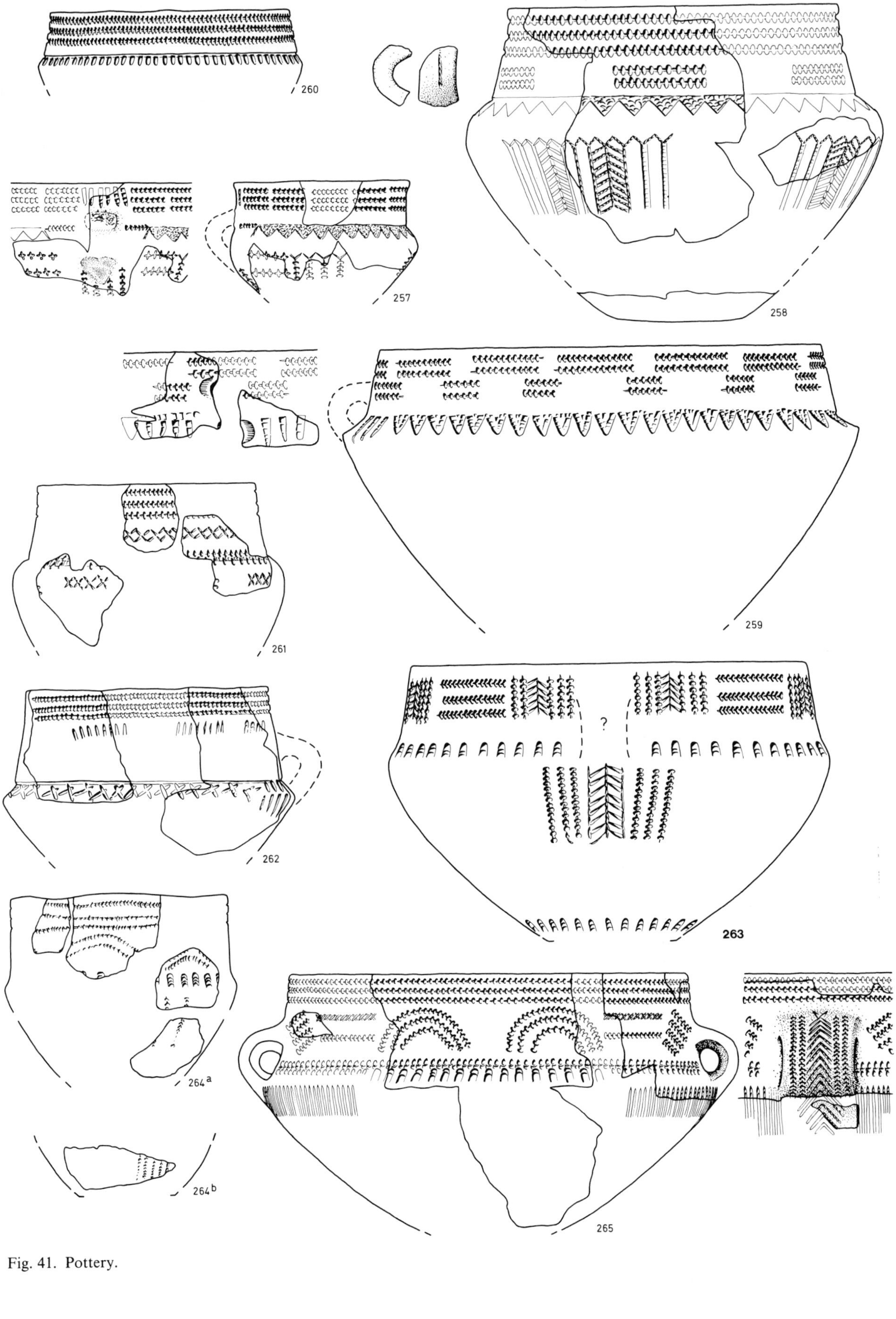

Fig. 41. Pottery.

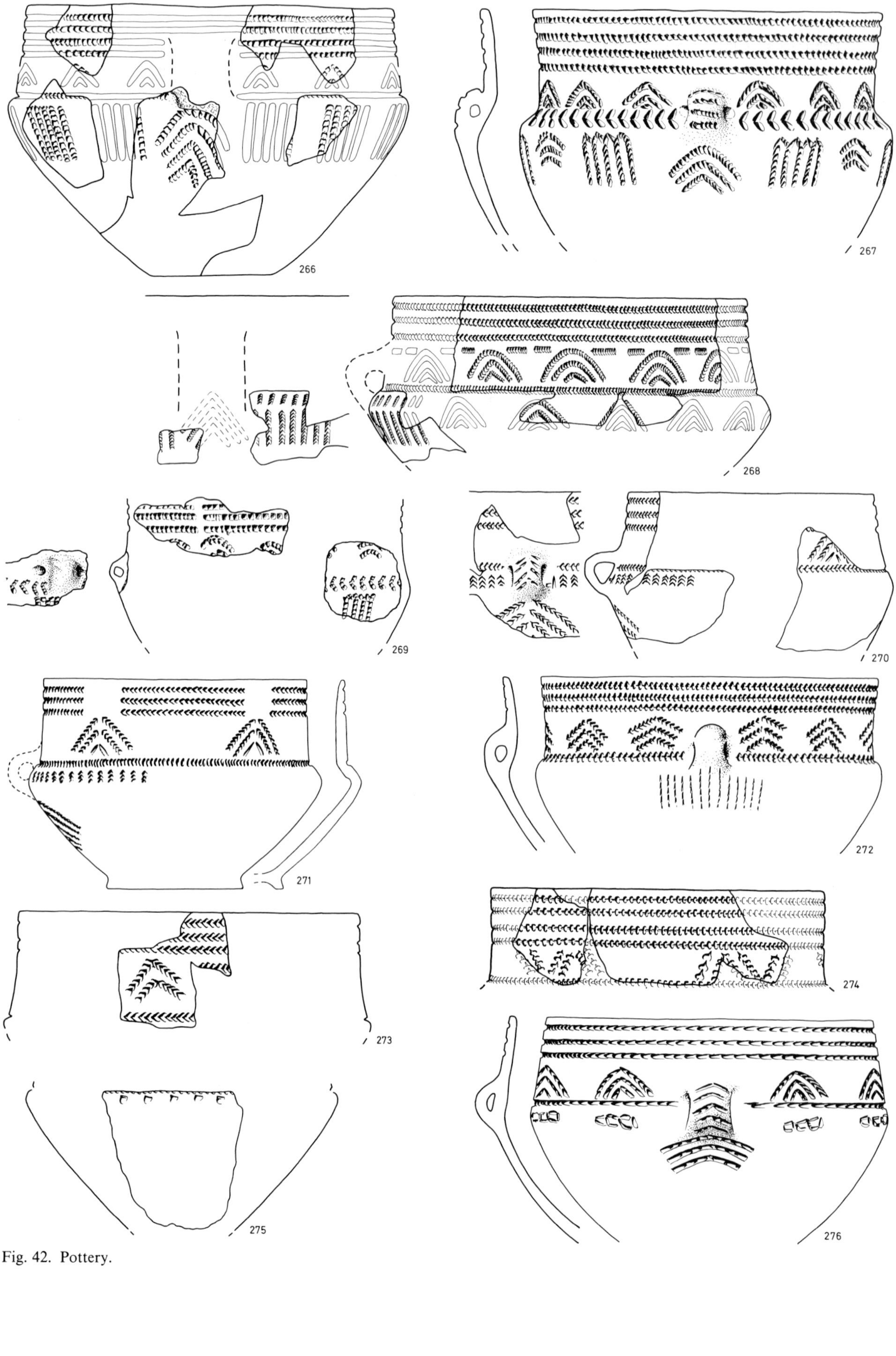

Fig. 42. Pottery.

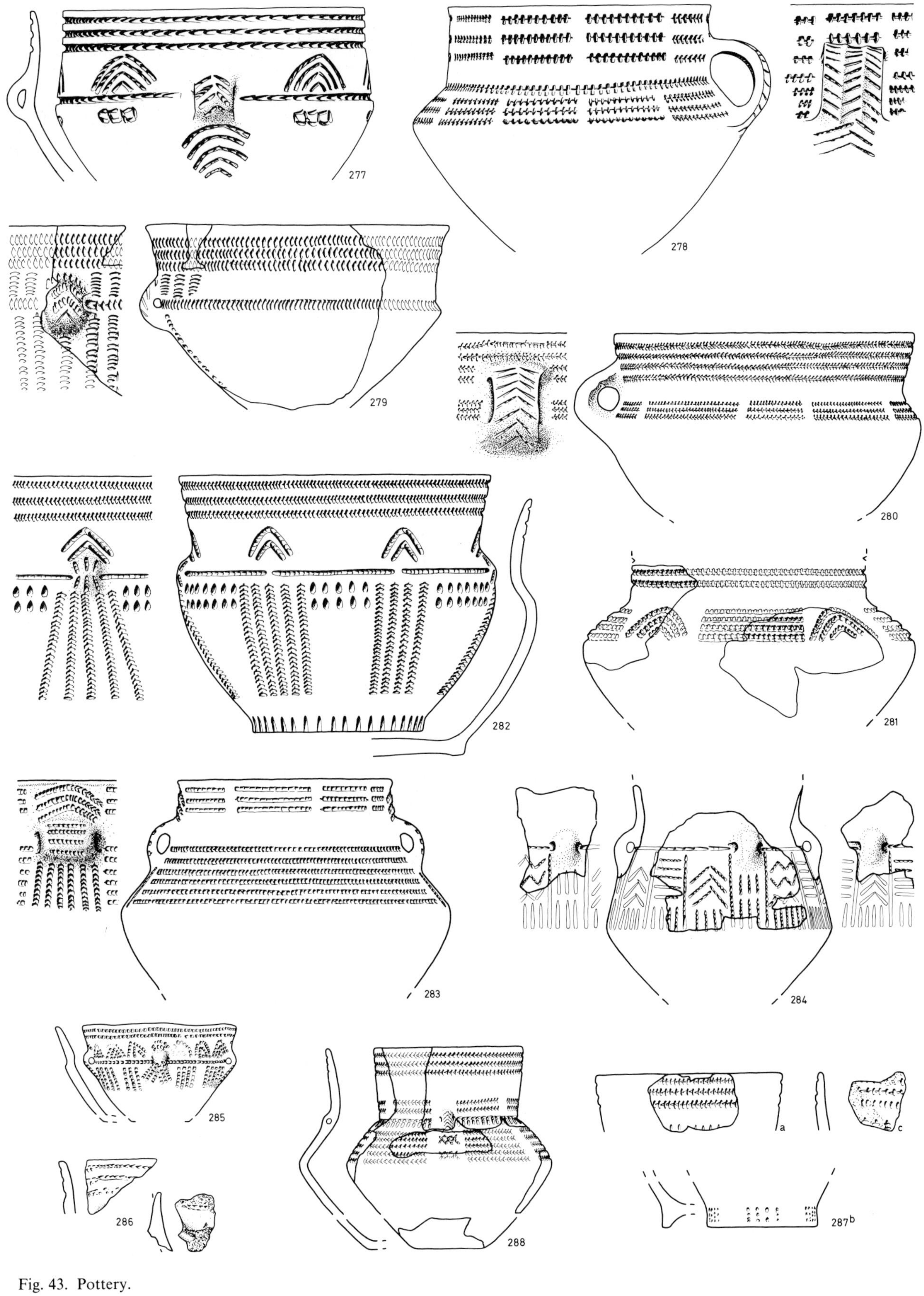

Fig. 43. Pottery.

Fig. 44. Pottery.

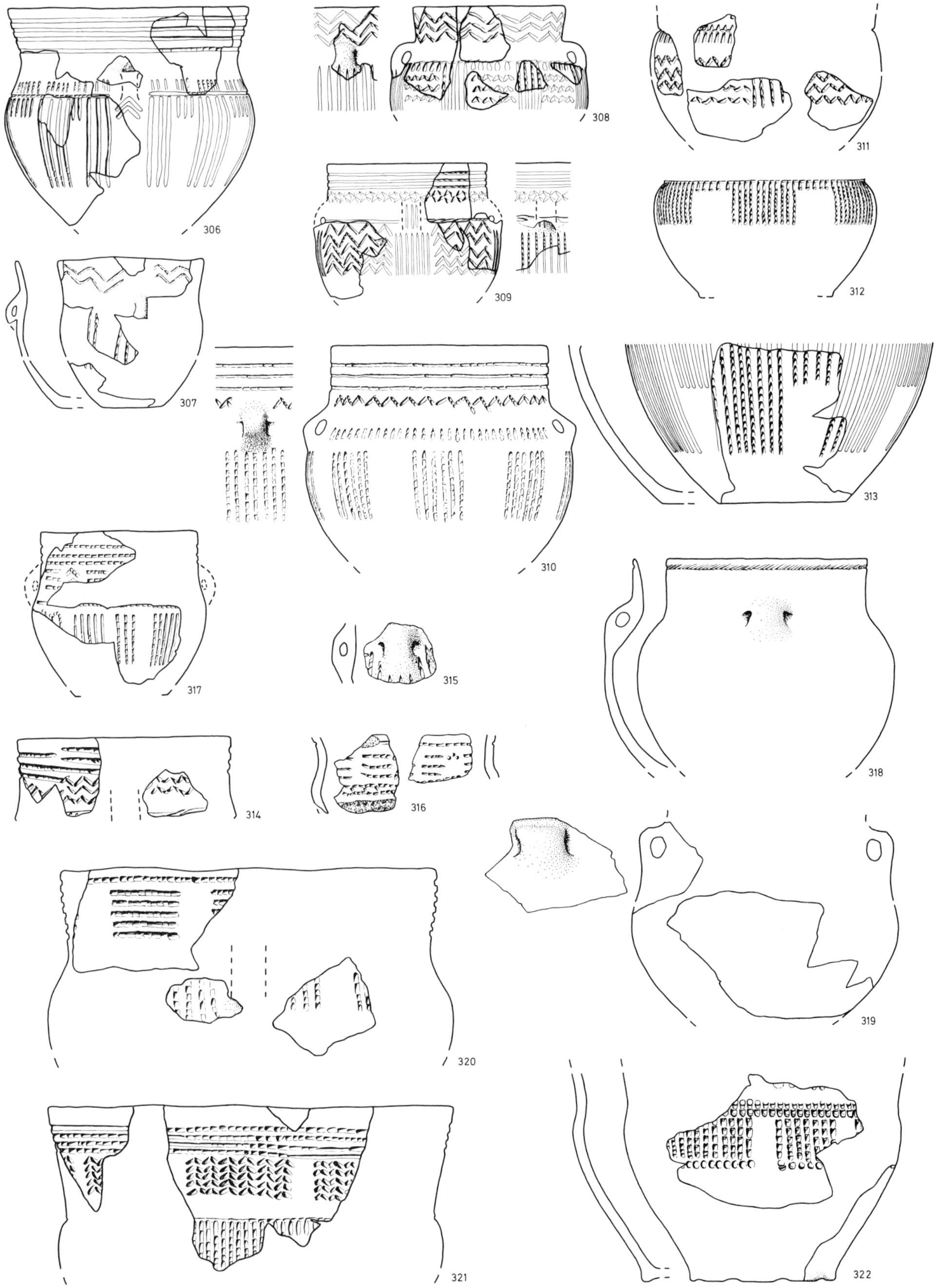

Fig. 45. Pottery.

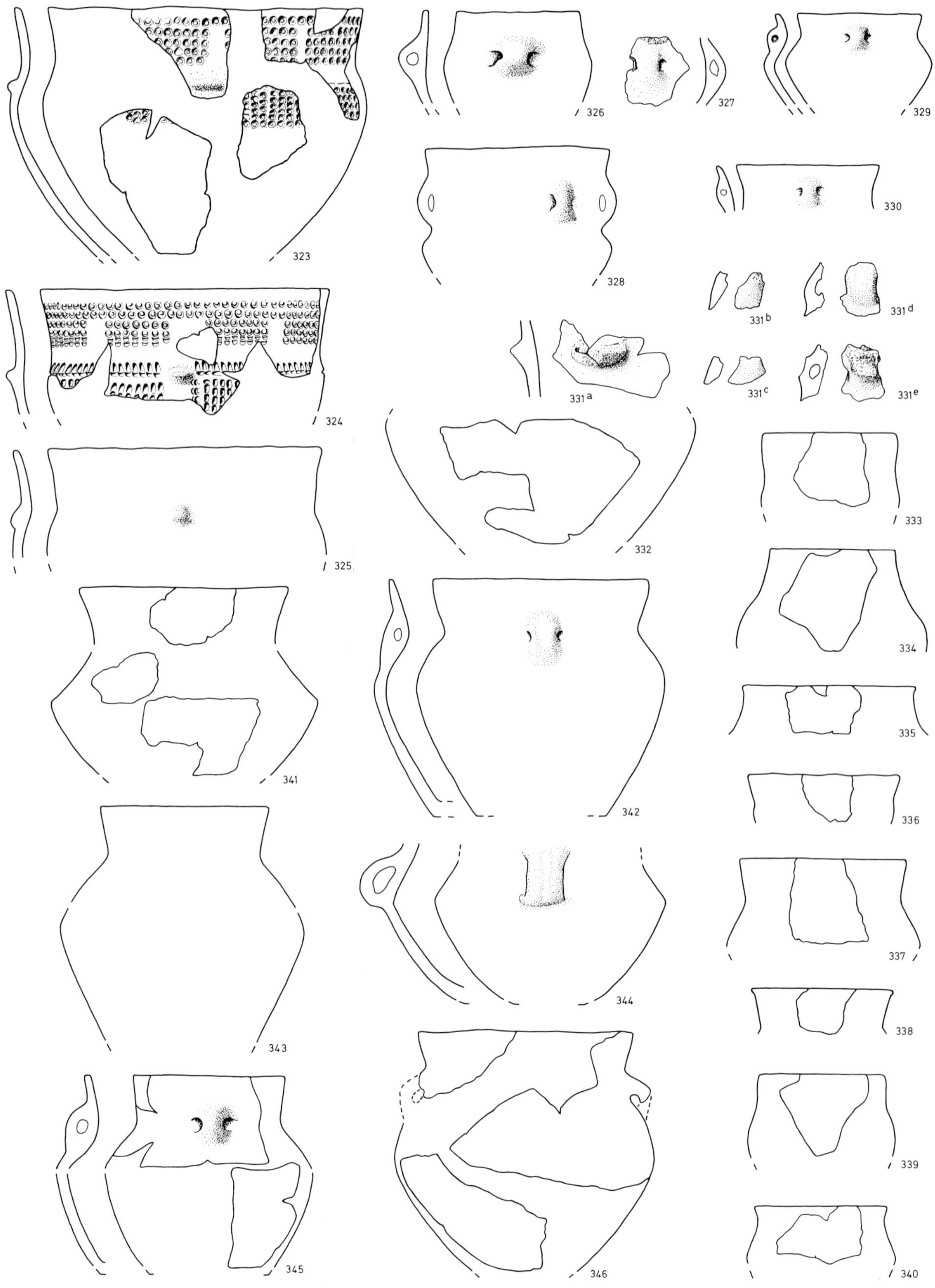

Fig. 46. Pottery.

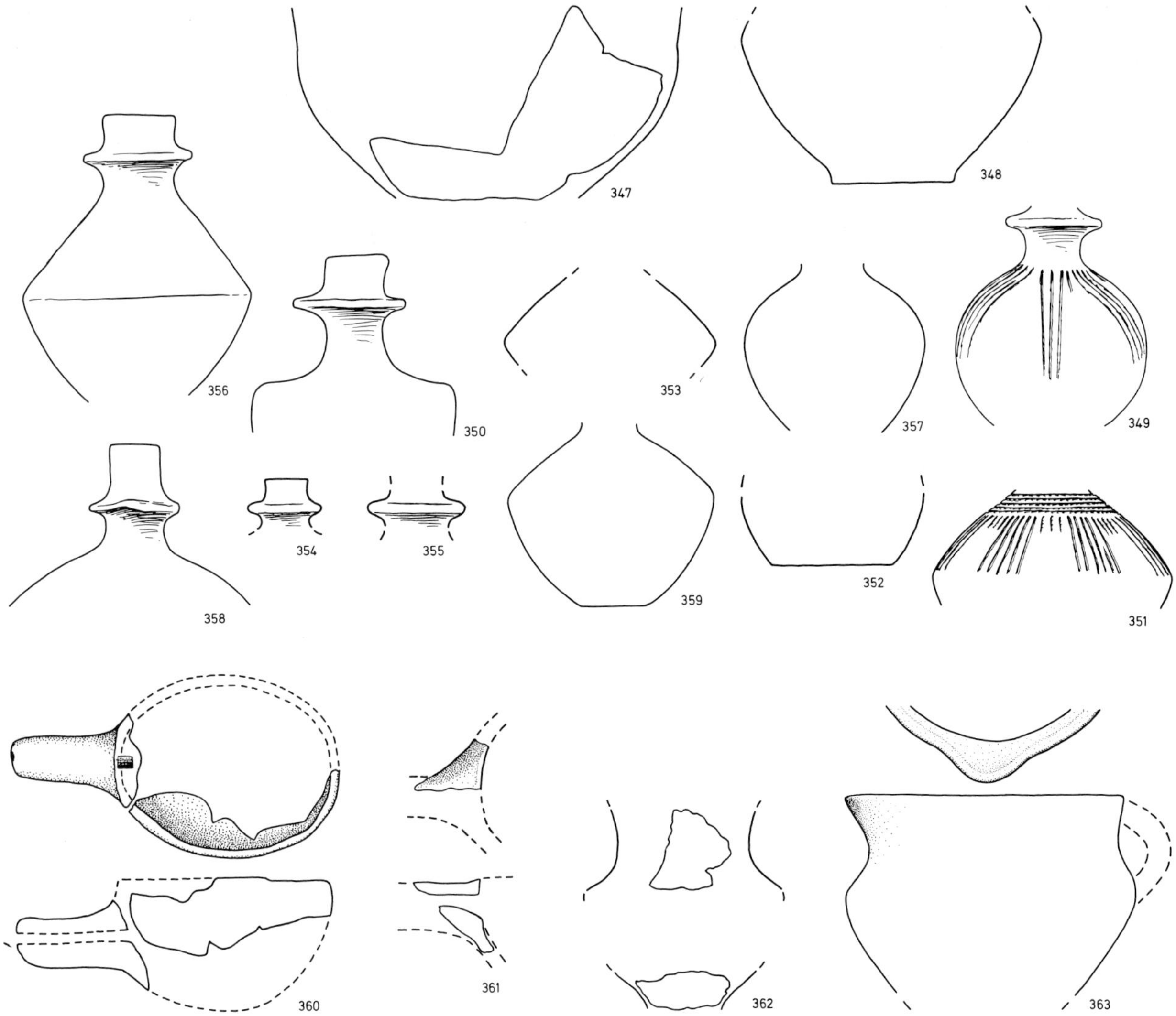

Fig. 47. Pottery.

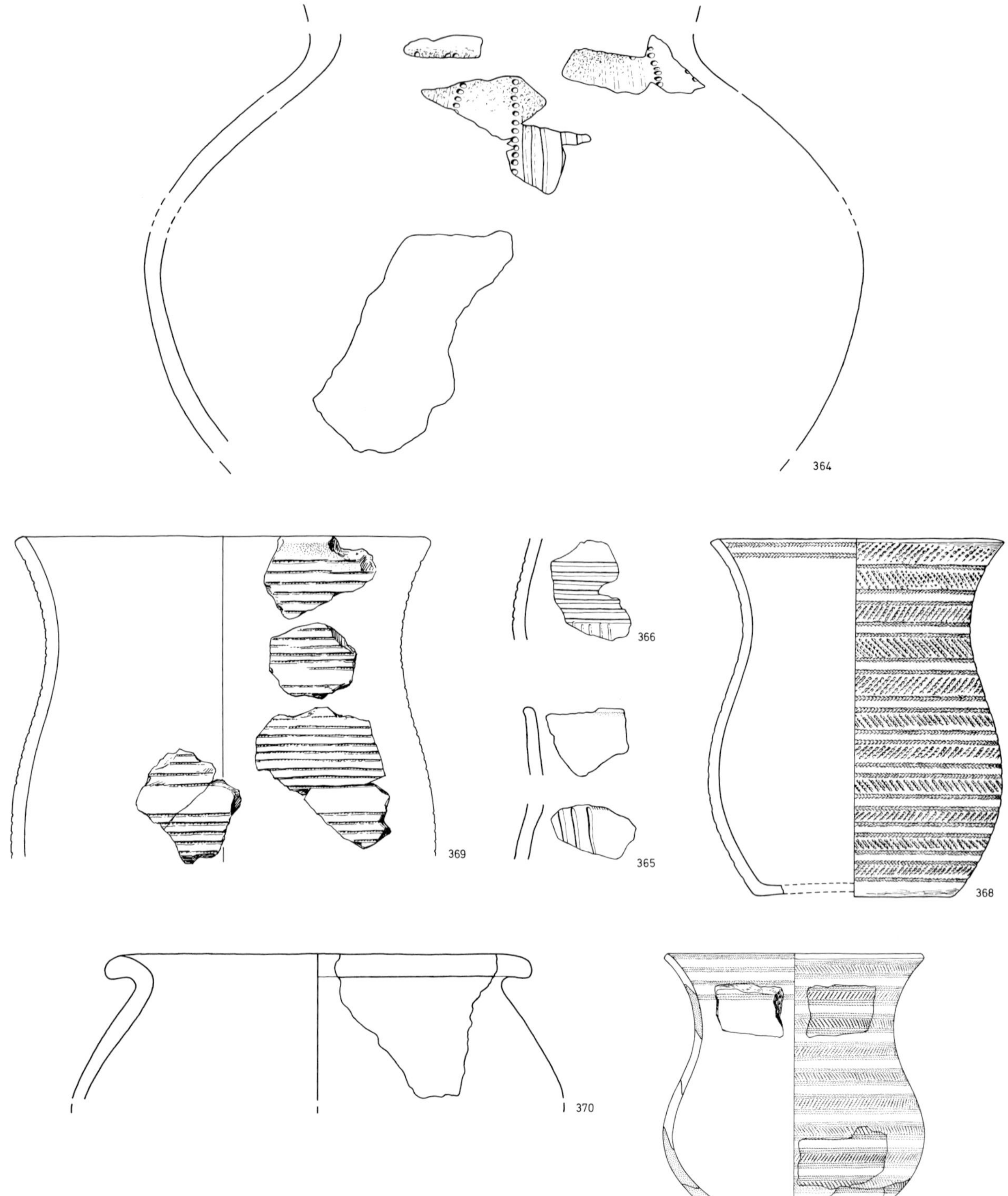

Fig. 48. Pottery.

Fig. 49. Pottery.

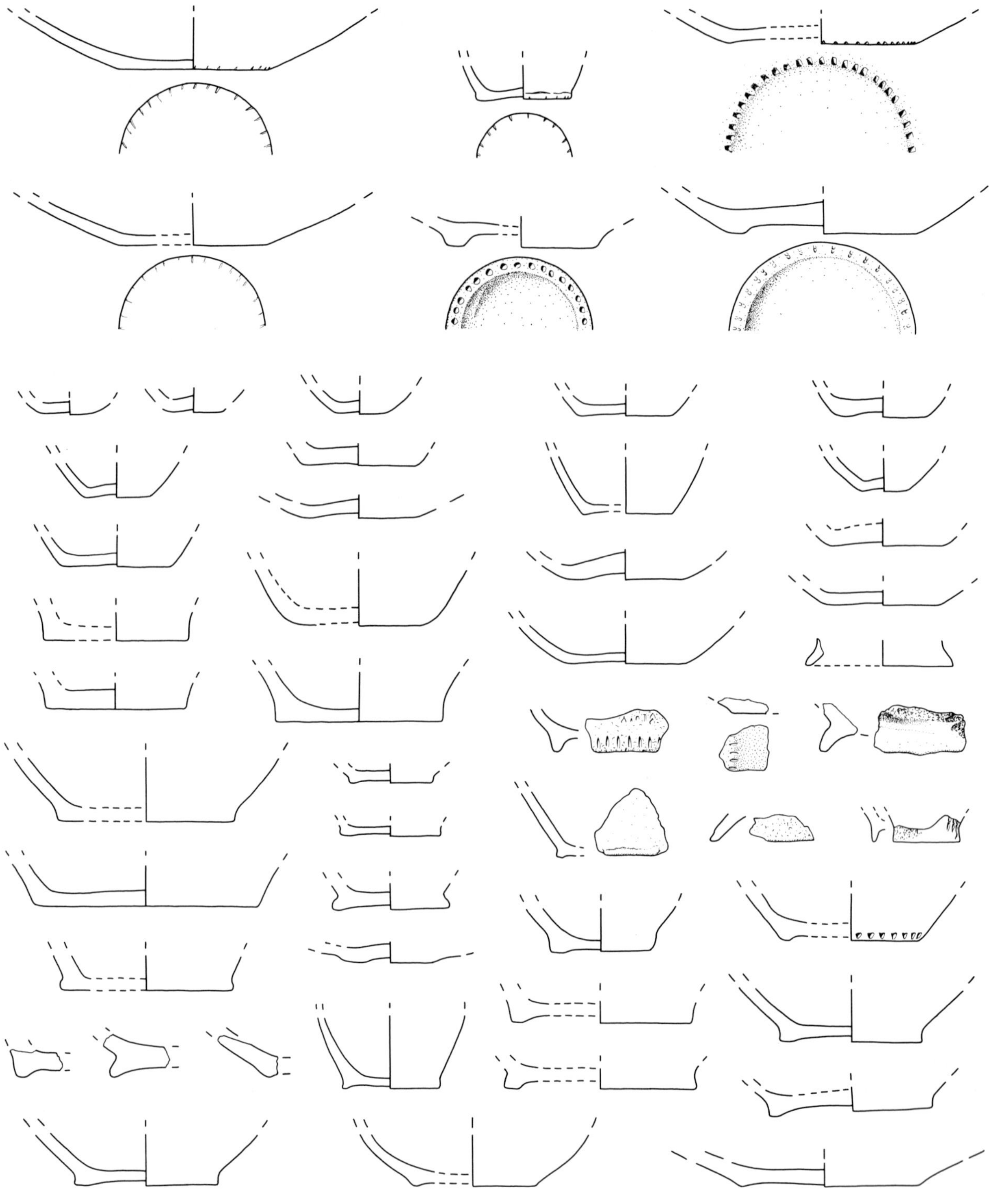

Fig. 50. Pottery.

THE TYPOCHRONOLOGY OF TRB WEST GROUP POTTERY

A.L. Brindley
Archaeological Survey OPW, Earl Court, Adelaide Road, Dublin, Ireland

ABSTRACT: In connection with the reconstruction of the large assemblage of pottery from the *hunebed* G2, near Glimmen, *gemeente* Haren, a survey of TRB West Group pottery was carried out. From this survey, a series of seven stages in the development of TRB pottery was identified, based on the development of form and decoration. These stages, 'horizons', are described in terms of the most popular motifs, techniques and forms at any one time. Horizon 1 shows distinct similarities to the pottery of the late Haassel-Fuchsberg group from which it is derived. The subsequent divergence and development of the West Group as a separate, regional entity is seen from Horizon 2 onwards.

The pottery belongs to two main traditions, one continuing through Horizons 1 to 4 and including most of the former Drouwen Style, and the second continuing through Horizons 4 to 7 and including most of the former Havelte Style. This development lasted approximately 550 years, beginning at about 3400 cal BC and ending at or around 2850 cal BC.

KEYWORDS: Northern Netherlands, Northwest Germany, Neolithic, TRB West Group pottery, typochronology.

1. INTRODUCTION

The reconstruction of the Glimmen G2 assemblage (Brindley, 1986) produced a representative collection of TRB pottery from all but the earliest stages of the development. This provided an opportunity to examine and review the typochronology outlined by Bakker (1979).

Van Giffen (1925/27) identified three styles of TRB pottery; the Drouwen, early Havelte and late Havelte styles. Later, Knöll (1959) published an exhaustive typological study from which he also identified three successive phases; Knöll 1, Knöll 1/2, Knöll 2. Although the material, sequences, and timespan of the two series are almost identical, the subdivisions are slightly different – the result of one being based ultimately on mutually exclusive assemblages and the other on typological considerations. This and Knöll's very detailed typological analysis led Bakker (1979: p. 48) to conclude that, during the 750 year long period which he believed was involved, further styles should be identifiable and that these should indicate chronological phases. He therefore used Knöll's typology (with slight amendment of two pilot series) as a basis on which to refine the Knöll and Van Giffen schemes. There is no denying that most of Knöll's typological explanations are reasonable and since then, radiocarbon dates have indicated that the general trend of his phases is correct.

Bakker discussed the Knöll typology in detail in chapter 3 of his book. The absence of any illustrations in this section is a very great hindrance to any appraisal of the discussion and indeed of the briefly outlined typochronology which he puts forward in the subsequent chapter. It is difficult to follow the lines of his argument and to apply the typochronology – both of which are open to criticism. At the onset and in personal discussion, Bakker has stated that this sequence is purely a typology but his use of the term 'phase' and his discussions, particularly of the dating of megalithic tombs, indicate that a chronology is understood and accepted by him. Secondly, the typology is heavily dependent on decoration, as Bakker felt that the profile would reflect the skill of the potter rather than its typological position. However, although individual examples may show a greater or lesser expertise, the level of competence on the whole usually does not create a problem when identifying vessels. It is more probable that working from photographs and drawings made an analysis of profile development impossible.

In practice, it proved very difficult to assign specific pots to Bakker's scheme. Bakker's typology consists of seven phases (A-G with both D and E subdivided) based on two pilot series, bowls/pails and jugs/tureens/amphorae (Bakker, 1979: pp. 61-75). Phase A material is scarce and the period was probably of short duration. For Bakker's Phases B, C and D1, bowls/pails are defined by the presence or absence of single decorative elements – zigzags, grooved lines or *tvaerstik* – and a general trend towards more rhythmic patterns. These bowls/pails are in fact all very similar and frequently occur together (Bakker, 1979: fig. 18). Similar problems arise with the second pilot series, this time from an over-emphasis of the chronological importance of triangles. It is more likely that Phases B, C and D1 represent a single phase, rather than subsequent and distinct chronological phases. Further problems of definition occur with Bakker's Phases D2 and E1. A large amount of material is assigned to these two phases, which Bakker himself states is difficult if not impossible to separate and which led him to suggest that a relatively long period of time might be involved. As with the definition of C and D1 material, the problems of definition arise from an over-emphasis of the importance of *tvaerstik* and in this case, the difficulties of reconciling the two systems, Van Giffen's and Knöll's, which diverge chiefly at this point.

In attempting to resolve some of these issues a series of seven horizons were identified, based on examination of a large number of assemblages, either in collections, publications or photographic archives (see Appendix) representing all stages of TRB development. The assemblages represent both long term and short term depositions. The horizons are described in terms of the most popular forms and decorative schemes at any one moment and the development of these. Within each horizon several typological changes may occur, but these are not necessarily chronologically significant. In general, typological developments are restricted at any one moment to a specific type, and secondly, the changes themselves are usually quite small and it is not difficult to imagine that some occurred spontaneously on separate sites. A horizon is only identified when changes can be noted across a wide spectrum. As with any sequence, the boundaries inevitably include some element of arbitrary judgement. In particular, within each horizon a general progression can be noted in the general direction of the overall sequence, while inevitably, pots with characteristics of two horizons will occur. In accepting the artificial constraints of an imposed system, allowances and judgements must sometimes be made.

2. THE TYPOCHRONOLOGY

In the following sections and the illustrations therewith, all examples are selected from groups of finds in the Netherlands, the Emsland (including the Hümmling) and Westphalia. Regional differences in decoration of TRB West Group pottery are recognizable even within this relatively small area. Although the typological development of the pottery here outlined is broadly valid for the whole of the West Group area, the citing of examples from Oldenburg and farther east is deliberately omitted because of the local variations that occur there. A good survey of the TRB West Group pottery in Oldenburg has recently been published by Fansa (1982).

Examples are cited from several large assemblages whose publications are widely available. For reasons of convenience the references consist only of site name and figure number. These large assemblages are: Emmeln (Schlicht, 1968), Gross Berssen (Schlicht, 1972), Wechte (Knöll, 1983) and Glimmen G2 (Brindley, 1986).

2.1. Horizon 1

Types: the full range of types has not been identified but include: bowls with lugs or perforations in pairs immediately below the rim, jugs with rounded bodies, funnel beakers with rounded bodies, lugged beakers, collared flasks, possibly lugged flasks (a single example is known from Glimmen G3 (Brindley, 1983: fig. 11); its early date is not in question but its precise date in relation to other material is not clear; lugged flasks may have occurred through a relatively short Horizon 1 and into Horizon 2). In addition to these, other lugged vessels seem to have occurred. The single sherd from a large pot with large handle at midbelly from Angelslo-Heemingeslag (fig. 3:6) seems to belong to a 4- or 8-handled amphora of the type found in Baalberge or Salzmünde contexts (Behrens, 1973: Abb. 26, 27, 29, resp. 33). An amphora of this type was also found in Oldendorf, Kr. Lüneburg, Grab II (Sprockhoff, 1952) together with pottery belonging to Laux's Group B, which may be equated to West Group Horizon 2 (see 4).

Decoration: horizontal elements immediately below the rim (e.g. maggot impressions or dots or oblique lines forming a rough zigzag). Vertical, decorated strips (chiefly ladders with horizontal, oblique or vertical hatching, zippers, maggot zippers) on remainder of body. Few bowls are sufficiently complete to allow a full analysis, but one of the three bowls from Exloo D30 and the bowl from Bronneger D21 show a regular and repeating series of what at first glance appear to be randomly arranged strips, for 2/3rd to 3/4th of their circumference (figs 3:1 and 4). Both jugs and funnel beakers usually have undecorated necks and decoration which covers shoulder and body in a single scheme. An exception is the jug from Emmen D43a which also has a narrow horizontal ladder below the rim (fig. 3:8, courtesy J. Molema).

Techniques: *Tiefstich*, incised lines, maggot impressions.

Motifs: narrow strips, either as ladders (horizontal hatching) or hatched (oblique or vertical hatching), zippers, dots.

Closed finds: Exloo D30, three bowls (fig. 3:1-3, one only recently recognized) found amongst the packing stones of the *hunebed* (Van Giffen, 1925/27, part II: p. 227), Angelslo-Heemingeslag (Bakker, 1979: p. 183, fig. B 1; new reconstruction this paper: fig. 3:5-6).

Other finds: Valthe, sherd of a lugged beaker (fig. 3:7). A very fine and well-preserved lugged beaker which is closely comparable to this is known from Neumünster (Struve, 1939: Abb. 1 and 2), Emmen D43a, sherds of jug (fig. 3:8), Emmen-Oude Roswinkelerweg (collection and information, E. Drenth), Bronneger D21 (figs 3:4 and 9), Rijs F1 (Van Giffen, 1925/27: Pl. 152: 5,8,9), Glimmen G3 (Brindley, 1983: fig. 11).

2.2. Horizon 2

Types: lugged bowls and pails, jugs with increasingly angular profile and handles with angular bend, angular tureens with wide and curving strap handles (rare, Tinaarlo), funnel beakers, collared flasks, undecorated amphorae type 1 (one example known, Hooghalen (Bakker, 1979: fig. B4:5)), lugged beakers (no known examples, but the type occurs in both Horizons 1 and 3), biberons, possibly lugged flasks (see above, Horizon 1).

Decoration: the horizontal elements below the rim tend now to consist of continuous or repeated elements, e.g. zigzag lines, sometimes doubled, or a band of short vertical lines, or a horizontal ladder. This zone is most frequently found on the bowls and pails but may also occur on jugs where it is usually narrow and separated from the shoulder decoration by an empty area (towards the end of the horizon a few vertical lines may occur in the zone, Tinaarlo, fig. 4:12, Hooghalen, fig. 4:14-15). On the bowls and pails, however, there is no empty area and the lower decoration starts immediately below the horizontal elements, and consists of vertical strips including ladders, zippers, hatching, and chevrons, interspersed with vertical lines (including *tvaerstik* lines). These strips are often combined to give wider strips of decoration. The strip itself may broaden and may be bordered by narrow ladder strips (e.g. Zeijen A and E, Bakker, 1979: figs B13:33, B14:20a). This combination of motifs leads eventually to the well-defined panels of Horizon 3. Although simple vertical lines in groups and fringes are now very common, strip decoration also occurs occasionally on the shoulders and bodies of funnel beakers (Hooghalen, fig. 4:19), and on the shoulders of the jugs where it is augmented by triangles or their equivalent. Apart from the *tvaerstik* line the most important addition to the decoration is the shoulder triangle, which appears first during this horizon, subsequently becoming perhaps the most characteristic motif of the following horizon. Its origin is unclear; it may have developed from the ladder. A jug from a stone cist at Gudendorf, Stadt Cuxhaven (Tempel, 1979) has shoulder decoration which may be an early mutant of these ladders. Its shoulder triangles are unusual in both form (very narrow and truncated) and hatching (horizontal). The overall impression suggests a connection with ladders rather than with the broader triangles which may have developed subsequently in line with the shorter shoulders of the tureens of the Düsedau horizon (see chapter 4). In West Group terms the Gudendorf stone cist can be dated to the transition of Horizons 1 and 2. The true hatched triangle occurs in conjunction with tureens with wide, angular shoulders and broad strap handles. It occurs in hatched form on tureens of the Düsedau Horizon of Altmark *Tiefstich*. Subsequently, probably shortly afterwards, it is replaced by the filled triangle which is a constituent of the Haldensleben Horizon (Preuss, 1980). In the West Group, hatched triangles are rare (e.g. Tinaarlo, fig. 4:12), but the variants – filled, pseudo triangles and shoulder stamps – all occur during Horizon 2, at least during the end part of it (filled: Hooghalen, fig. 4:14; pseudo triangles: Hooghalen, fig. 4:15; shoulder stamp: Zeijen, fig. 4:11).

Techniques: chiefly *Tiefstich* or simple lines. Maggot impressions (Elspeet, Bakker, 1979: fig. B10:25); Glimmen G2, No. 126; Hooghalen (fig. 4:20), these do not seem to occur after this horizon. The earliest examples of *tvaerstik* date from this horizon. Sherds from a pail or bowl found in a pit at Emmen D43 show a combination of ladder and single *tvaerstik* line. Two jugs (Exloo D30, Van Giffen, 1925/27, part II: fig. 9:22; Weerdinge, fig. 4:10) have *tvaerstik* lines on the shoulder. The horizontal ladder pattern on the Weerdinge strap handle is comparable to that on the early Horizon 3 tureen from Glimmen G2 (No. 246) which on other grounds has been compared to jugs (Brindley, 1986). This handle decoration may also be compared to decoration on the strap handles of tureens from Issendorf (Tempel, 1972: figs 4-6). Zigzags are executed as a continuous incised or *Tiefstich* line, or as discrete, incised or maggot impressed strokes.

Motifs: strips as for Horizon 1 but increasing in width and including chevron zigzag and zigzag and interspersed with vertical lines. Narrow vertical ladders play an increasingly important role in defining wide strips of ornament. Horizontal zigzags, true hatched and filled triangles, pseudo-triangles, shoulder stamp.

Closed finds: Diever (Bakker, 1979: fig. B3, except No. 3 which was found outside the cist and is later), Zeijen (Bakker, 1979: fig. B12-15, the relatively broad strips on both the jug (B15:44) point to a late stage in this horizon), Hooghalen (Bakker, 1979: fig. B4-B5).

Other material: Weerdinge (fig. 4:10), Barger-compascuum (fig. 4:2, information O.H. Harsema),

Bronneger D21/22, Glimmen G2 (Nos 113, 238-240), Exloo D30 (Van Giffen, 1925/27, part II: fig. 9:22), Elspeet (Bakker, 1979: fig. B6:3), Rijs F1 (fig. 4:24).

2.3. Horizon 3

Types: lugged bowls, pails, tureens with angular profiles and strap handles, funnel beakers, lugged beakers, amphorae type 1, collared flasks, biberons.

Decoration: within this horizon, the decoration diversifies further and in certain instances becomes more specific to individual forms. Immediately below the rim, a horizontal zone develops, usually consisting of vertical lines bordered by horizontal elements. The horizontal elements are usually zigzag or related motifs or simple lines, the lower border may be a broad and shallow grooved line. On bowls/pails, the band of vertical lines is usually continuous, but occasionally the lines are grouped. On tureens, the lines are usually in small, well-spaced groups but are sometimes replaced by pointed multiple arcs or by narrow, horizontal blocks of two lines (e.g. Glimmen G2 Nos 258, 259). Vertical lines occur frequently on both sides of the handle, and on the handle itself, extending down onto and sometimes over the shoulder. Zigzags, spaced verticals and pointed multiple arcs occasionally occur also on the necks of funnel beakers.

On the bowls/pails, below the horizontal zone the strips develop into relatively large panels often defined by either broad shallow grooves or broader *Tiefstich* lines pendant from the basal line of the horizontal zone (the grooves may also penetrate the upper horizontal zone). The panels form fairly regular, vertical zones, filled with vertical lines, chevron zigzags, vertical chevrons (particularly common below the lugs), zippers and in the upper portions of the widest panels, horizontal lines and zigzags.

The decoration on the shoulders of tureens also includes these panels, but the major decorative element of the tureens is the triangle or a related motif. The triangle occurs in various forms, true triangles (i.e. outlined and filled) pseudo-triangles (i.e. approximately triangular impressions) and as shoulder stamp. All these occur throughout Horizon 3, but only the shoulder stamp occurs frequently in Horizon 4. Decoration occasionally moves over the shoulder onto the upper body. A decorative panel, usually incorporating a chevron or multiple arc often occurs below the handle itself. Simple vertical lines are the most common form of decoration on the handles.

Decoration on funnel beaker bodies is similar to that of the preceding horizon, except that the decorated strips seem to have disappeared entirely.

Lugged beakers, having comparatively long necks, include a horizontal zone below the rim, and two vertical zones, one on the lower part of the neck and the second covering the shoulder and body. These zones consist of alternating panels. Lugged beakers are not particularly common but examples occur at Glimmen G2 and Exloo D30 where a second has been recognized recently (fig. 5:29-31), Huntedorf (Schirnig, 1979: Abb. 5) and Tannenhausen (information J.A. Bakker). Amphorae do not appear to be numerically significant until Horizon 4. A very finely decorated example was found in a flat grave at Werlte, Kr. Aschendorf-Hümmling (Schlicht, 1968: Abb. 7). The neck, which has a pseudo-*tvaerstik* line immediately below the rim, is incomplete and its decorative scheme is unreconstructable, but the shoulder shows a well-developed panel decoration. Other amphorae show chiefly *tvaerstik* technique but include elements such as vertical lines on the neck and panels on the shoulder and body which are more at home in Horizon 3 contexts (Bottrop-Kirchhellen, fig. 5:26; and Emmeln, fig. 5:24).

Typological developments within this horizon: several important typological developments contribute to the diversification in decoration, both in motif and technique. One of these is the progressive movement from true zigzag (executed by a single continuous movement) and zigzag by discrete opposed strokes, by way of separate strokes which are no longer discrete but which cross at each extremity, to strokes which cross to such a degree that they are termed skating motif. All of these variations occur within this horizon on vessels which are otherwise identical (compare Glimmen G2 Nos 113, 120, 118, 114). The *Tiefstich* technique also varies considerably through the use of variously shaped stylus heads, several different impressions often appearing side by side on a single vessel. The most important of these is the pointed nib- or quill-like stylus giving a deep-tipped triangular impression. This results in a line of close set impressions looking remarkably similar to *tvaerstik*. In fact, this form of *Tiefstich* is executed along a guideline which may occasionally protrude at the end of a line, or even more rarely, may still be visible behind the pointed impressions. However, the execution of these impressions normally destroyed the guidelines entirely (see Glimmen G2, Nos 111, 116, 251). A second variant of this technique involves the use of a pointed stylus to create a line of opposed, chevron-like deeply indented stabs. This is also executed along a guideline which is subsequently destroyed by the deep indentations of the stylus tip (e.g. Glimmen G2, No. 248). The typology of the panel defining lines or grooves is demonstrated most clearly on two large, well-decorated pails from Bronneger D21. The first (fig. 5:2) represents an undeveloped example of Horizon 3 decoration – the layout of the lower zone is emphatically vertical, with the few panels outlined by ladder strips. The second (fig. 5:4) shows a well-developed example of the same horizon, with wide panels and ladders replaced by very broad and shallow *Tiefstich* lines. The association of ladder strips with strips of other types which occurs from Horizon 1 onwards is highly significant for the ultimate development of the defined panels. Two additions to the motifs range, the multiple arc and the horizontal block, are significant in their continued importance through the following horizons. The multiple arc in its pointed form occurs sporadically during Horizon 3. Its use is confined chiefly to the necks of tureens (Glimmen G2, No. 251). It seems to have developed out of the single standing triangles at the base of the neck, examples of which are also known from early Horizon 3 tureens (i.e. the Seeste tureen, Bakker, 1979: fig. 31:1. The marked angularity and panel decoration of its shoulder are more easily reconciled with similar features not only of Horizon 3, but also of the previous horizon. *Tvaerstik*, presumably the main reason for Bakker's somewhat later dating of this vessel (to Phase D1) is an infrequent but well-established element of both Horizons 2 and 3). In its rounded form, the multiple arc becomes one of the most popular motifs of Horizon 4. Two line horizontal blocks also occur sporadically during Horizon 3, chiefly on the neck of tureens (Glimmen G2, Nos 258, 259) and at the top of the vertical panels on bowls and pails (Fig. 5:5). The motif has a long life, occurring throughout Horizon 4 where its lines have increased to between three and eight or nine, and well in to Horizon 5 where it is an equally important constituent of both the Heek-Emmeln and the Uddelermeer-Anlo Styles. In the former, it may include up to thirteen lines.

Techniques: *Tiefstich* (see paragraph above) is the most common technique employed. The size of the stylus and the form of the *Tiefstich* line vary, contributing to the general range in decoration. True *tvaerstik*, where the guideline is a dominating visual part of the technique, increases in popularity during this horizon. Early examples include Glimmen G2 Nos 245 (vertical), 246 (horizontal), pails from Bronneger D21 (vertical, fig. 5:2) and Mander O2 (information Bakker), and a bowl from Emmen D43a (fig. 5:11, courtesy J. Molema).

Motifs: panels, zigzags, filled triangles, pseudo-triangles, shoulder stamp, pointed multiple arcs, two-line horizontal blocks, chevrons (usually chevron zigzag). Narrow strips decline.

Bakker places considerable emphasis on typological changes to separate his Phases B and C, including in the first only true zigzag and using as a major indicator of the second the presence of a broad shallow groove at the base of the upper decoration (Bakker, 1979: p. 64 and p. 66). Broad shallow grooves, zigzags and other related motifs occur on otherwise identical bowls and pails. These two elements are insufficient to categorize vessels which are largely identical into two separate and distinct typochronological phases. In the second of his pilot types, strip patterns are given as the shoulder decoration type for Phase B and the hatched triangle as the type for Phase C (Bakker, 1979: fig. 29(B), p. 66). However, strip/panel decoration occurs well into Phase C (Bronneger D21, Bakker, 1979: fig. 30:2), while triangles occur on the jugs from Hooghalen which is otherwise a B assemblage. Within the Glimmen G2 assemblage, it has been argued that the tureens Nos 245 and 246 have proportions which are closer to jugs than to tureens, have triangles and shoulder stamp but no hatched triangles on the shoulders, and also include *tvaerstik* decoration. The vessels therefore show examples of triangle decoration at an early stage of development, in conjunction with *tvaerstik* technique. Individually, these elements belong to three separate phases in Bakker's typochronology (shape Phase B, triangles Phase C, *tvaerstik* Phase D).

Assemblages: many *hunebeds* contain large quantities of this material, including Emmeln, Gross Berssen, Glimmen G2, Bronneger D21/22.

2.4. Horizon 4

Types: lugged bowls, non-lugged bowls, pails (more squat than formerly and probably only during the early part of the horizon), tureens (including two-handles tureens and spouted tureens) with various profiles but with a marked tendency towards progressively contracting shoulders and smaller lug handles, funnel beakers with angular shoulders, lugged beakers, tureen-amphorae, amphorae type 1, collared flasks, biberons.

The decoration on the lugged bowls of the previous horizon breaks down and splits into three fairly distinct groups which are matched on both the tureens and the amphorae. The lugged bowls (and non-lugged bowls with similar ornament) form a typological link between the bowl series of Horizons 3 and 4. The upper zone remains distinct and fairly well-defined, although the middle part contracts sharply. The band of verticals often consists of little more than a line of short stab marks, or is replaced by a line of lozenges, 'C' motif or zigzag, or is simply left vacant. The lower boundary line consists most characteristically of a well-marked line of chevrons along which the lugs are situated, when they occur. Below this, vertical chevrons (below the lugs) divide the lower body into otherwise undefined areas, either left empty or partly filled with vertical lines or a combination of horizontal and vertical lines. Bowls of this type occur at Glimmen G2 (Nos 127, 128); Landersum (Bakker, 1979: fig. B15:8, 12b, 15); Heek-Ammerter Mark, Kr. Borken (information W. Finke, Münster); Emmeln (e.g. Nos 64, 159, 161), Gross

Berssen (e.g. Nos 44, 45). Bowls of this type belong to the first part of the horizon.

A second group of bowls shows the band of verticals being replaced by blocks of horizontal lines, occasionally incorporating multiple arcs. The lower border line (and sometimes the upper one as well), disappears. Blocks of vertical lines, interspersed with or replaced by blocks of horizontal lines replace the panels on the body. A third zone, repeating this pattern may occur, and in general the decoration covers a large portion of the bowl. Block decoration is widespread, occurring not only on bowls and pails, but also on tureen-amphorae and amphorae type 1.

A third group of bowls incorporates a zigzag (often multiple) in the mid area of the former upper zone which is associated with chevron panels or blocks on the lower body. Multiple arcs rarely occur on bowls.

On the tureens and amphorae type 1 the decoration follows similar lines, allowing for the motifs which are specific to the tureens themselves. Thus there is a group of tureens with horizontal lines on the neck and a very small, usually empty zone just above the shoulder (Glimmen G2, Nos 260, 280). This contracted zone may also include elements such as the 'C' motif, or lozenges which may then be repeated on the narrow shoulder (Glimmen G2, No. 261). The neck may also include horizontal blocks of more than two lines, the number of which seems to increase with the largest blocks of up to eight occurring at the end of the sequence. Tureens with shoulder decoration of this type usually have broad shoulders to incorporate horizontal bands of several lines or blocks of horizontal lines alternating with vertical blocks (Glimmen G2, No. 278). The decoration may derive ultimately from panel decorated shoulders. The width of the shoulder is related more to the necessity to accommodate the block decoration than to the angular and broader shoulders of the previous horizon which in fact contract.

The most widespread motifs on the necks of the tureen are the zigzag and the multiple arc usually in conjunction with narrow shoulder stamp. Tureens with this decoration develop very narrow, in some cases almost non-existent, shoulders, sometimes marked by a deeply indented line, and small lugs, becoming increasing more like tureen-amphorae. The lugs are now usually decorated with chevron motif, in place of the vertical lines. The multiple arc has its origin in the occasional pointed arcs of the Horizon 3 tureens and amphorae, and does not occur after Horizon 4. The zigzag also occurs sporadically on tureens of Horizon 3 (e.g. Gross Berssen No. 10, in association with both pointed multiple arcs and shoulder triangles). It continues into Horizon 5. Multiple arcs may occur with horizontal blocks and lozenge lines (Landersum, Bakker, 1979: fig. B15:14), and zigzags may occur with horizontal chevron and lozenge (Emmeln No. 194). Late examples of the multiple arc occur on a tureen-amphora at Glimmen G2 (No. 282) and on small tureens from Ekelberg (Bakker, 1979: fig. B17:15) and Mesum (Bakker, 1979: fig. B18:9). Multiple arcs and zigzags occur as alternative motifs in some assemblages. At Glimmen G2, the multiple arc predominates, with only one possible example of a zigzag (No. 264). However, both are present at Wechte, Emmeln and Gross Berssen (a.o.) where intermediate forms occur, probably due to the presence of both motifs on the sites rather than to a typological link between the two (Wechte Nos 303, 210, 211, 221).

The tureen-amphorae develop out of tureens. As they frequently include zigzags and block patterns on the necks but rarely multiple arcs, it is probable that they developed from tureens with these motifs only. This suggests that tureens with multiple arcs may have disappeared at a slightly earlier stage.

Amphorae type 1 also show these three types of ornament. Some amphorae have basically horizontal line decoration, but may include chevrons below the lugs and some lozenge motif in the otherwise empty band (Glimmen G2, Nos 288, 289). Others show horizontal blocks, occasionally interspersed with multiple arcs on the neck and horizontal blocks with or without vertical blocks on the shoulder, these last sometimes extending onto the body (Glimmen G2, No. 292).

Motifs: block patterns occur throughout Horizon 4, on vessels of all types. Blocks of vertical lines or multiple arcs (or a combination of the two) occur frequently below the shoulders of tureens. Shoulder stamp, 'C' motif, lozenge (this and horizontal chevron are very reliable indicators of early Horizon 4), vertical chevron, multiple arc, zigzag (often multiple and larger than the zigzag of Horizon 2 and 3).

Techniques: *tvaerstik* of various forms executed with a variety of different stylus heads, *Tiefstich*, occasionally hollow stamp impressions.

Assemblages and other finds: the assemblage from Drouwen D19 includes a large representative group of Horizon 4 material. Landersum (Bakker, 1979: fig. B15), Oldenburg-Ofenerdiek (Knöll, 1952: Taf. 15:1-2), Oldenburg-Donnerschwee, Grambergs Gründe (Knöll, 1952: Taf. 15: 12, 15, 16), Ekelberg (Bakker, 1979: figs B16; B17), Sleen-Kruidhaarsveld (Van Giffen, 1937); Rheine-Mesum (*Neujahrsgruss Münster*, 1985: pp. 28-29 and Bild 13). The assemblage from Mesum (Bakker, 1979: fig. B18), if a closed find (see Bakker, 1979, p. 193), represents a late Horizon 4/early Horizon 5 group, with elements of both.

2.5. Horizon 5

Types: bowls (sometimes globular in shape, frequently with small unpierced lugs), tureen-amphorae, amphorae type 2, some funnel beakers, funnel beaker cups, collared flasks.

Decoration: during this horizon, the decoration becomes more simplified. Two distinct styles are identifiable – the Heek-Emmeln Style which predominates in the eastern part of the West Group and the Uddelermeer-Anlo Style which predominates in the Northern and Central Netherlands. The two styles are confidentally assigned to a single horizon because of the various motifs and techniques which they share. The most important of these are the zigzag motif and the block pattern, both of which derive from the previous horizon and form the basis of all subsequent decoration.

Heek-Emmeln Style

Types: within this style, the most important forms are the bowl with/without low, narrow lugs and the tureen-amphorae with sinuous curved profile and small lugs. Amphorae type 2 do not seem to occur.

Decoration: bowls have 2-3 horizontal lines below the rim. These always occur with block pattern, but may be omitted where a band of continuous large multiple zigzag or the related oblique blocks (see Wechte No. 266) is present. Vertical blocks occur on the lower body. Initially these continue almost to the base, but become shorter during the horizon. The decoration on the tureen-amphorae (which replace the tureens) is essentially similar, with the frequent addition of a line of *Tiefstich*, dots or stabs at the base of the neck and the occasional occurrence of a band of large zigzag on the shoulder, in addition to the vertical panels which occur on the body. Small blocks of zigzag or chevron related ornament occur below the lugs. Zigzag decoration occurs frequently on the necks of funnel beakers (e.g. Glimmen G2 No. 48).

Motifs: the large zigzag is derived from the zigzag of Horizon 4, but usually incorporates 4 or more lines. The horizontal blocks incorporate as many as 13 lines. The combined effect of both motifs is vertical, in comparison to that of the Uddelermeer-Anlo Style which is more horizontal in effect. The oblique block, a combination of the zigzag and the block pattern, also has an ancestory in Horizon 4.

Technique: chiefly fine pointed *Tiefstich*.

Assemblages: this style is represented in the following assemblages – Emmeln, Heek (Finke, 1983), Gross Berssen, Wechte, Gittrup, Ostenwalde (Fansa, 1978), S. of Hunneschans (Bakker, 1979: fig. B20). It also occurs in small quantities in many other assemblages, including Beekhuizerzand (Modderman, Bakker & Heidinga, 1976), Baalder Es (Bakker, 1979: fig. 34:1), Glimmen G2, and other *hunebeds*. Late within this style, and already showing elements of Horizon 6 in form and decoration: Haltern-Westruper Heide (Knöll, 1959: Taf. 41:14-16).

Uddelermeer-Anlo Style

Types: the most important forms within this style are bowls (with/without small round knobs), tureen-amphorae with cylindrical necks, small round shoulder and a distinct junction between neck and shoulder where the small lugs are situated, and including towards the later part of the horizon, large examples with heavy, horizontally pierced lugs (e.g. Harderwijk), and amphorae type 2 (derived from amphorae type 1, having cylindrical necks and spherical bodies).

Decoration: the bowls are essentially similar to those of the Heek-Emmeln Style, but the large zigzag is replaced by a band of close-set multiple zigzag. The horizontal blocks do not incorporate the very large numbers of lines which occur within the Heek-Emmeln Style. The vertical blocks on the lower body are frequently interspersed with 2-3 lines of close-set zigzag or stabs. Ultimately, these disappear, giving way to a continuous band of short verticals. These frequently terminate in 'tear' shaped impression. The tureen-amphorae have similar decoration, but with one important addition – in addition to blocks of vertical or horizontal lines and close-set zigzag, block patterns may also be executed in hollow stamp. Amphorae type 2 do not include block patterns on the neck. Two forms of funnel beakers occur: slack profiled funnel beakers with a line of fine dots beneath the rim are present at Beekhuizerzand (Modderman, Bakker & Heidinga, 1976: fig. 9, Nos 7-13); several small funnel beakers at G2 bear resemblances to the smaller amphorae type 2 in fabric, form and decoration to such an extent that the presence or absence of lugs was taken as the defining characteristic (e.g. compare No. 64 and Nos 298, 306, 307).

Motifs: close-set small zigzag in bands and blocks, linear blocks, hollow-stamp blocks. The close-set zigzag is a regional variation of the larger zigzag motif of the Emsland-Westfalen area where it occurs from Horizon 3 onwards. The block pattern has a well established ancestry in Horizon 4.

Techniques: fine *Tiefstich*, hollow stamp, grooved lines (rare: Glimmen G2 Nos 305, 306).

Assemblages: G2, Anlo (Waterbolk, 1960), Uddelermeer and Ugchelen (Bakker, 1979: figs B19-20). Some material in this style also occurs at Emmeln. Late within this style, and already showing elements of Horizon 6 in form and decoration: Harderwijk (Manssen, 1980; this paper: fig. 7:-22-23).

The Uddelermeer-Anlo Style as a whole can be characterized as horizontal when compared to the Heek-Emmeln Style. It is a regional offshoot of this latter style and probably started somewhat later. It occurs in most areas at least sporadically, although it is absent from the Heek-Gittrup area.

2.6. Horizon 6

This material, identified first by Kat-van Hulten (1947), is essentially a further typological step beyond Horizon 5 in the direction of Horizon 7.

Types: only a limited amount of material is known from this period which on the basis of the undecorated material from both Horizons 5 and 7, can be assumed to include a large proportion of undecorated material. The main types include large amphorae derived from the tureen-amphorae of the preceding horizon and large, flaring, straight-walled pails also derived from tureen-amphorae with shoulders which have become so subtle as to have disappeared entirely. These two types have characteristically large horizontally pierced lugs. Other types include necked bowls, also developing out of tureen-amphorae (e.g. fig. 8:7-8), bowls, collared flasks and funnel beaker cups.

Decoration: the decoration is restricted to the necks and to a narrow area along the shoulder. The neck decoration may occur either with small breaks or as a continuous band. No empty area occurs in the vicinity of the base of the neck. Initially the decoration covers most of the neck, but subsequently contracts towards the base. Similarly, the shoulder decoration contracts upwards to the base of the neck. The neck decoration consists almost entirely of close-set zigzag, related motifs or horizontal lines. The shoulder decoration is usually in very narrow, discrete blocks or a narrow band, chiefly contracted versions of the decoration styles occurring during the previous horizon. Narrow, horizontal lugs with vertical lines occur (fig. 8:7-9), from which develop the slashed cordons of the following horizon.

Technique: pointed *Tiefstich*.

Material: Havelte D53 (collection B.A.I.: fig. 8:6,7 and 9; the most extensive and important collection of Horizon 6 material, with good examples of all typological developments within this horizon), Glimmen G2 (fig. 8:4-5), Nottuln (Eckert, 1986: Abb. 11), Angelslo, grave 4 (fig. 8:1), Harderwijk (Manssen, 1980: this paper fig. 8:2).

2.7. Horizon 7

The material in this horizon may be identified with Bakker's Phase G.

Types: large amphorae and/or large, flaring, straight-walled pails, with large lugs, were still produced. The rim fragment with paired handles from settlement pit 7 at Angelslo (Bakker & Van der Waals, 1973: fig. 7) and the lower part of a high, straight-walled pot from flat grave 13 at Angelslo (Bakker & Van der Waals, 1973: fig. 8) belong to these types. Other types include necked bowls which are ultimately derived from tureen-amphorae via Horizon 6, simple bowls with/without lugs including so-called cordoned bowls, funnel beaker cups, collared flasks. The necked bowls show considerable variation in proportions and may occur with lugs. The lugs may be vertically or horizontally pierced or take the form of bosses.

Decoration: the decoration on these vessels is extremely limited, confined to the shoulder and consisting of narrow, often discontinuous, slashed cordons, and narrow panels of small stabs and opposed strokes. Decoration has disappeared from the neck area entirely. Decoration is a feature of the North-East Netherlands and the German coastal area (Bakker & Van der Waals, 1973). There are some indications that regional variations in form also occur. Amongst the undecorated material are funnel beaker cups, including some fairly broad examples, bowls and collared flasks.

Motifs: small slashes, stabs and strokes, applied, narrow cordons.

Assemblages: Denekamp (Bakker & Van der Waals, 1973: fig. 1), Angelslo (Bakker & Van der Waals: figs 6-9), Spier (Bakker & Van der Waals: fig. 10:4), Valthe (Musch, 1970), Drouwen, (Van Giffen & Glasbergen, 1964: the so-called pre-Drouwen finds), Glimmen G2 (e.g. Nos 174, 175, 213, 237).

3. RECONCILING THE HORIZON SERIES WITH THE WORK OF KNÖLL AND BAKKER

The most important typological studies of West Group TRB pottery are those of Knöll (1959) and Bakker (1979), both of which must be acknowledged and discussed in any new approach to the subject. Knöll's detailed typological analysis and his large photographic catalogue provide a corpus of information not otherwise easily accessible. Bakker's extensive study of TRB material, his discussion of Knöll's typology and application of his conclusions to problems of the West Group, are pioneering works in a field which had, during the Van Giffen era, concentrated on the excavation of TRB monuments and sites in terms chiefly of structure and morphology, largely ignoring any analysis of the artefacts either as sources of cultural documentation or as dating tools. By high-lighting aspects of applied typochronology, Bakker opened up new avenues of research. However, perhaps inevitably, it has become necessary to upgrade the tool itself, in this case, the typochronology.

The weaknesses of Bakker's typochronology are most apparent in:

1. The overlap in many of his small find groups

and assemblages, which often include at least two, and sometimes even three, phases (Bakker, 1979: fig. 18). The fact that the find groups overlap, indicates that the sequence is basically correct but that the relevant stages have been incorrectly identified.

2. Emphasis on minor typological features to define phases B, C and D1 on the one hand, and lack of defining characteristics for Phases D2, E1 and E2 on the other (Bakker, 1979: pp. 48-50, 61-75).

Throughout his study, Bakker records his debt to Van Giffen, but above all, to Knöll and the typology which he worked out. Indeed, it is absolutely impossible to assess Bakker's work without Knöll's text and illustrations. Bakker's phases are essentially a reworking and refinement of the Knöll typology. This was achieved by identifying each step of Knöll's very detailed typological argument of each pot type (e.g. bowls, shouldered pots etc.), with a letter. Two 'pilot types' (Bakker, 1979: p. 48) were then identified as being diagnostically useful and the lettered steps for these were brought together to form a series of stages or phases. Bakker (1979: p. 51) summarized his revision as shown in table 1.

Unfortunately, the absence of any illustrations in the text at this point, makes his arguments extremely difficult to follow. The main discrepancies between the series of horizons offered here and that proposed by Bakker arise through:

- The absence of specific pots amongst Knöll's large, but not fully comprehensive catalogue.
- Knöll's stages which heavily influenced Bakker and which he retained within his new series of phases.
- Knöll and Bakker's reliance on the upper zone as a) typologically pre-eminent at the expense of the lower zone, b) the basis on which most of the sequence depends. Knöll's sequence ends when the upper zone breaks down and disappears.
- Bakker's over-reliance on the importance of technique at the expense of individual motif and overall pattern. (When the impressions are filled with paste, the technique employed to execute the impressions becomes of little visual importance).

Table 1. Correlation of TRB West Group phases according to Bakker, Van Giffen and Knöll (columns 1, 2 and 4), and the steps in the typological development of two pilot types, worked out by Knöll and Bakker (column 3).

A	= Drouwen A	= K+R+H	Knöll 1
B	= Drouwen B	= L+T+U1+I	
C	= Drouwen C	= M+V+U2+J	Knöll ½
D1	= Drouwen D1	= N+W+U3	
D2	= Drouwen D2	= O+X	
E	= Early Havelte	= P+Y	Knöll 2 (or Havelte)
F	= Middle Havelte	= ∋	
G	= Late Havelte	= ⊢◯+ꟼ	

In correlating the three systems (Knöll's, Bakker's and the Horizon series) it is clear that

- Knöll's three stages are too broad for the timescale and typology involved. He paid little attention to the end phases of the TRB and to material which he considered of poor quality or degenerate and therefore late.
- Bakker's typochronology leans heavily on the typological explanation behind Knöll's three stages, leading him to define seven phases and two subphases. These phases are somewhat shakily established on the basis of minor typological changes and on cross-associations of only two pilot types in a number of closed and semi-closed finds.

Horizon 1. This horizon is identical with Bakker's Phase A and the material included within Knöll's Stage 1 which is identified by Bakker with the letters K, H, R. However, Knöll also placed in this stage material considered by Bakker to constitute a separate Phase B.

Horizon 2 consists basically of jugs of Bakker's Phase B and bowls considered by Knöll and Bakker to be 'incompletely decorated' (U1, U2 and T) and therefore typologically useless. This misconception arises from Knöll's and Bakker's belief that the upper zone decoration was of greater importance than that of the lower zone in showing typological change, and that this upper zone must consist of three elements, namely a middle 'band' rigorously defined on either side by horizontal lines. As the descriptions of the lower portions of bowls from Horizons 1-3 show, the lower zone is of equal value for the purposes of definition: Horizon 1 – narrow vertical strips; Horizon 2 – broader, usually combined strips, with intervening vertical lines; Horizon 3 – broad, well-defined panels.

Knöll was influenced by the theory of the degradation of potting traditions through time and therefore placed sparsely or 'incompletely' decorated bowls in the latest stages of the TRB. Bakker realized that this material was considerably earlier in date, but continued to be influenced by Knöll into accepting that these zones were incompletely decorated. He divided them amongst several different phases or stated that they were unclassifiable. Most of these bowls should be viewed as the necessary typological step between those of Horizon 1/Bakker Phase A/Knöll 1 and the bowls with developed well-defined horizontal upper zones and well-defined lower panels. Bakker's Phase B is artificially composed of jugs of direct typological descent from Phase A and bowls which are typologically developed from the 'incompletely decorated' bowls which in reality belong to the following phase, i.e. pilot types of two different steps. The main evidence put forward by Bakker for the existence of Phase B, despite some difficulties, is the Hooghalen find group.

Horizon 3 consists of Bakker's Phases B, C and D1 material/part of Knöll 1 and most of Knöll 1/2, with the exception of the jugs which Bakker placed in his Phase B. This material shows the normal range of expected variations within any single horizon. These variations, however, Bakker considered to have chronological and typological significance. This was partially caused by the initial composition of Phase B; in associating the 'wrong' bowls with jugs of Phase B, Bakker was forced to form his Phase C by giving phase status to bowls which differed only on minor typological grounds. Knöll had already conveniently indicated a break between his stages 1 and 1/2, which Bakker therefore adopted. But Knöll's sequence, composed of a range of types, appears to move smoothly from stage to stage through these bowls, because his much broader groups also placed less emphasis on detail. Bakker's recognition that the jugs constitute a separate phase led him to elevate one of Knöll's minor typological steps in the bowl series into a phase of its own, which in turn put the typological series under stress and unfortunately linked material from two different stages of development. Knöll's broader groups also placed less emphasis on the importance of the *tvaerstik* technique, which occurs only as one aspect of a large diverse group of material. Bakker used *tvaerstik* to define further phases. Material within Horizon 3 shows the increasing importance of this technique, but the overall decoration of the pots remains unchanged. The broader terms of both the Knöll and Horizon system allow for such an eventuality.

Horizon 4 consists of Phases Bakker D2/Knöll 1/2 and Bakker E1/Knöll 2. This section of Knöll's catalogue unfortunately includes few examples of either multiple zigzags or multiple arcs. Bakker's (1979: pp. 66-72) typochronology is at its weakest at this point. D1 material is described as similar to the preceding Phase C material except for the use of *tvaerstik*; D2 which may be distinguished from D1 on the basis of new motifs, is described as being practically indistinguishable from E1 which itself differs only from E2 in terms, once more, of technique. With such very small differences, at least as described in the text, there seems to be no justification for separate phases at all. For Bakker, however, who wished to refine Knöll's stages, this was too much, and he therefore adopted Knöll's break between 1/2 and 2 to divide D2 from his E1. The sub-divisions 'D1', 'D2 and E1', and 'E2' are in fact the more important divisions (D1 going to Horizon 3, D2 and E1 together forming Horizon 4 and E2 constituting a very homogenous Horizon 5). However, Bakker was heavily influenced by Knöll in choosing to divide his D2 from E material but apparently ignored the fact that within the much broader stages envisaged by Knöll some overlap could be accommodated, which in a narrower phase system would lead to practically identical phases if adequate care was not given to the detailing of the phases in the first place. Knöll's division of his Stage 1/2 from his Stage 2 bowls (what Bakker terms Knöll's X and Y bowls) was based once again on his reliance on the importance of the upper zone, which is crucial because this division is based on the breakdown of the upper zone. X pots (Bakker's annotated step) are those where the upper zone is still defined between horizontal border lines (below the rim and at the base of the upper zone). Y pots are those where these lines break into sections or even disappear ... degeneration. In using this feature, Knöll paid insufficient attention to other motifs and patterns appearing while this process was taking place, and which make it difficult to identify D2 from E1 in Bakker's division. In fact, there are no real grounds for dividing D2 from E1. Together they constitute Horizon 4, which may be summed up as the horizon during which the evolving strip-panel-zone patterns break down and a new decorative scheme replaces it, allied to new pot forms evolving out of the existing range. This development may be traced by placing more emphasis on the totality of the decoration and individual motifs, rather than on the degeneration of the upper zone.

Horizon 5 consists of E2 material. Knöll dealt very cursorily with material after Horizon 4/Phases D2-E1/Knöll 2, seeing it as degraded, undecorated and therefore late. The finely decorated material of this horizon and the subsequent one was unknown to him. Bakker, however, for no very clear reason, considered his E2 material to be part of a larger E Phase. Once more, he subdivided the phase on the grounds of technique, this time, the finely pointed *Tiefstich*, but stated in his description that otherwise the two subphases were the same. In this case, this is not true. Bakker's own text (1979: pp. 70-71) clearly describes E2 material rather than E Phase material (E1 and E2). Subsequently he states "to E1 was assigned the pottery which as regards shape and arrangement of ornamentation was applied in *tvaerstik* lines or with a heart stamp, hollow stamp or double stamp" (Bakker, 1979: p. 72). The vessels which he illustrates for his two subphases are, however, quite dissimilar and in any case, some of the E2 patterns cannot be executed in *tvaerstik* (e.g. the very dominant bands of closely set small zigzag). There is no justification in placing the two subphases together. Horizon 5/E2 material is fairly homogenous, with good typological antecedents in Horizon 4 and a well-defined progression through the following phases. It occurs on several settlement sites without the admixture of other material.

Horizons 6 and 7. Knöll's typology is obviously

inadequate in its handling of the later material, some of which only became available after he had completed his work. Horizon 6/Phase F was first identified by Kat-van Hulten (1947) working on material with Van Giffen. This was greatly enlarged by the work of Bakker and Van der Waals (1973) on both this and the subsequent Horizon 7.

4. TRB DEVELOPMENTS IN THE ADJACENT AREAS OF SOUTH DENMARK AND NORTH GERMANY

In the areas adjacent to the TRB West Group, related and parallel developments in the pottery took place. The Haassel-Fuchsberg Group of the late Early Neolithic C occurs in Denmark and north Germany, extending as far south as the river Elbe. In a small area of north-east Lower Saxony, it extended south of this river (Schwabedissen, 1979: Abb. 5). The earliest relevant material belongs to the latest stages of this group. Of particular importance are the bowls with decoration of large bar chevron. Other relevant elements are the jugs with very slightly flaring undecorated necks, slightly angular profiled thick handles, and rounded bodies with strip decoration. These are comparable to the jugs of Horizon 1 (e.g. Bronneger, fig. 3:9). Collared flasks and funnel beakers with vertical lines on the body also occur. The assemblage from Bistoft belongs to the end of the Haassel-Fuchsberg Group, already incorporating elements of the Troldebjerg development. The assemblage is important not only because of its radiocarbon dates, but also because of the close similarities this assemblage bears to Horizon 1 material (in particular, the lugged flask and the two bowls with a line of dots below the rim and simple, single ladders on the body, Johansson, 1981: Tafel 13 and 15; Meurers-Balke, 1978-79: Abb. 1). It is quite clear that the earliest TRB pottery arrived in the area of the West Group during the transitional stage of the Haassel-Fuchsberg and Troldebjerg Groups. Horizon 1 material includes informal lines of dots etc. below the rim and simple strip ornament which occur during this transition.

The subsequent developments are indicated by the Troldebjerg Group in Denmark and Schleswig-Holstein (Schwabedissen, 1979: Abb. 8). The Troldebjerg Group includes (amongst others) bowls with a band of short verticals below the rim and body ornament derived from the bar chevron. Sherds of a bowl from Emmen-Oude Roswinkelerweg belong to this type (collection E. Drenth). The strips may now be doubled or trebled and incorporate various elements including *tvaerstik* lines, in addition to simple hatching (e.g. the lugged beaker from Neumünster, Schwabedissen, 1979: Abb. 8:1). The jug from Tinnum/Sylt (Schwabedissen, 1979: Abb. 8:3) shows the development of an angular profile with broad strap handle and large shoulder decorated with vertical ladders. This stage of development is represented in the West Group by material of Horizon 1 and the earlier part of Horizon 2.

South of the Elbe, the Haassel-Fuchsberg Group occurs only in the Uelsen-Lüneberg area (Schwabedissen, 1979: Abb. 5). It was succeeded by Altmark *Tiefstich* which is not immediately derivable from the Haassel-Fuchsberg Group. It may either have arrived as a developed entity or it evolved by a short-lived but as yet unrecognized step from the Haassel-Fuchsberg Group. Preuss (1980) identified two phases of Altmark *Tiefstich.*

– Düsedau Horizon, characterized by (amongst others) standing triangles below the rims of the bowls, true hatched triangles on the shoulders of the tureens and *tvaerstik* lines. The material from this horizon, including the material from Issendorf, Kr. Stade (Tempel, 1972: Abb. 4-8), may be equated with West Group Horizon 2 where all the elements occur, although rarely. Elements of both horizons are combined on a bowl from Achim, Kr. Verden, with standing triangles below the rim and vertical ladders and chevrons in the lower zone (Schünemann, 1972: Abb. 20a). The single vertical *tvaerstik* line is also of interest. Both Preuss (1980: p. 90) and Schwabedissen (1979: p. 158) have indicated that the Düsedau Horizon is contemporary with the Troldebjerg Group.

– Haldensleben Horizon, characterized by (amongst others) bowls with two decorative zones, the upper consisting of a band of short verticals, and tureens with filled shoulder triangles. This horizon is comparable to the West Group Horizon 3.

In most of the area in which Altmark *Tiefstich* occurred, it was succeeded by the Walternienburg-Bernburg Groups. However, within the Uelsen-Lüneburg area Laux distinguished a local and according to him continuous development of five steps (Laux, 1979: pp. 59-82).

– Group A includes the material of the Haassel-Fuchsberg Group;

– Group B belongs to the Düsedau Horizon of Altmark *Tiefstich*;

– Group C is the Haldensleben Horizon of Altmark *Tiefstich*;

– Group D includes material which is comparable to the West Group Horizon 4;

– Group E has no specific similarities to West Group material. Laux unconvincingly suggested that Group E may show influences from the Bernberg Group.

5. CHRONOLOGY AND DATING OF THE TRB WEST GROUP

The chronology and dating are based not only on the available dates from the West Group itself but on dates from earlier TRB material from south Denmark and Schleswig-Holstein, too.

The earliest relevant dates are from the Satrup Group. According to Schwabedissen eighteen dates are known for this group, which fall between 5050 and 4720 BP (no standard deviation given; Schwabedissen, 1979: Abb. 12). The Fuchsberg Group which follows Satrup is apparently well-dated to the period c. 4750-4550 BP. Schwabedissen (1979: p. 156) mentions 'eine Reihe von ^{14}C Daten zwischen 4700 and 4550 BP' from the settlements of Fuchsberg itself and Sachsenwald. Elsewhere in the same paper (Abb. 12) seven dates are indicated between 4740 and 4555 BP (without standard deviation) for the Fuchsberg Group. These are probably the dates from the two settlement sites; the figures mentioned in the text seem to be incorrect.

Eleven dates are available from the Fuchsberg settlement at Toftum (Madsen, 1977; Madsen & Petersen, 1982-83: p. 117, note 53). Charcoal samples: K-2985 4500±85 BP, K-2986 4690±90 BP, K-2987 4690±90 BP, and K-2988 4730±90 BP, mean c. 4650 BP. Marine shells: K-2978 4500±85 BP, K-2979 4620±70 BP, K-2980 4520±85 BP, K-2981 4690±70 BP, K-2982 4550±90 BP, K-2983 4590±90 BP, and K-2984 4530±85 BP, mean c. 4570 BP.

No details of contexts are given in the publication. The dates fall into two groups, those from charcoal samples being on average 80 years earlier than those from shells. Two main problems are connected with the dating of shell samples.

a) *Reservoir effect*, which is brought about by the mixing of ^{14}C depleted deep ocean water (i.e. below c. 200 metres) with surface water. This mixing results in a lower ^{14}C content in the surface water in comparison to that in the atmosphere (Olsson, 1980; Tauber, 1983). Marine materials thus affected take on an own age in ^{14}C years. In the North Atlantic and adjacent seas this age is about 400 ^{14}C years.

b) *Isotopic fractionation*: Due to isotopic fractionation marine carbonate (shell) contains more ^{14}C than the surrounding sea-water and seems to be younger, therefore. This effect also amounts to about 400 ^{14}C years in the North Atlantic and adjacent seas (Olsson, 1980; Tauber, 1983).

Because of the coincidental cancellation of the two effects no corrections are applied to radiocarbon dates derived from marine shells in a number of laboratories, including Copenhagen (Tauber, 1983: p. 371). However, in the Baltic Sea, the reservoir effect is reduced because there is less mixing with deep ocean water. It amounts to c. 340±30 years on the Swedish west coast, and c. 320±30 years on the Danish east coast (Olsson, 1980). Toftum lies in east Jutland. It is therefore very likely that the shell dates from Toftum are c. 80 years too young in comparison with contemporaneous terrestrial material, like charcoal. Taking this in account, the discrepancy between the dates disappears. After correction the shell dates have an average age of c. 4650 BP, similar to that of the charcoal samples.

In addition, several dates are known from the Fuchsberg settlement near Sarup (Andersen, 1980; see also Bakker, 1979: p. 144). These dates are from the remains of three posts in the palisade trench (A 307): K-2630 4600±90 BP, K-2631 4620±90 BP, K-2632 4760±90 BP (mean 4660±55 BP), from charcoal in a flat grave (A 310): K-2629 4690±90 BP, and from charred grain in a pit with pottery (A212): K-2628 4580±70 BP.

The settlement of Bistoft LA 11 may be considered as belonging to the last stages of the Fuchsberg Group. There are no indications that this settlement (Johansson, 1981) was of long duration. This is not contradicted by the fact that in the peatbog beside the settlement site a large fragment of a decorated bowl, formally of MN1 type, was found at a higher level than fragments which can be attributed to EN C (Meurers-Balke, 1978-79). According to Johansson (letter to A.E. Lanting, 10.5.1983) definite EN C sherds were also found at different levels. He is convinced that the site was visited only once by a single group of people, and was used as an extraction camp during the summer. The pottery should be treated therefore as a closed find. If the fragments illustrated by Johansson (1981: Taf. 15:13) and Meurers-Balke (1978-79: Abb. 1:5) had been found in a West Group context, they would have been attributed to Horizon 1. Two pieces of charred wood from the settlement have been dated: KI-921 4700±65 BP and KI-1231 4610±120 BP.

The dates from Sarup and Bistoft have been included in figure 10. When all these dates are combined they indicate absolute dates of 3800-3550 cal BC for the Satrup Group and 3550-3400 cal BC for the Fuchsberg Group. The dates from Bistoft which are slightly older than the Sarup grain date are in accordance with the calibration curve at this point; around 3400 cal BC radiocarbon ages are greater than around 3500-3450 cal BC.

The earliest West Group material – Horizon 1 – is comparable to the decorated pottery from Bistoft LA11 (Johansson, 1981: Tafel 15; Meurers-Balke, 1978-79: Abb. 1). Therefore Horizon 1 must have begun at or around 3400 cal BC, with radiocarbon dates of about 4700 BP. So far, no material from this horizon has been dated. The following dates for younger West Group material are known.

Odoorn: GrN-2226 4590±80 BP. Charcoal from

a flat grave or pit in front of the entrance to *hunebed* D32. According to Van Giffen (1961) this feature was intersected by the socket of one of the orthostats. This interpretation is at least questionable. The intersecting pit seems to be due to recent stone digging, not an original socket. Two funnel beakers were found in the grave/pit, which cannot be more closely dated than to Horizons 1-4.

Odoorn D32a: GrN-12609 4550±40 BP. Charcoal from pit next to east end stone of destroyed *hunebed* D32a. This pit may be connected with the construction of the tomb (Taayke, 1985). The oldest pottery in D32a belongs to Horizon 3.

Odoorn D32c: GrN-13184 4630±60 BP. Charcoal from under in situ stone packing of destroyed *hunebed* D32c (cf. Taayke, 1985). Oldest pottery belongs to Horizon 3.

Gittrup F. 707: GrN-12263 4490±60 BP. Charcoal from flat grave with Horizon 5 pottery (see *Neujahrsgruss Münster* 1983: pp. 36-39).

Harderwijk-Beekhuizerzand: GrN-7746 4520±70 BP. Charcoal fragments from filled-in bed of stream, in layer with sherds of Horizon 5 pottery (Modderman et al., 1976).

Heek F. 4: GrN-9202 4520±35 BP. Charcoal concentration in flat grave with Horizon 5 pottery (Finke, 1983).

Heek F. 27: GrN-11764 4400±60 BP. Charcoal from flat grave with Horizon 5 pottery (cf. Finke, 1983).

Heek F. 38a: GrN-11766 4480±60 BP. Charcoal from flat grave with Horizon 5 pottery (cf. Finke, 1983).

Anlo: GrN-1824C 4410±60 BP. Charcoal from settlement pit with pottery belonging to Horizon 5 (Waterbolk, 1960).

Glimmer Es: GrN-6156 4380±40 BP. Charcoal from pit with Horizon 6 pottery next to destroyed *hunebed* G2 (Brindley, 1986).

Nottuln: GrN-12414 4240±60 BP. Charcoal from layer with Horizons 6 and 7 pottery in upper part of filled-in ditch of Michelsberg fortified settlement (Eckert, 1986: Abb. 11-14).

Angelslo: GrN-4200 4415±65 BP and GrN-4201 4380±75 BP, mean 4400±50 BP. Charcoal from settlement pit 5 with Horizon 7 pottery (Bakker & Van der Waals, 1973: fig. 6).

Angelslo: GrN-5767 4315±60 BP. Charcoal from settlement pit 7 with Horizon 7 pottery (Bakker & Van der Waals, 1973: fig. 7).

Angelslo: GrN-2370 4145±100 BP. Charcoal from flat grave 3 with cremated bone and Horizon 7 pottery (Bakker & Van der Waals, 1973: fig. 5).

Angelslo: GrN-5070 4100±30 BP. Charred sticks from flat grave 14 with cremated bone and Horizon 7 pottery. In the filling, a sherd of Horizon 6 pottery and a sherd of Corded Ware pottery were also found (Bakker & Van der Waals, 1973: fig. 9).

Six dates are not included in figure 10.

Angelslo: GrN-5103 4355±45 BP. Fragments of charcoal found in the filling of a tree-fall pit, together with sherds of Horizon 4 pottery. There is no certainty that the charcoal and sherds were associated.

Harderwijk-Beekhuizerzand: GrN-7745 4165±35 BP. Fragments of charcoal from filling of pit in which sherds of Horizon 5 pottery were also found. In an adjacent pit of the same type a Late Bronze Age pot was found. It is possible that the dated pit was also dug in the Late Bronze Age but filled with soil containing TRB settlement refuse. The charcoal appears to have been mixed with about 20% of Late Bronze Age charcoal.

Heek F.5:	GrN-11762	5030±70 BP
Heek F.25:	GrN-11763	4980±60 BP
Heek F.34:	GrN-11765	4890±80 BP
Gittrup F.440:	GrN-12262	4920±70 BP

These four samples are from charcoal in flat graves with good Horizon 5 pottery. The dates are much older than expected. This is probably due to the use of mature wood. In this part of Westphalia, where previous Neolithic settlement is lacking, such wood must have been widely available.

The dates for Horizon 6 and 7 material show clearly that the latest TRB West Group material should be dated to around 2850 cal BC. The available radiocarbon dates correspond clearly with the steep part of the wiggle between 2930 and 2850 cal BC, where radiocarbon dates drop from c. 4400 BP to 4080 BP. On archaeological grounds there is no reason to suppose that Horizon 7 continued beyond c. 2850 cal BC, although this cannot be excluded on the basis of the radiocarbon dates alone.

This indicates that the West Group lasted about 550 years. Horizons 1 and 2 seem to have lasted only a short time, together probably about 100 years. Horizons 6 and 7 appear to have been fairly short also, possibly lasting about 100 years too. The remaining 350 years must encompass Horizons 3, 4 and 5. From the amount of material and the variations which occur within each horizon, one can suggest a period of around 100 years for Horizon 3, about 150 years or slightly more for Horizon 4 and about 100 years or slightly less for Horizon 5.

When evaluating the radiocarbon dates for the West Group the source of the samples should not be forgotten. With the exception of GrN-5070, the dated material consisted of fragments of charcoal which may have had an age of their own of several decades. This may explain why two of the Horizon 7 dates are older than expected. In the wiggle section of the calibration curve between 2930 and 2850 cal BC, an age differential of several decades could mean an increase in radiocarbon age of 100-200 years.

According to the calibration curve of Pearson et al. (1986), between c. 3350 and 2930 cal BC, differences between the radiocarbon age of the sample and the event should not be apparent unless very small standard deviations are involved. The age differential of the sample, therefore, should have no great influence during most of this period. However, the age differential of samples which fall in the steep part of the curve between 3350 and 3400 cal BC, will have a marked influence on the result. An age differential of several decades can result in differences of 100-150 radiocarbon years.

6. ACKNOWLEDGEMENTS

I wish to thank S.W. Jager, J.M. Smit and H.R. Roelink for illustrations of pottery in the collections of the B.A.I. and the Provinciaal Museum of Drenthe; H.R. Roelink for the tables and for mounting the tables and figures; J.H. Zwier for the ^{14}C figure. I am most grateful to J.A. Bakker for showing me the material from the *hunebeds* D26 and O2, and to W. Finke for his hospitality during several visits to examine material from his excavations at Heek, Gittrup, Heek-Ammerter Mark and Mesum. I am grateful to A.E. Lanting for pointing out the Toftum dates to me and to B. Kamlag for reading the manuscript, and checking numbers. For discussion along the way and especially for seeing this paper through press, my best thanks to J.N. Lanting.

7. REFERENCES

ANDERSEN, N.H., 1980. Sarup. Befaestede neolitiske anlaeg og deres baggrund (with English summary). *Kuml*, pp. 63-103.

BAKKER, J.A., 1979. *The TRB West Group. Studies in the chronology and geography of the makers of hunebeds and Tiefstich pottery*. Amsterdam.

BAKKER, J.A., 1982-83. Het hunebed G1 te Noordlaren. *Groningse Volksalmanak*, pp. 115-199.

BAKKER, J.A. & J.D. VAN DER WAALS, 1973. Denekamp-Angelslo. Cremations, collared flasks and a Corded Ware sherd in Dutch final TRB contexts. In: G. Daniel & P. Kjaerum (eds.), *Megalithic graves and ritual*. Papers presented at the III Atlantic Colloquium, Moesgård 1969. Moesgård, pp. 17-50.

BEHRENS, H., 1973. *Die Jungsteinzeit im Mittelelbe-Saale-Gebiet*. Berlin.

BRINDLEY, A.L., 1983. The finds from hunebed G3 on the Glimmer Es, mun. of Haren, prov. of Groningen, the Netherlands. *Helinium* 23, pp. 209-236.

BRINDLEY, A.L., 1986. Hunebed G2: excavation and finds. *Palaeohistoria* 28, pp.

ECKERT, J., 1986. Ein mittel- und jungneolithischer Siedlungsplatz bei Nottuln, Kreis Coesfeld. Bericht über die Ausgrabungen 1983-1984. *Ausgrabungen und Funde in Westfalen-Lippe* 4, pp. 39-63.

FANSA, M., 1978. Die Keramik der Trichterbecherkultur aus dem Megalithgrab I von Ostenwalde, Kreis Aschendorf-Hümmling. *Neue Ausgrabungen und Forschungen in Niedersachsen* 12, pp. 33-77.

FANSA, M., 1982. *Die Keramik der Trichterbecherkultur aus den Megalith-und Flachgräbern des oldenburgischen Raumes* (= Göttinger Schriften zur Vor- und Frühgeschichte 20). Neumünster.

FINKE, W., 1983. Ein Flachgräberfeld der Trichterbecherkultur bei Heek, Kreis Borken. Vorbericht über die Ausgrabungen 1978-1981. *Ausgrabungen und Funde in Westfalen-Lippe* 1, pp. 27-32.

GIFFEN, A.E. VAN, 1925/27. *De hunebedden in Nederland*. Text: 2 vols. Atlas: 2 vols. Utrecht.

GIFFEN, A.E. VAN, 1937. Twee grafvelden uit de hunebeddenen bekertijd in het Kruidhaarsveld bij Sleen, gem. Sleen. *Nieuwe Drentse Volksalmanak* 55, pp. 74-77.

GIFFEN, A.E. VAN, 1961. Een vlakgraf van de trechterbekercultuur, gesneden door een standkuil van hunebed D32 te Odoorn (Dr.). *Helinium* 1, pp. 39-43.

GIFFEN, A.E. VAN & W. GLASBERGEN, 1964. De vroegste faze van de TRB-cultuur in Nederland. *Helinium* 4, pp. 40-48.

JOHANSSON, L., 1981. Bistoft LA 11. Siedlungs- und Wirtschaftsformen im frühen Neolithikum Norddeutschlands und Südskandinaviens. *Offa* 38, pp. 91-129.

KAT-VAN HULTEN, J.C., 1947. Het hunnebeddenaardewerk en zijn stijlen. In: *Een kwart eeuw oudheidkundig bodemonderzoek in Nederland*. Gedenkboek A.E. van Giffen, directeur van het Biologisch-Archaeologisch Instituut der Rijksuniversiteit te Groningen, 1922 - 17 juni - 1947. Meppel, pp. 203-219.

KNÖLL, H., 1952. Neue und alte tiefstichkeramische Flachgräber aus Oldenburg. *Germania* 30, pp. 303-313.

KNÖLL, H., 1959. *Die nordwestdeutsche Tiefstichkeramik und ihre Stellung im nord- und mitteleuropäischen Neolithikum*. Münster.

KNÖLL, H., 1983. *Die Megalithgräber von Lengerich-Wechte (Kreis Steinfurt)*. Münster.

LAUX, F., 1979. Die Grosssteingräber im nordostlichen Niedersachsen. In:
H. Schirnig (ed.), *Grossteingräber in Niedersachsen*. Hildesheim, pp. 59-82.

MADSEN, T., 1977. Toftum ved Horsens. Et 'befaestet' anlaeg tilhørende tragtbaegerkulturen (with English summary). *Kuml*, pp. 161-184.

MADSEN, T. & J.E. PETERSEN, 1982-83. Tidligneolitiske anlaeg ved Mosegården. Regionale og kronologiske forskelle i tidlig-neolitikum (with English summary). *Kuml*, pp. 61-120.

MANSSEN, W.J., 1980. A flat grave cemetery of the Funnel Beaker culture near Harderwijk. *Helinium* 20, pp. 105-113.

MEURERS-BALKE, (end-und)J., 1978/79. Pollenanalytische Untersuchungen zu früh- und mittelneolithische Moorfunden von Bistoft, Kreis Schleswig-Flensburg. *Kölner Jahrbuch* 16, pp. 35-39.

MODDERMAN, P.J.R., J.A. BAKKER & H.A. HEIDINGA, 1976. Nederzettingssporen uit Midden-Neolithicum (TRB), Late Bronstijd en Middeleeuwen in het Beekhuizer Zand onder Harderwijk, prov. Gelderland (met bijdragen van C.C. Bakels en J. Ypey). *Analecta Praehistorica Leidensia* 9, pp. 39-74.

MUSCH, J.E., 1970. Een vondst van Laat-Havelte aardewerk ten noorden van Valthe, gem. Odoorn. *Nieuwe Drentse Volksalmanak* 88, pp. 213-219.

OLSSEN, I.U., 1980. Context of ^{14}C in marine mammals from northern Europe. *Radiocarbon* 22, pp. 662-675.

PEARSON, G.W., J.R. PILCHER, M.G.L. BAILLIE, D.M. CORBETT & F. QUA, 1986. High-precision ^{14}C measurement of Irish oaks to show the natural ^{14}C variations from AD 1840 to 5120 BC. *Radiocarbon* 28, pp. 911-934.

PREUSS, J., 1980. *Die altmärkische Gruppe der Tiefstichkeramik(end-und). Berlin.*

SCHIRNIG, H., 1979. *Das 'Huntedorf' bei Lembruch am Dümmer. In: H. Schirnig (ed.), Grossteingräber in Niedersachsen.* Hildesheim, pp. 235-238.
SCHLICHT, E., 1967. Drei nichtmegalithische Gräber der Trichterbecherkultur im Emsland. *Die Kunde* NF 18, pp. 16-29.
SCHLICHT, E., 1968. *Die Funde aus dem Megalithgrab 2 von Emmeln, Kreis Meppen* (= Göttinger Schriften zur Vor- und Frühgeschichte 9). Neumünster.
SCHLICHT, E., 1972. *Das Megalithgrab 7 von Gross Berssen, Kreis Meppen* (= Göttinger Schriften zur Vor- und Frühgeschichte 12). Neumünster.
SCHÜNEMANN, D., 1972. Das Megalithgrab im 'Düvelshagen' bei Völkersen, Kr. Verden und die Trichterbecherkultur in Kreis Verden. *Die Kunde* NF 23, pp. 8-43.
STRUVE, K.W., 1939. Der Trichterbecherfund von Neumünster. *Offa* 4, pp. 83-91.
SCHWABEDISSEN, H., 1979. Zum Alter der Grossteingräber in Norddeutschland. In: H. Schirnig (ed.), *Grossteingräber in Niedersachsen.* Hildesheim, pp. 143-160.
SPROCKHOFF, E., 1952. Ein Grabfund der nordischen Megalithkultur von Oldendorf, Kr. Lüneburg. *Germania* 30, pp. 164-174.
TAAYKE, E., 1985. Drie vernielde hunebedden in de gemeente Odoorn. *Nieuwe Drentse Volksalmanak* 102, pp. 125-144.
TAUBER, H., 1983. ^{14}C dating of human beings in relation to dietary habits. In: W.G. Mook & H.T. Waterbolk (eds.), *Proceedings of the first international symposium on* ^{14}C and archaeology, Groningen 1981 (= PACT 8). Strasbourg, pp. 365-375.
TEMPEL, W.-D., 1972. Erdgräber der Trichterbecherkultur in der Gemarkung Issendorf, Kreis Stade. *Neue Ausgrabungen und Forschungen in Niedersachsen* 7, pp. 46-59.
TEMPEL, W.-D., 1979. Flachgräber der Trichterbecherkultur. In: H. Schirnig (ed.), *Grossteingräber in Niedersachsen.* Hildesheim, pp. 111-116.
WATERBOLK, H.T., 1960. Preliminary report on the excavations at Anlo in 1957 and 1958. *Palaeohistoria* 8, pp. 59-90.

APPENDIX: Assemblages utilized for this study

The Horizons 1-7 were identified after a survey of the assemblages given in table 2.

Most of the important flat graves and other groups of short timespan have been published by Bakker (1979), Knöll (1952) and Schlicht (1967; 1968). The finds from the recently excavated flat graves of Heek, Kr. Borken, Heek-Ammerter Mark, Kr. Borken, Münster-Gittrup and Rheine-Mesum, Kr. Steinfurt, were examined in Münster, courtesy of W. Finke.

Hunebed inventories

F1	Rijs	Coll. F.M.
G2	Glimmer Es	Coll. G.M.
G3	Glimmer Es	Coll. G.M.
G5	Heveskesklooster	B.A.I. inf. J. Schoneveld
D9	Annen	Coll. P.M.D. inf. D.J. de Groot
D14	Eexterhalte	Coll. B.A.I.
D21/22	Bronneger	Coll. B.A.I.
D26	Drouwenerveld	I.P.P. inf. J.A. Bakker
D28	Buinen	Coll. B.A.I.
D30	Exloo	Coll. P.M.D.
D32a	Odoorn	B.A.I. inf. E. Taayke
D32c	Odoorn	B.A.I. own research
D32d	Odoorn	B.A.I. inf. B. Kamlag
D40	Emmen	Coll. B.A.I.
D43a	Emmen	B.A.I. inf. J. Molema
D53	Havelte	Coll. B.A.I.
D54b/c	Hooghalen	Coll. B.A.I. inf. H. Bouman
O1	De Eeze	Coll. B.A.I.
O2	Mander	Coll. O.K.T./R.M.T. inf. J.A. Bakker
Spr. 981	Rheine	Coll. W.M.A. Münster inf. J. Eckert

Hunebed inventories, seen only in photos

D19	Drouwen	courtesy I.P.L.
D43	Emmen	courtesy I.P.P.

Hunebed inventories, seen only in publication

G1	Noordlaren	Bakker, 1982/83
-	Emmeln 2	Schlicht, 1968
Spr. 861	Gross Berssen 7	Schlicht, 1972
-	Lengerich-Wechte I&II	Knöll, 1983
Spr. 835	Ostenwalde I	Fansa, 1978

	TYPE	HORIZON
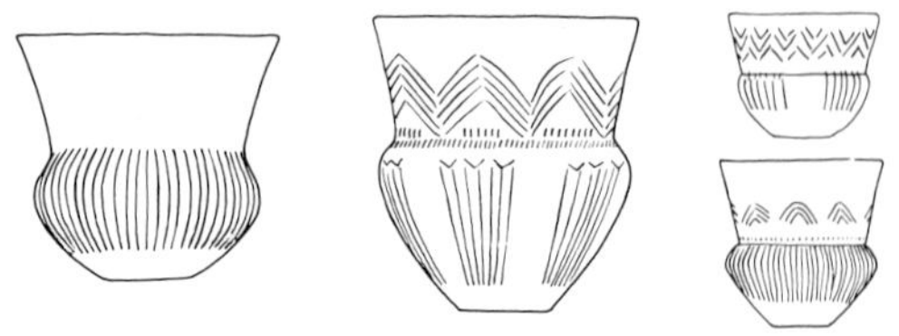	FUNNEL BEAKER	1-5
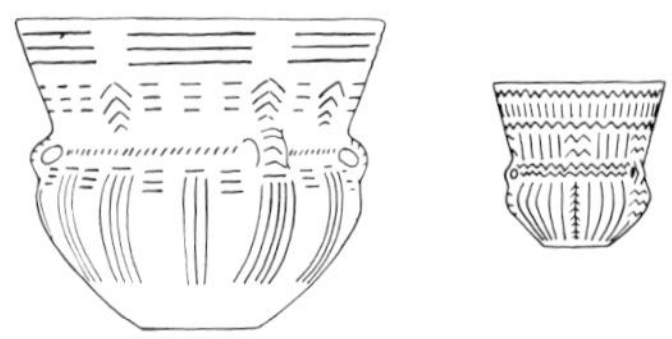	LUGGED BEAKER	1-4
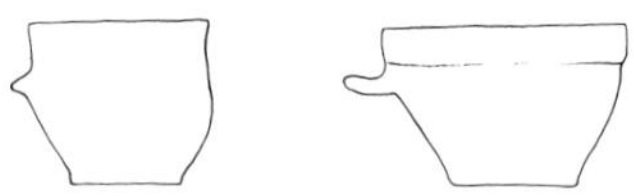	FUNNEL BEAKER CUP	5-7
	JUG	1-2
	TUREEN	3-4
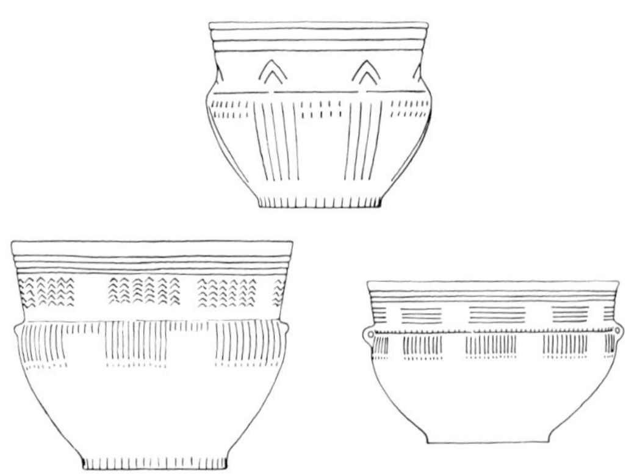	TUREEN-AMPHORA	4-5

Fig. 1. Table of most common pot types.

	TYPE	HORIZON
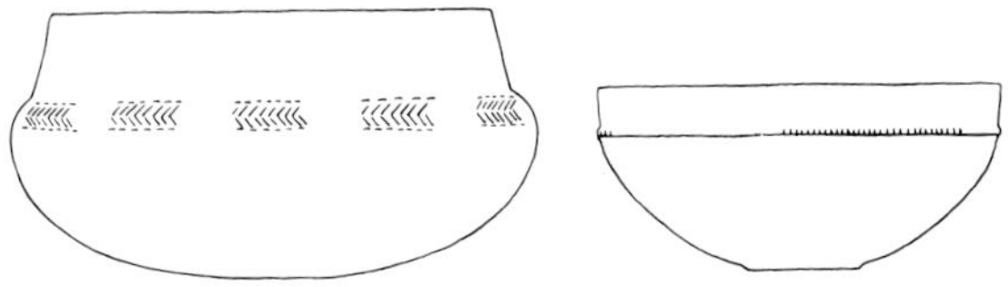	NECKED BOWL	7
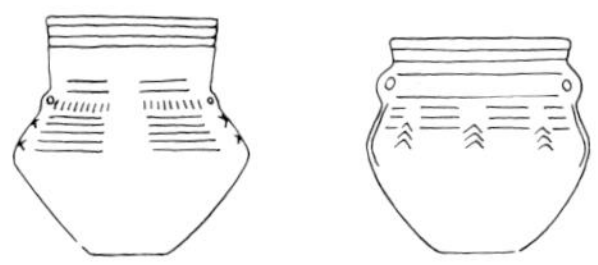	AMPHORA TYPE 1	3–4
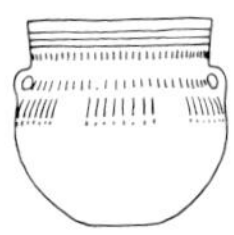	AMPHORA TYPE 2	4–5
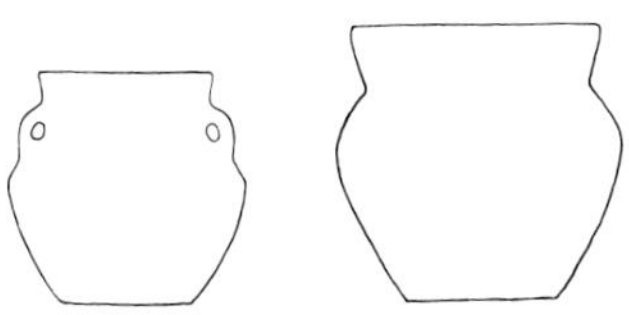	LUGGED OR NON-LUGGED SHOULDERED VASE	2–7
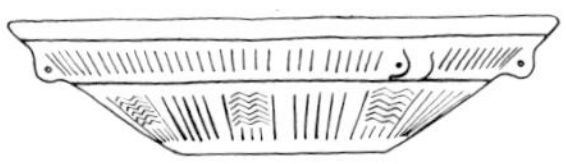	LUGGED BOWL	1–3
	PAIL	2–4
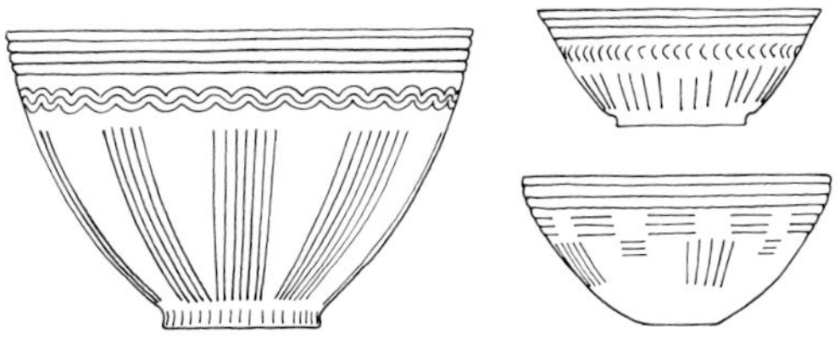	NON-LUGGED BOWL	decorated 3–5 undecorated 2–7
	COLLARED FLASK	1–7
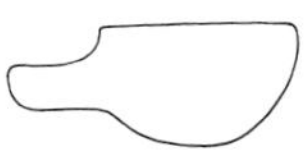	BIBERON	2–4

Fig. 1 (cont.).

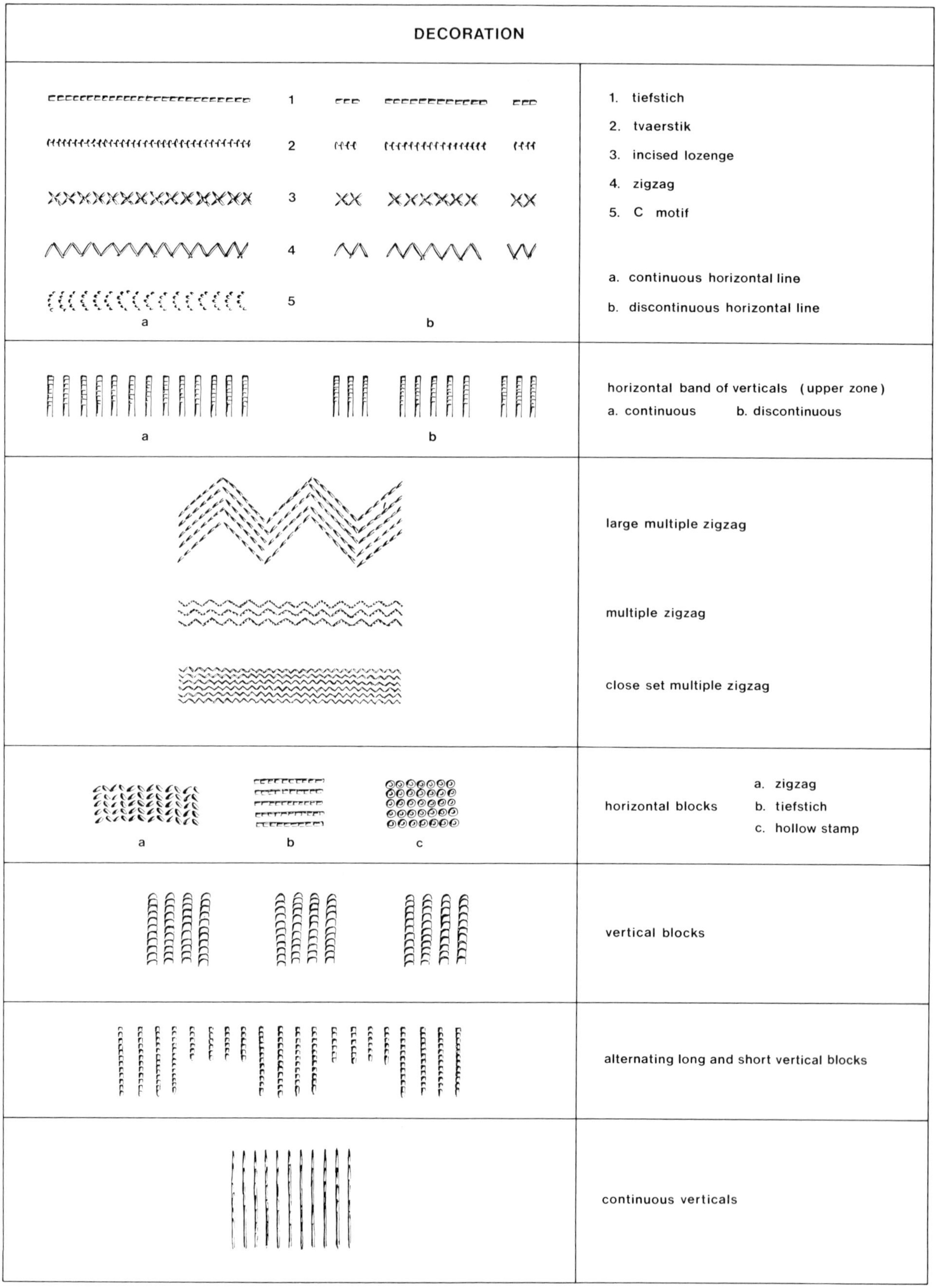

Fig. 2. Table of main decorative elements, layout and applied features.

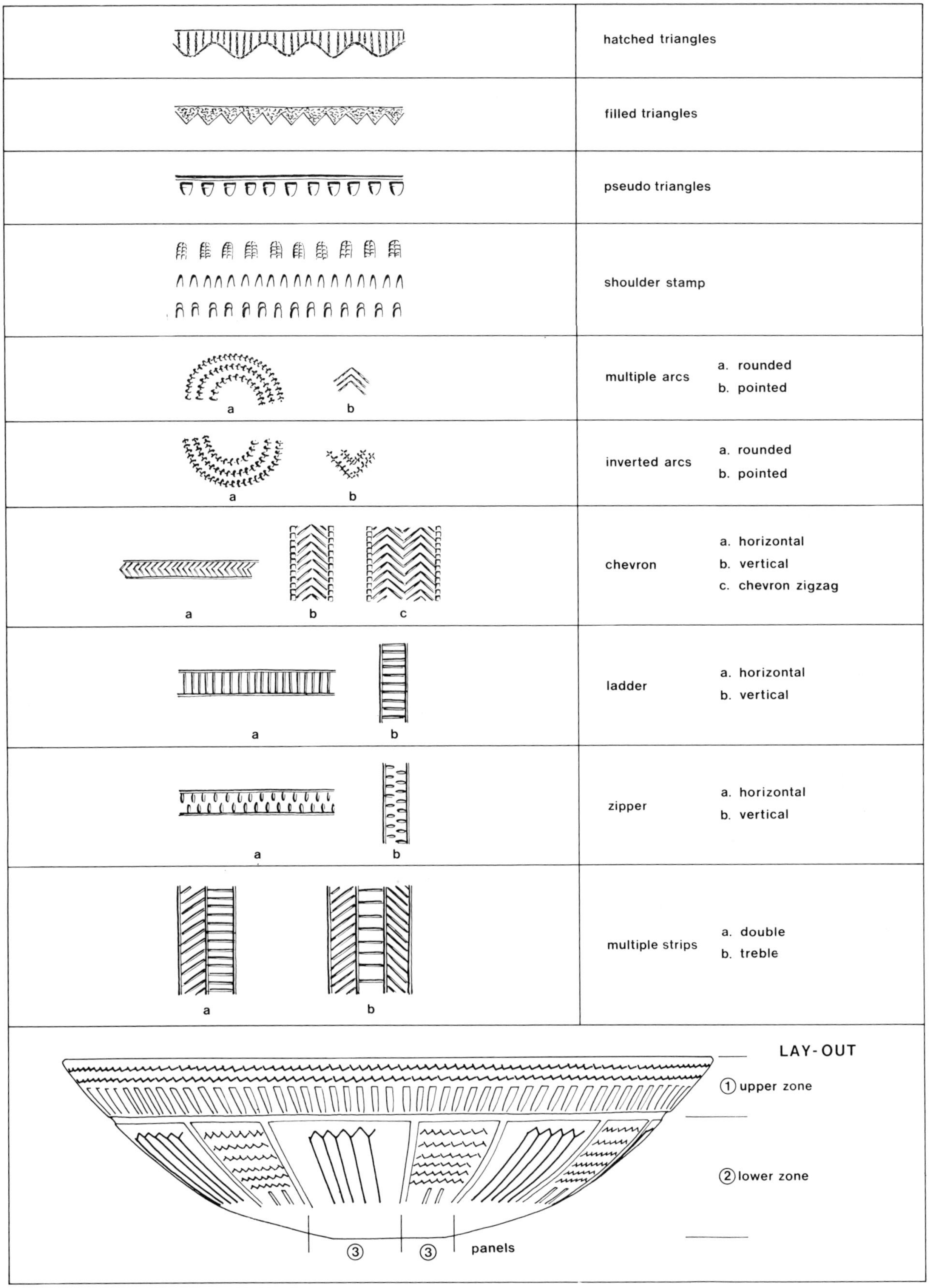
hatched triangles
filled triangles
pseudo triangles
shoulder stamp
multiple arcs
a. rounded
b. pointed
a
b
inverted arcs
a. rounded
b. pointed
a
b
chevron
a. horizontal
b. vertical
c. chevron zigzag
a
b
c
ladder
a. horizontal
b. vertical
a
b
zipper
a. horizontal
b. vertical
a
b
multiple strips
a. double
b. treble
a
b
LAY-OUT
① upper zone
② lower zone
③ ③ panels

Fig. 2 (cont.).

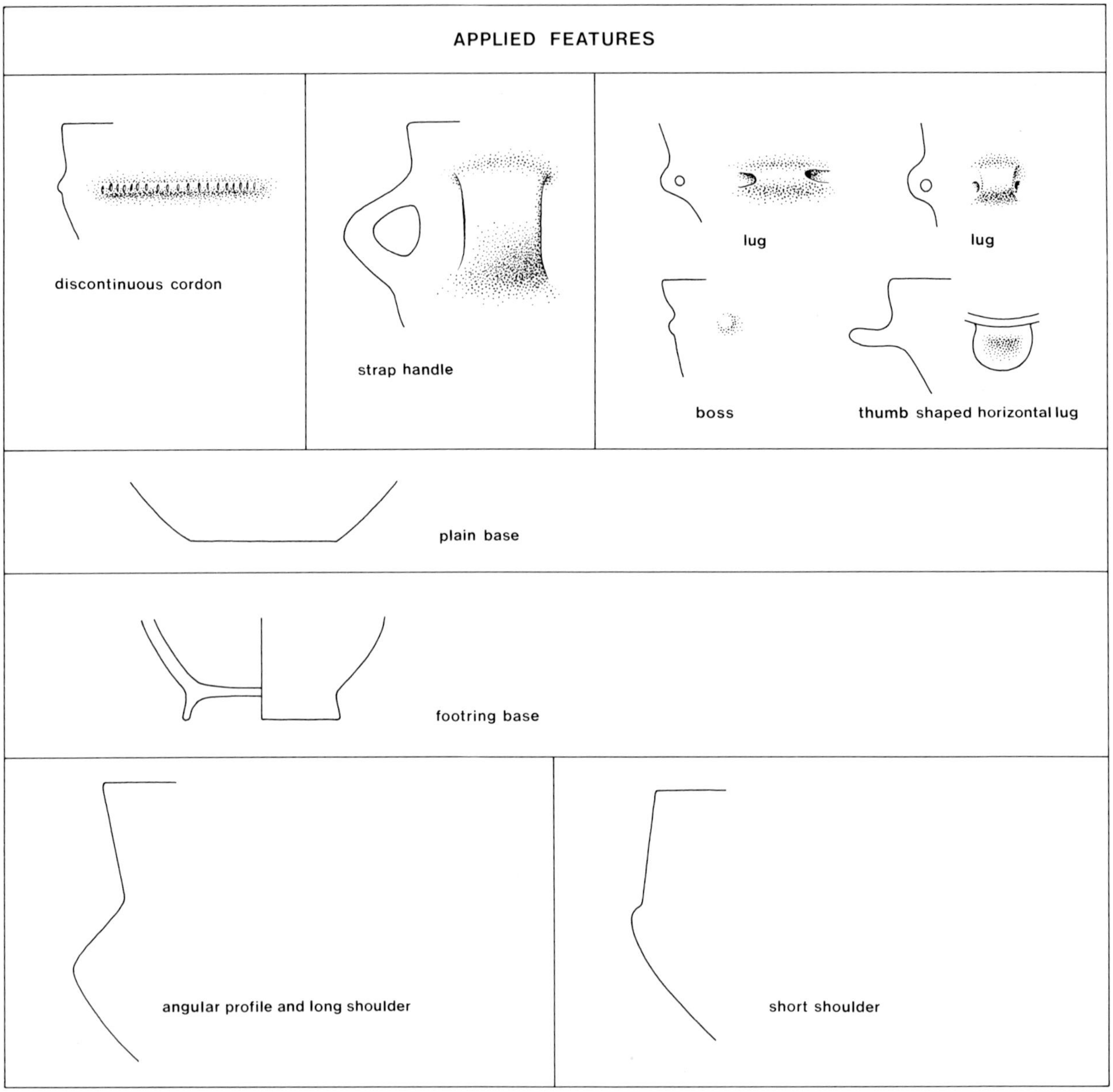
APPLIED FEATURES
discontinuous cordon
strap handle
lug
lug
boss
thumb shaped horizontal lug
plain base
footring base
angular profile and long shoulder
short shoulder

Fig. 2 (cont.).

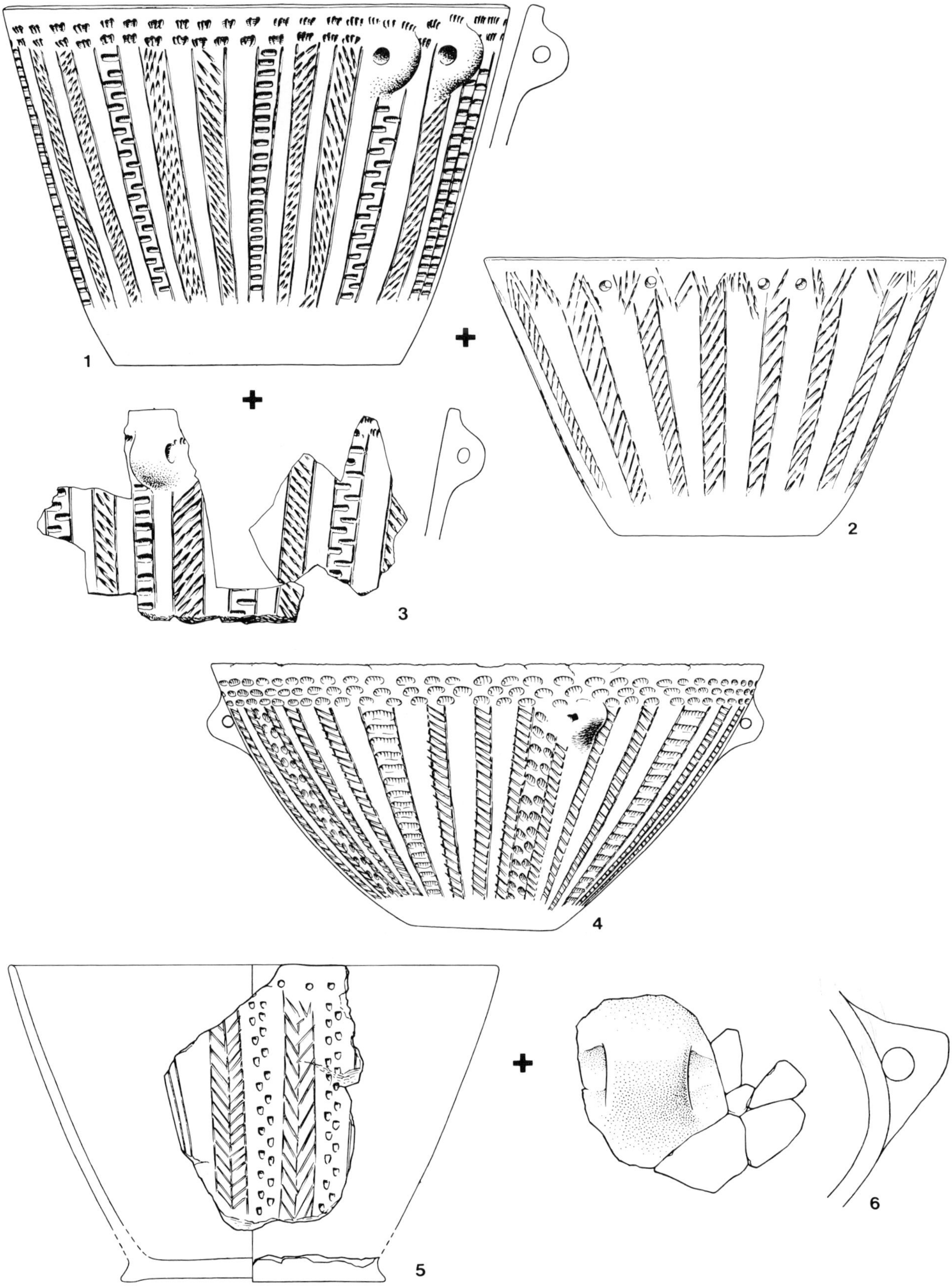

Fig. 3. Horizon 1. 1-3. Exloo, D30; 4 and 9. Bronneger, D21; 5-6. Angelslo-Heemingeslag 1963; 7. Valthe 1897; 8. Emmen, D43a; 10. Glimmer Es, G3 (after Brindley, 1983).

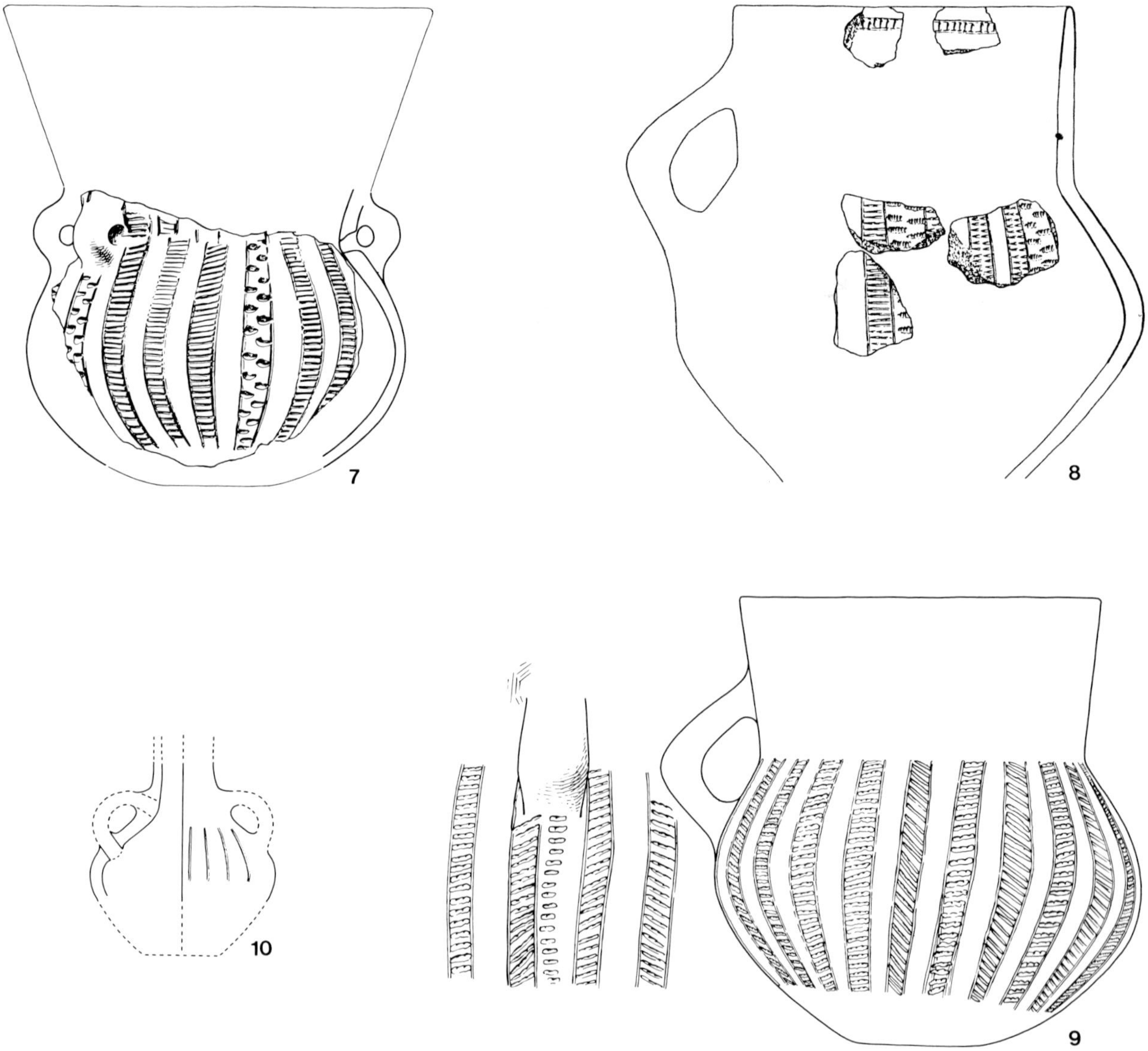

Fig. 3 (cont.).

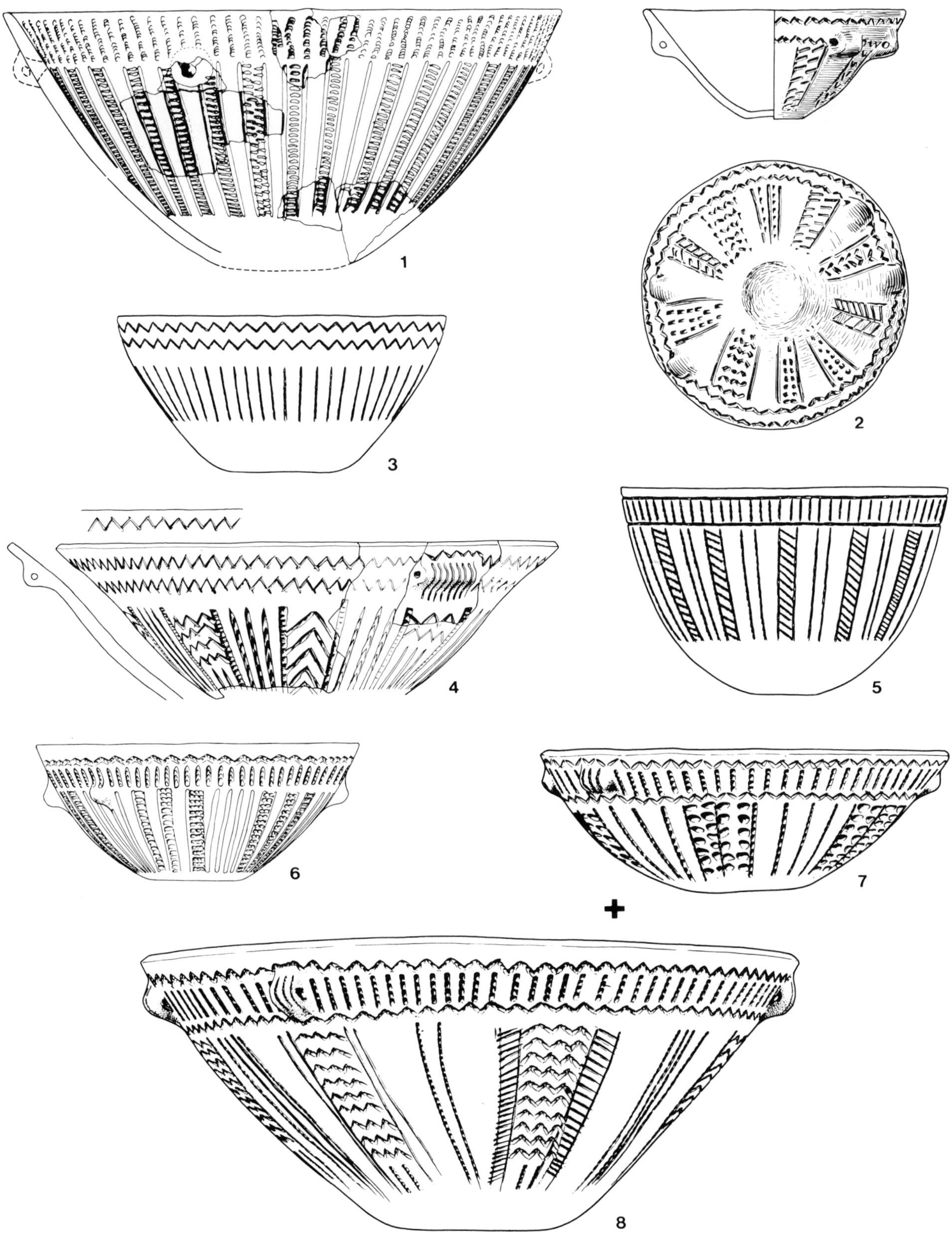

Fig. 4. Horizon 2. 1. Elspeet (after Bakker, 1979); 2. Bargercompascuum, peat find (private collection); 3 and 11. Zeijen, flatgrave e (after Bakker, 1979); 4. Glimmer Es, G2; 5. Diever, stone cist (after Bakker, 1979); 6. Bronneger, D21; 7-8 and 14-22. Hooghalen, stone cist(?) (after Bakker, 1979, but No. 18 after Brindley, 1983); 9. Hooghalen, D54b; 10. Weerdinge, peat find; 12. Tinaarlo, near D6e/f (after Bakker, 1979); 13. Zeijen, flatgrave a (after Bakker, 1979); 23. Zeijen, flatgrave(?) f (after Bakker, 1979); 24. Rijs, F1 (redrawn after van Giffen, 1925/27).

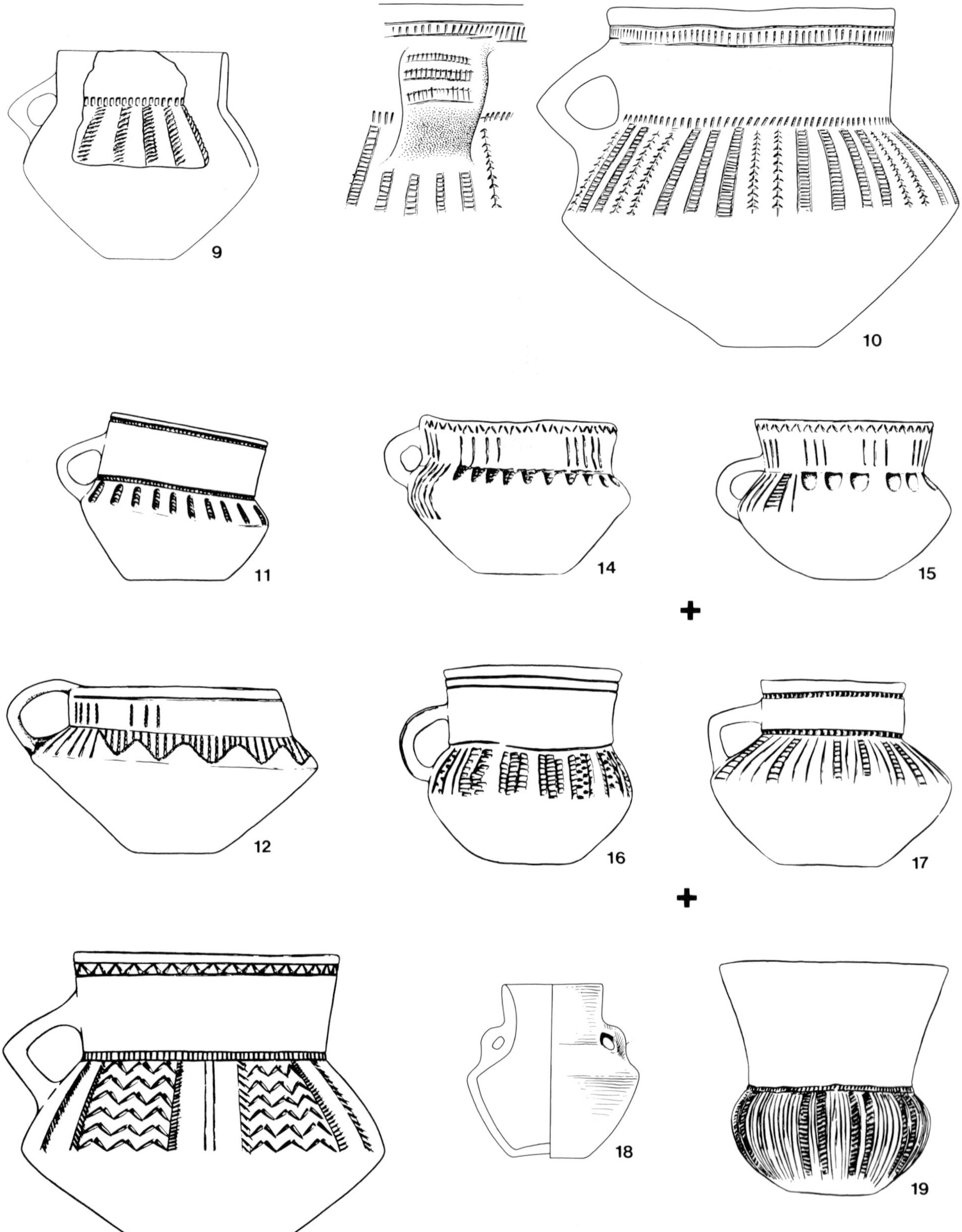

Fig. 4 (cont.).

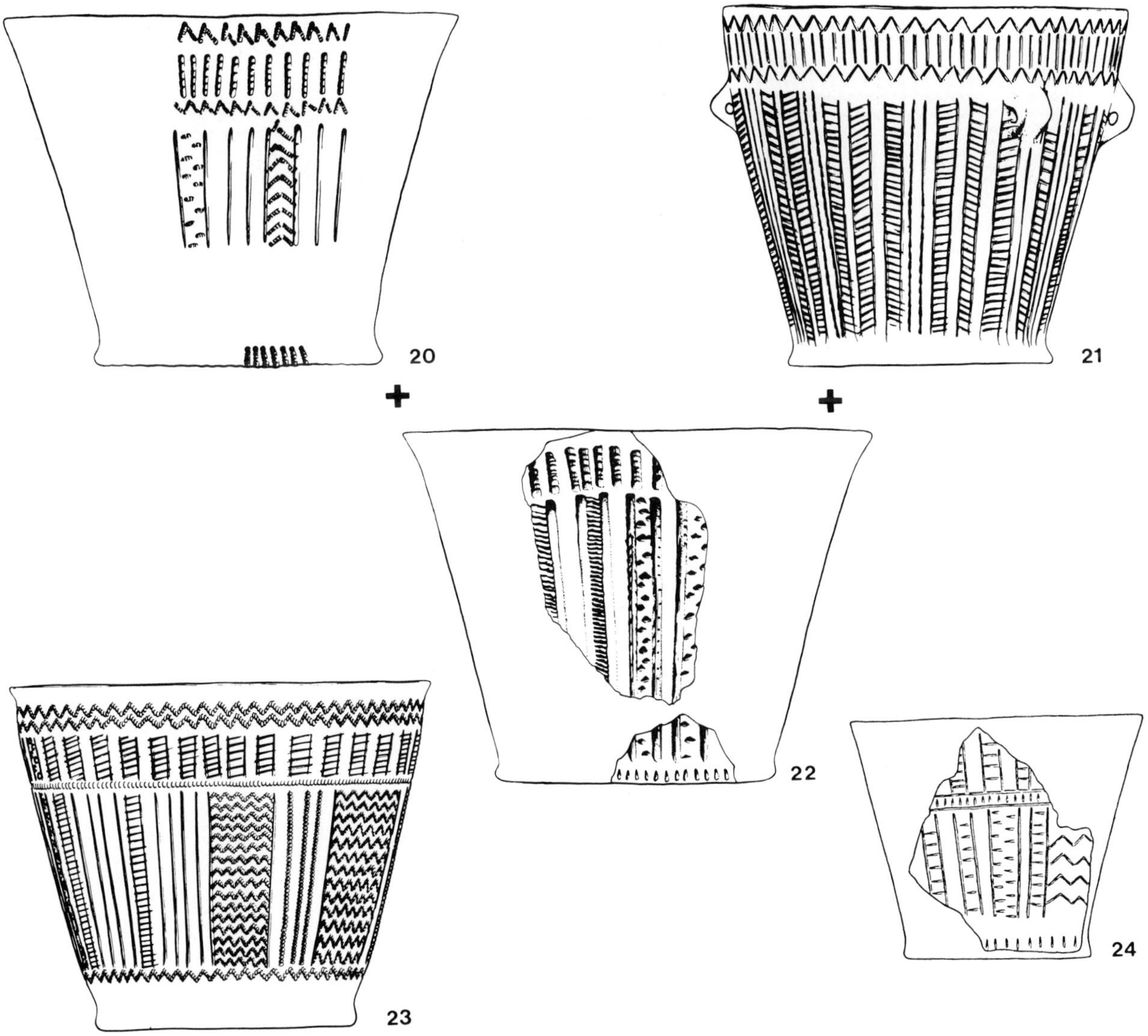

Fig. 4 (cont.).

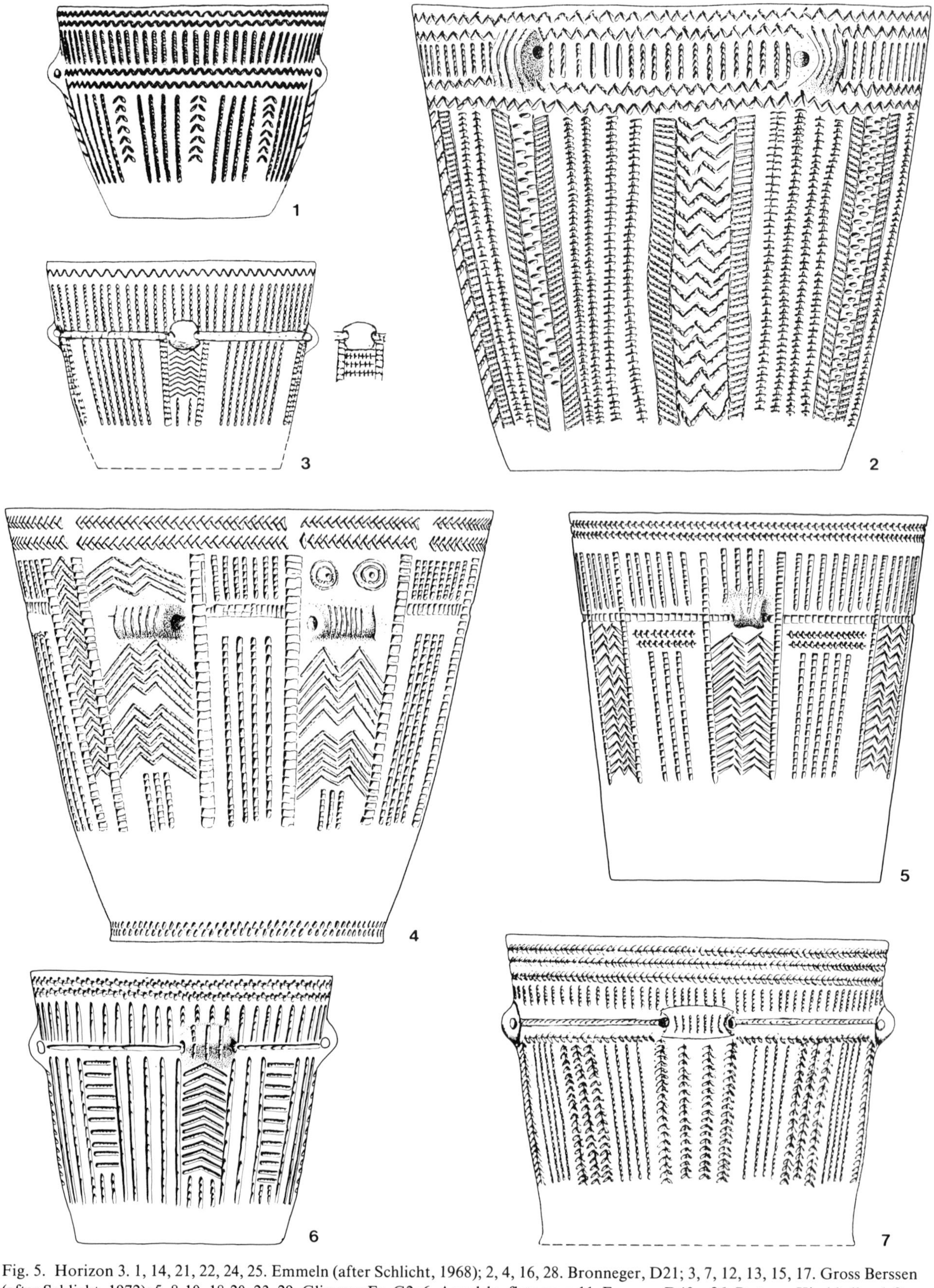

Fig. 5. Horizon 3. 1, 14, 21, 22, 24, 25. Emmeln (after Schlicht, 1968); 2, 4, 16, 28. Bronneger, D21; 3, 7, 12, 13, 15, 17. Gross Berssen (after Schlicht, 1972); 5, 8-10, 18-20, 23, 29. Glimmer Es, G2; 6. Angelslo, flatgrave; 11. Emmen, D43a; 26. Bottrop-Kirchhellen (after AFWL 3, 1985); 27. Angelslo, flatgrave; 30 and 31. Exloo, D30.

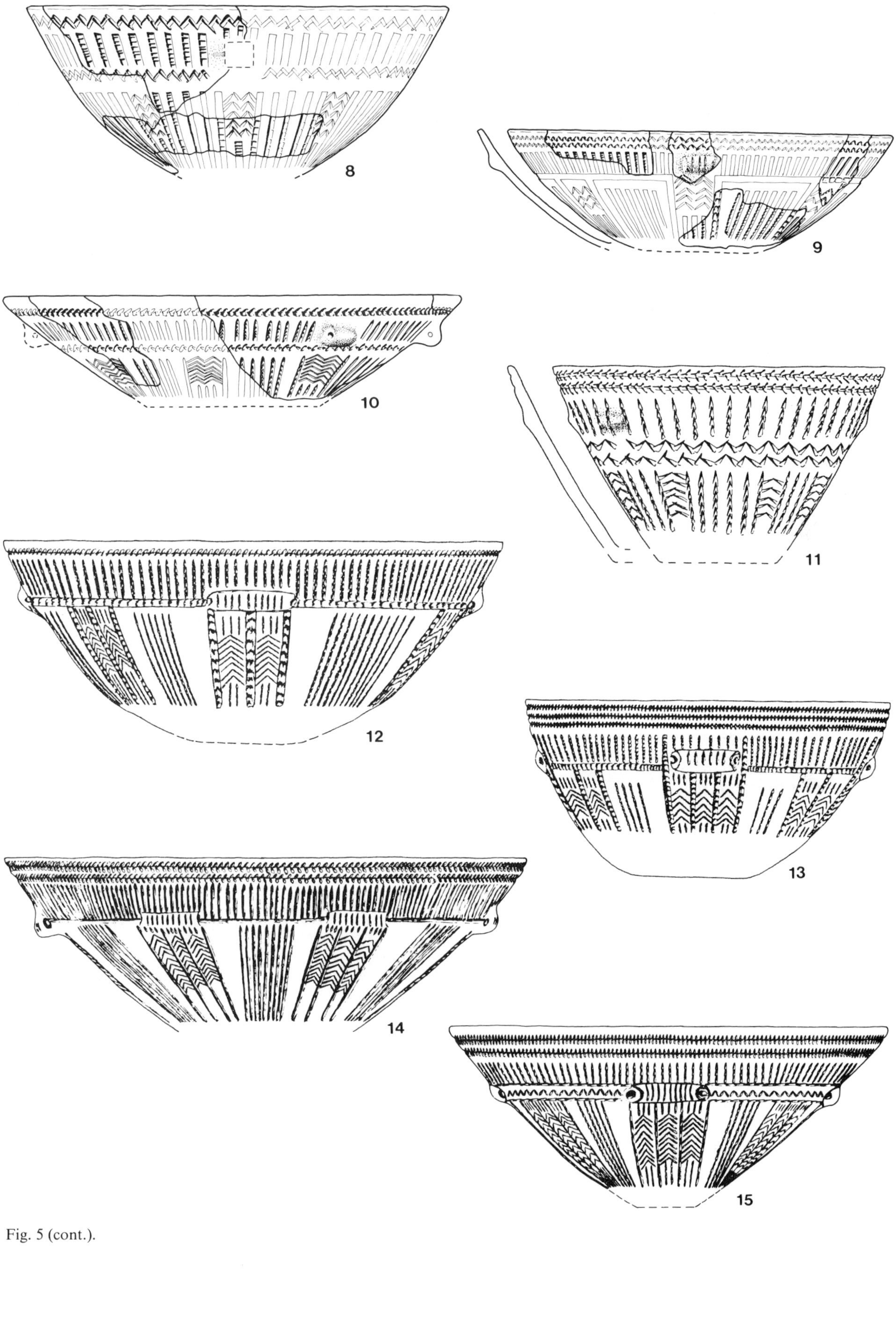

Fig. 5 (cont.).

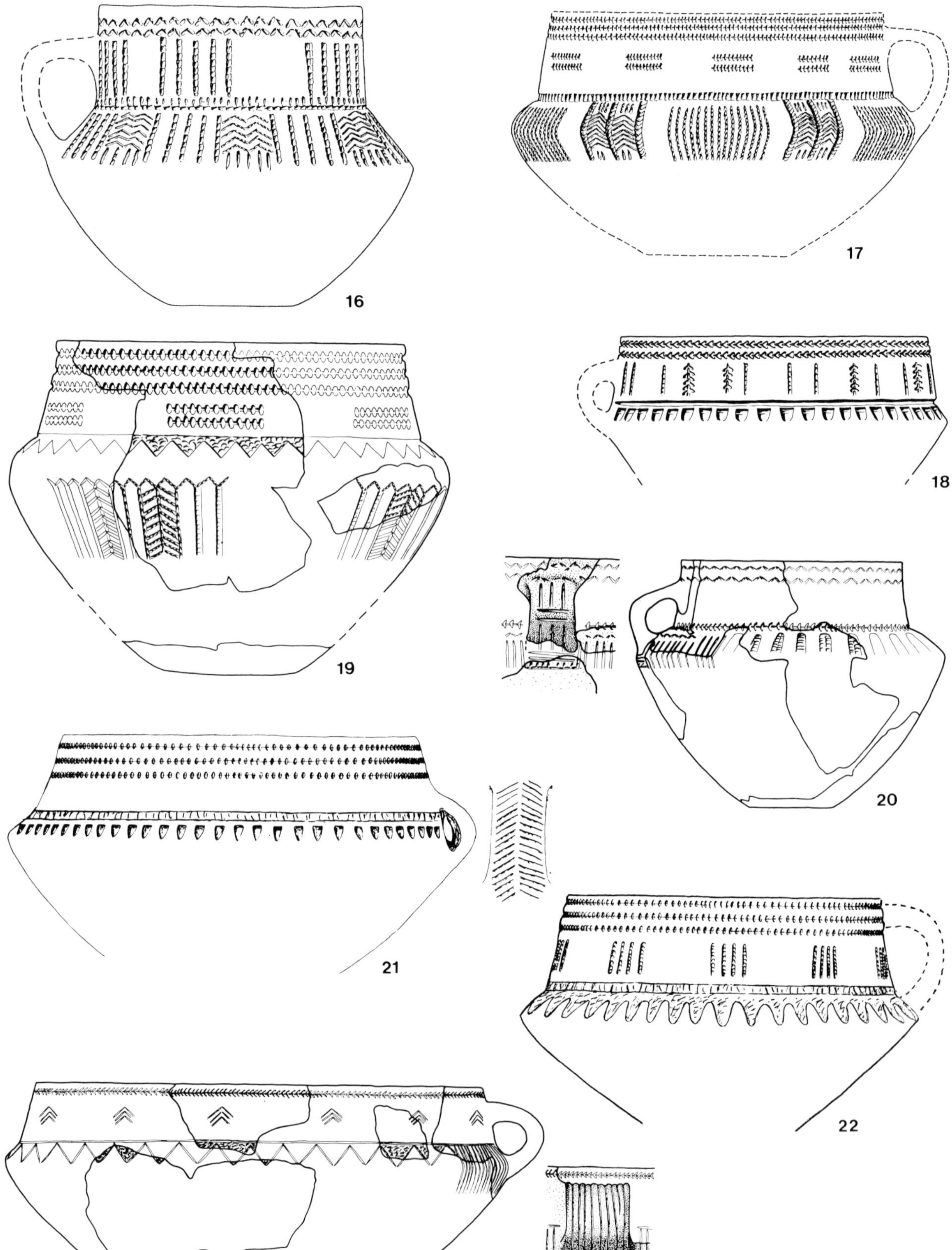

Fig. 5 (cont.).

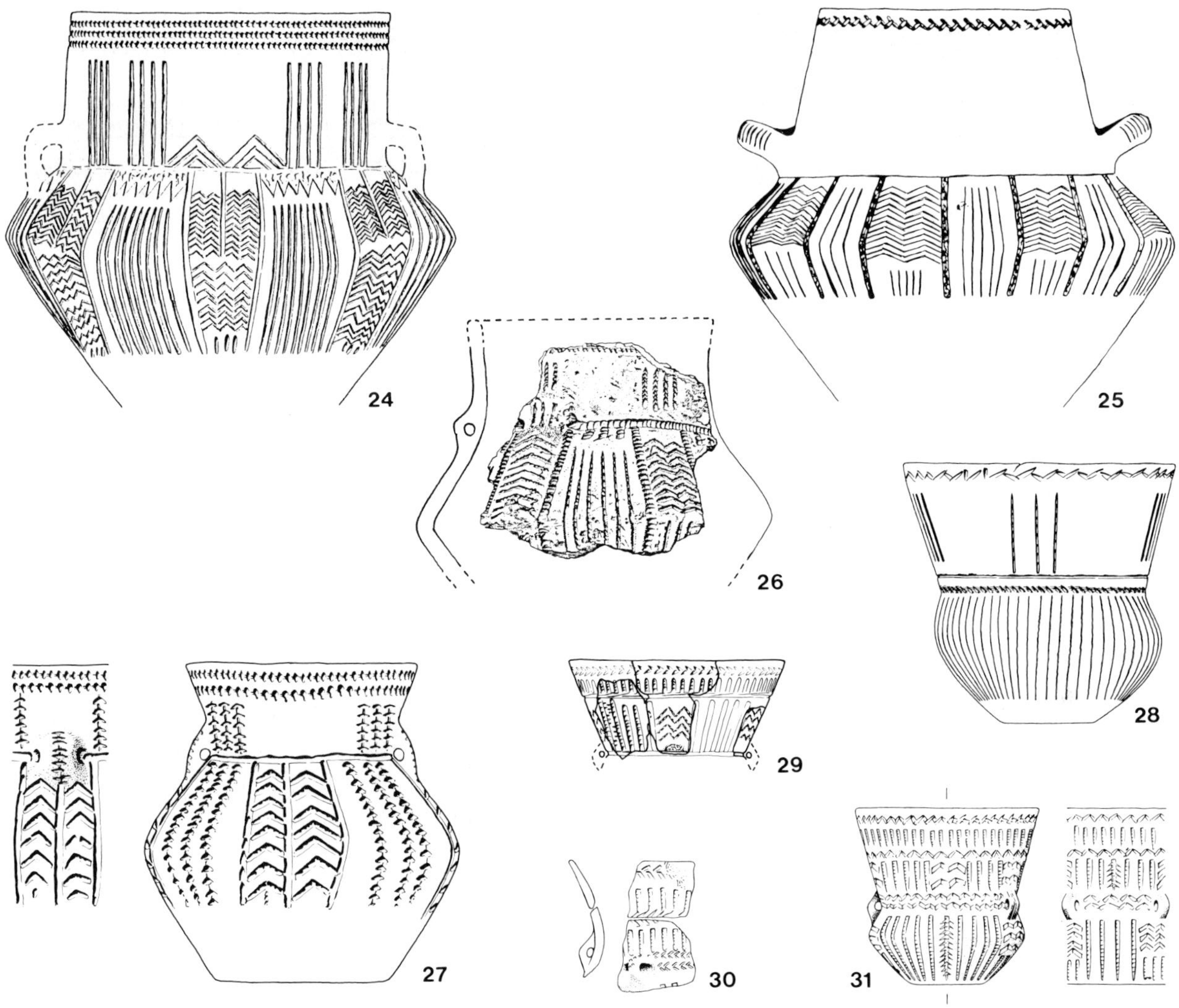

Fig. 5 (cont.).

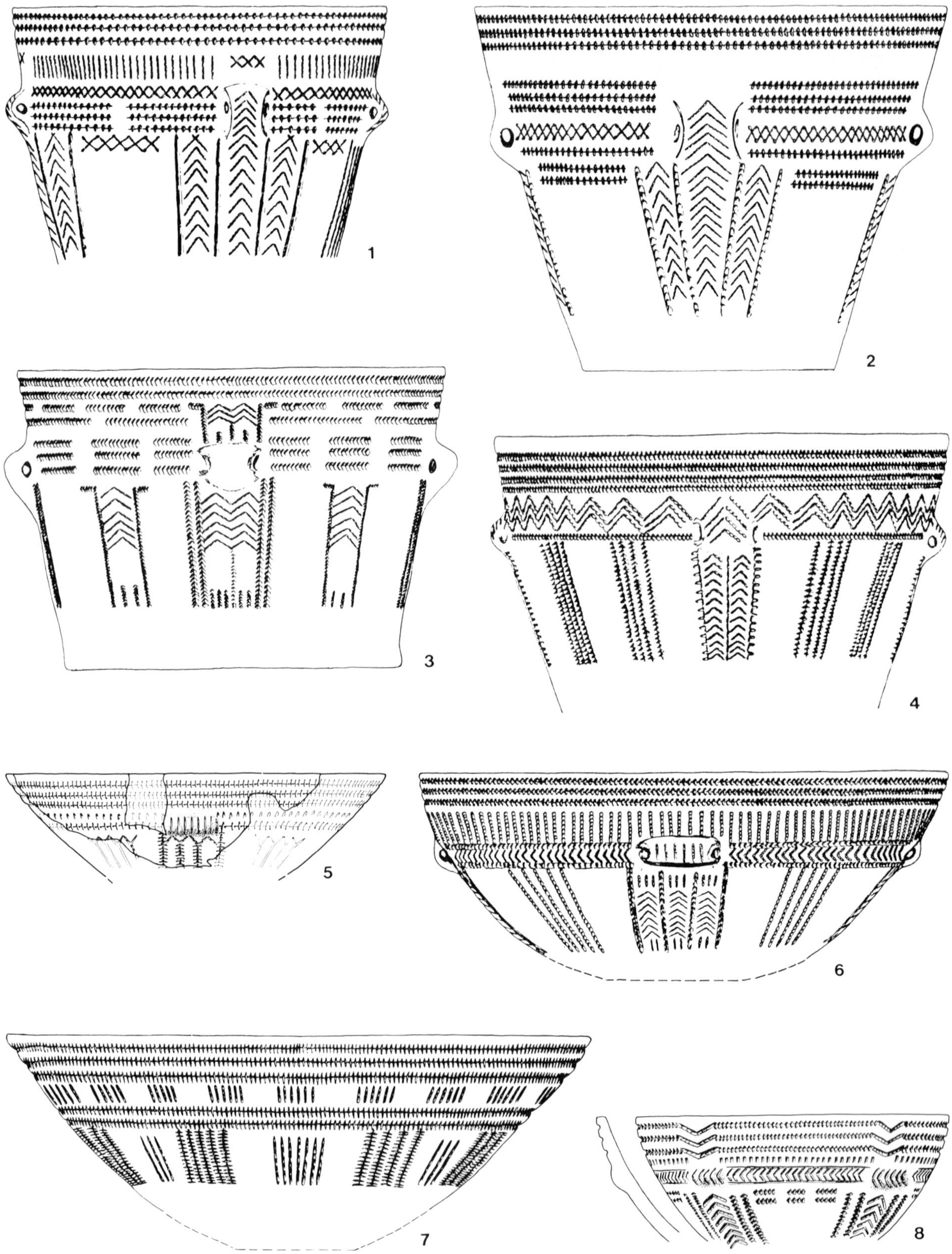

Fig. 6. Horizon 4. 1-4, 19, 20, 22-24, 26, 28, 29. Emmeln (after Schlicht, 1968); 5, 8, 11, 13-18, 21, 25, 27, 30-32, 35, 36. Glimmer Es, G2; 6, 7, 9. Gross Berssen (after Schlicht, 1972); 10. Veele, flatgrave(?); 12. Hankampsveld, flatgrave(?); 33. Ekelberg, flatgrave e; 34. Ekelberg, flatgrave (33 and 34 after Bakker, 1979).

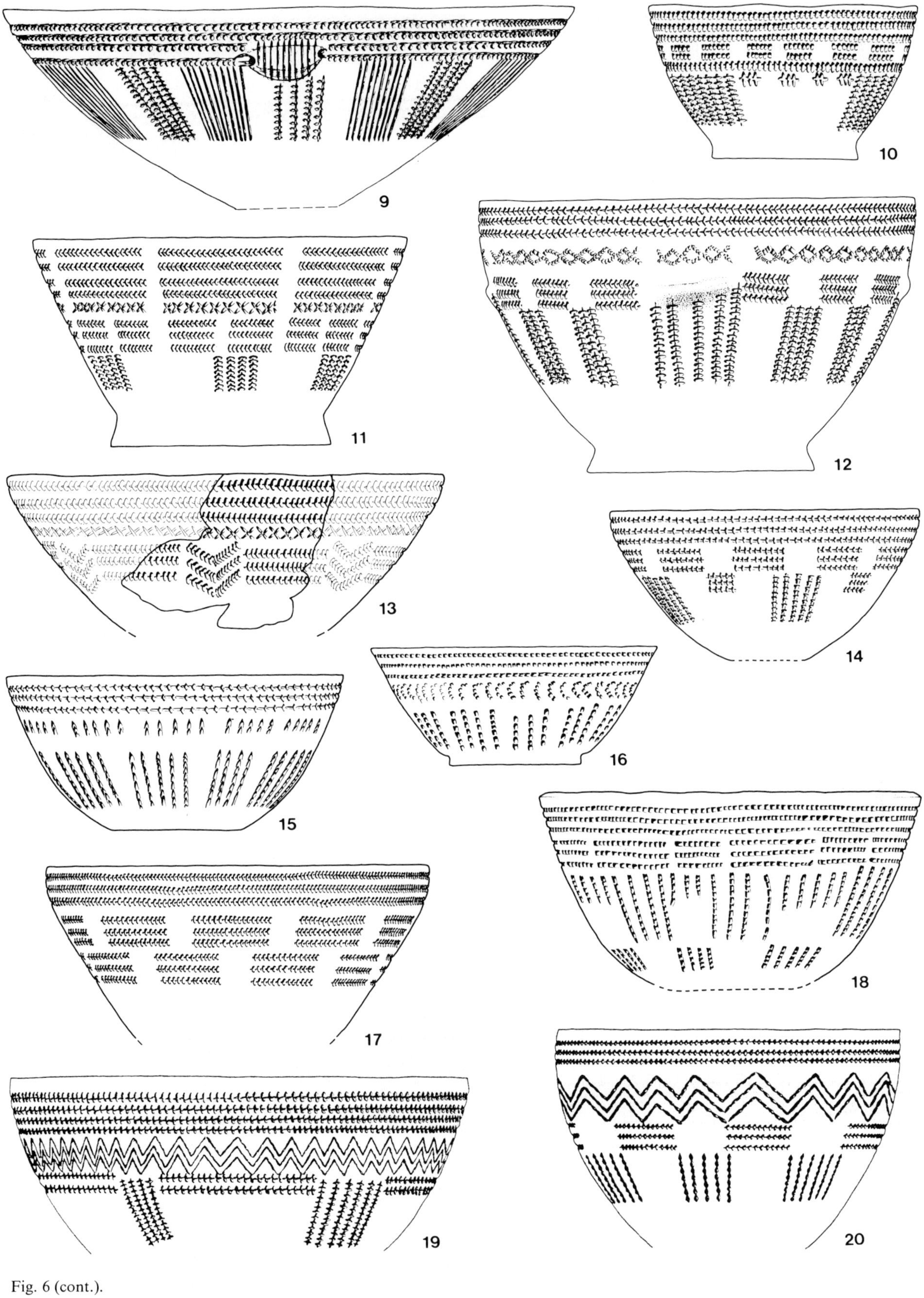

Fig. 6 (cont.).

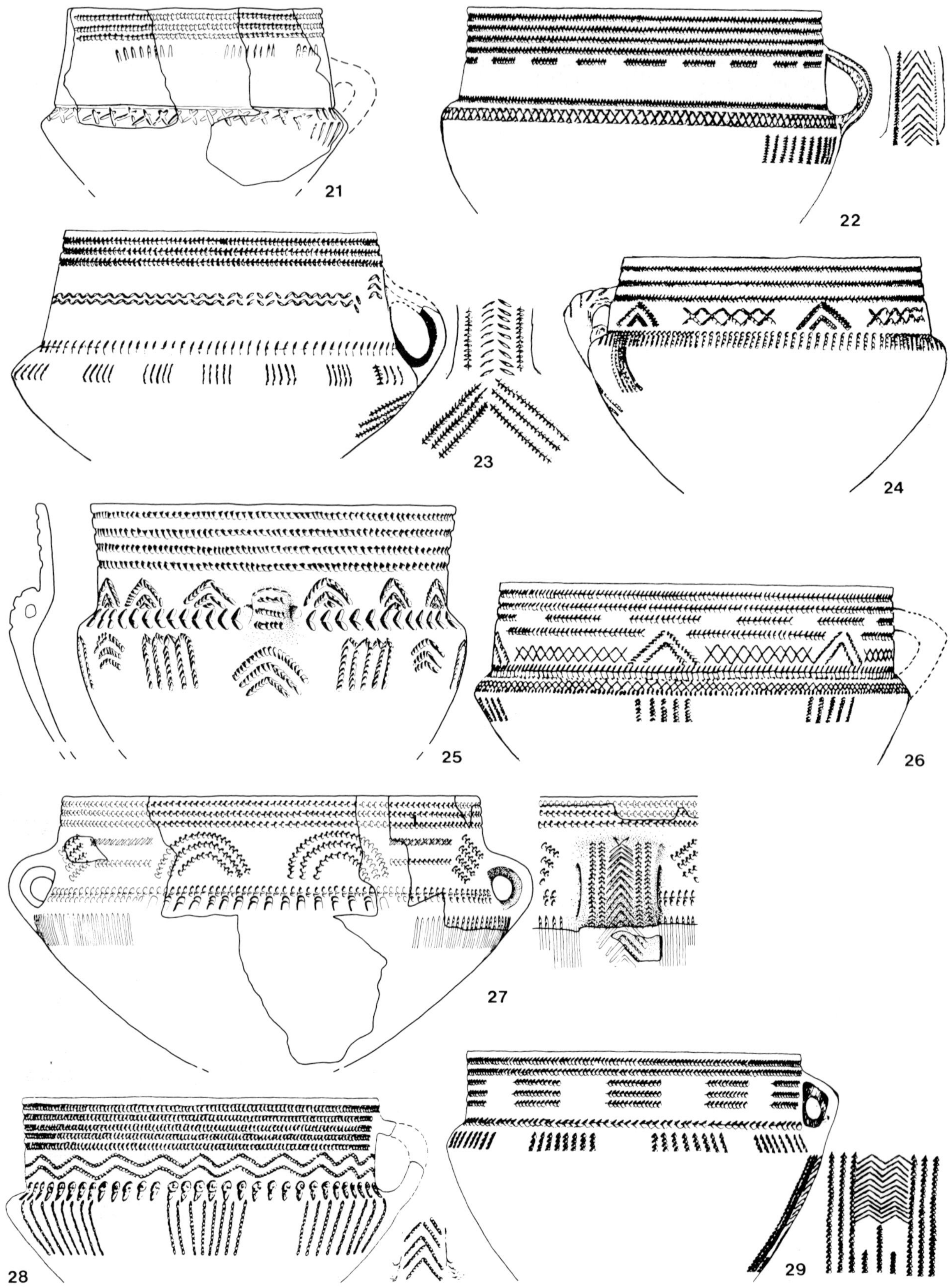

Fig. 6 (cont.).

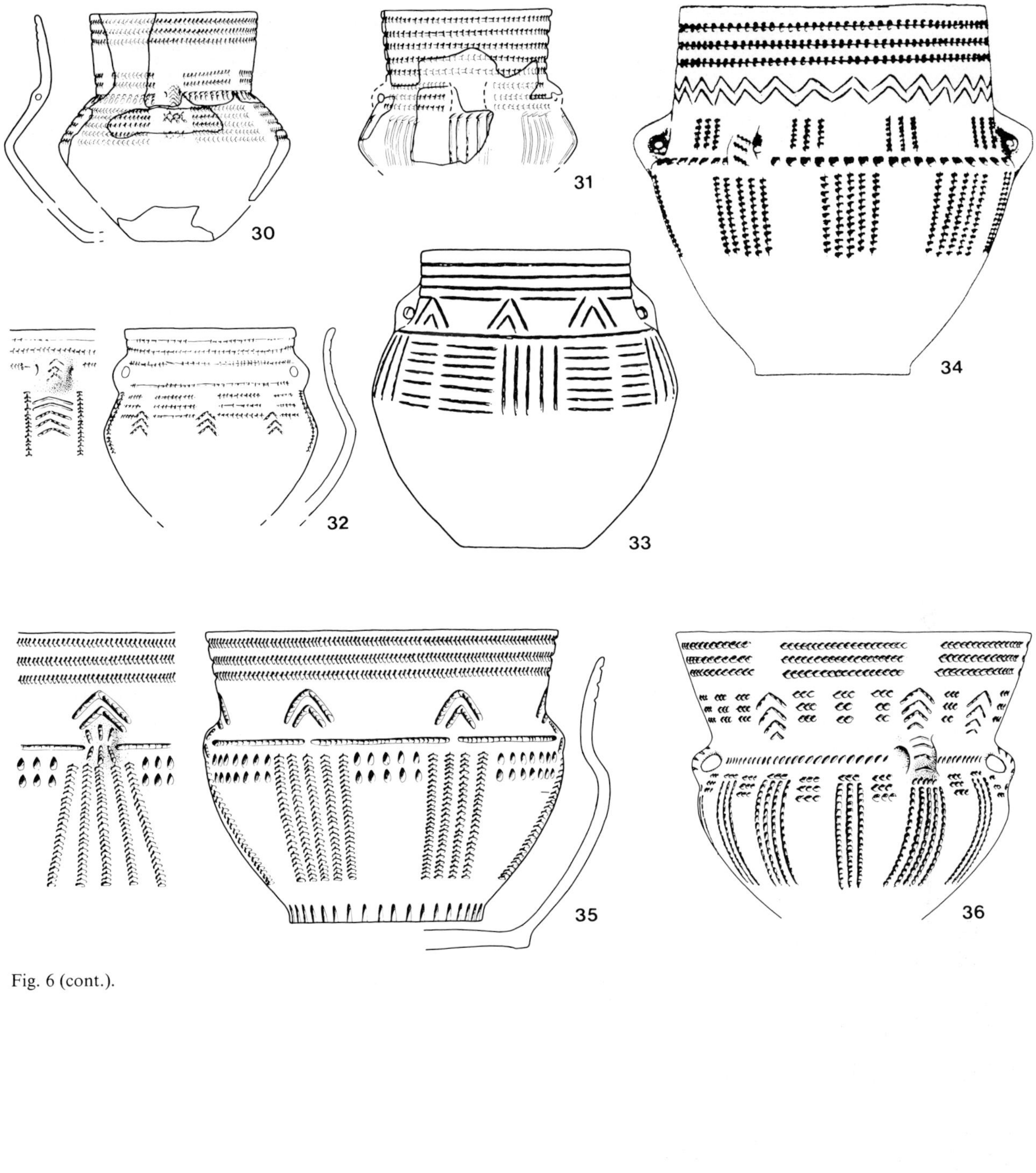

Fig. 6 (cont.).

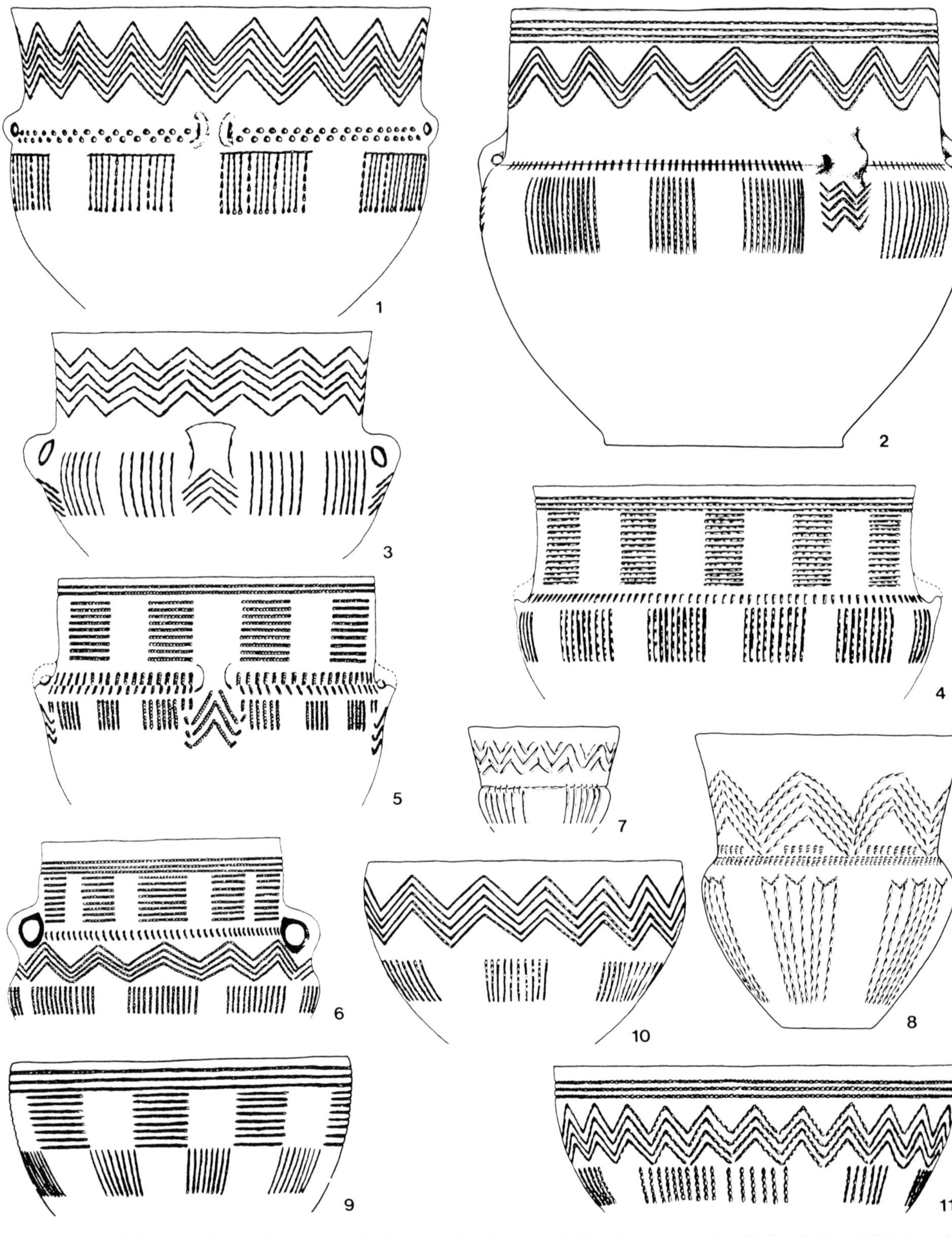

Fig. 7. Horizon 5. Emmeln-Heek Style: 1, 3-6, 10, 11. Emmeln (after Schlicht, 1968); 2. Uddelermeer, hut 3 (after Bakker, 1979); 7 and 8. Glimmer Es, G2; 9. S. of Hunneschans, settlement pit (after Bakker, 1979). Uddelermeer-Anlo Style: (22, 23 and 26 typologically late) 12 and 24. Anlo, settlement (redrawn after Waterbolk, 1960); 13. Uddelermeer, hut 8 (after Bakker, 1979); 14. Steenwijkerwold, flatgrave?; 15. Uddelermeer, grave 2 (after Bakker, 1979); 16-18. Glimmer Es, G2; 19. Ugchelen, flatgrave (after Bakker, 1979); 20. Uddelermeer, grave 1 (after Bakker, 1979); 22-23. Harderwijk (after Manssen, 1980); 25. Uddelermeer, grave 4 (after Bakker, 1979); 26-27. Havelte, D53. Funnel beaker cups and undecorated funnel beakers, found with both styles: 28. Uddelermeer, grave 2 (after Bakker, 1979); 29. Glimmer Es, G2.

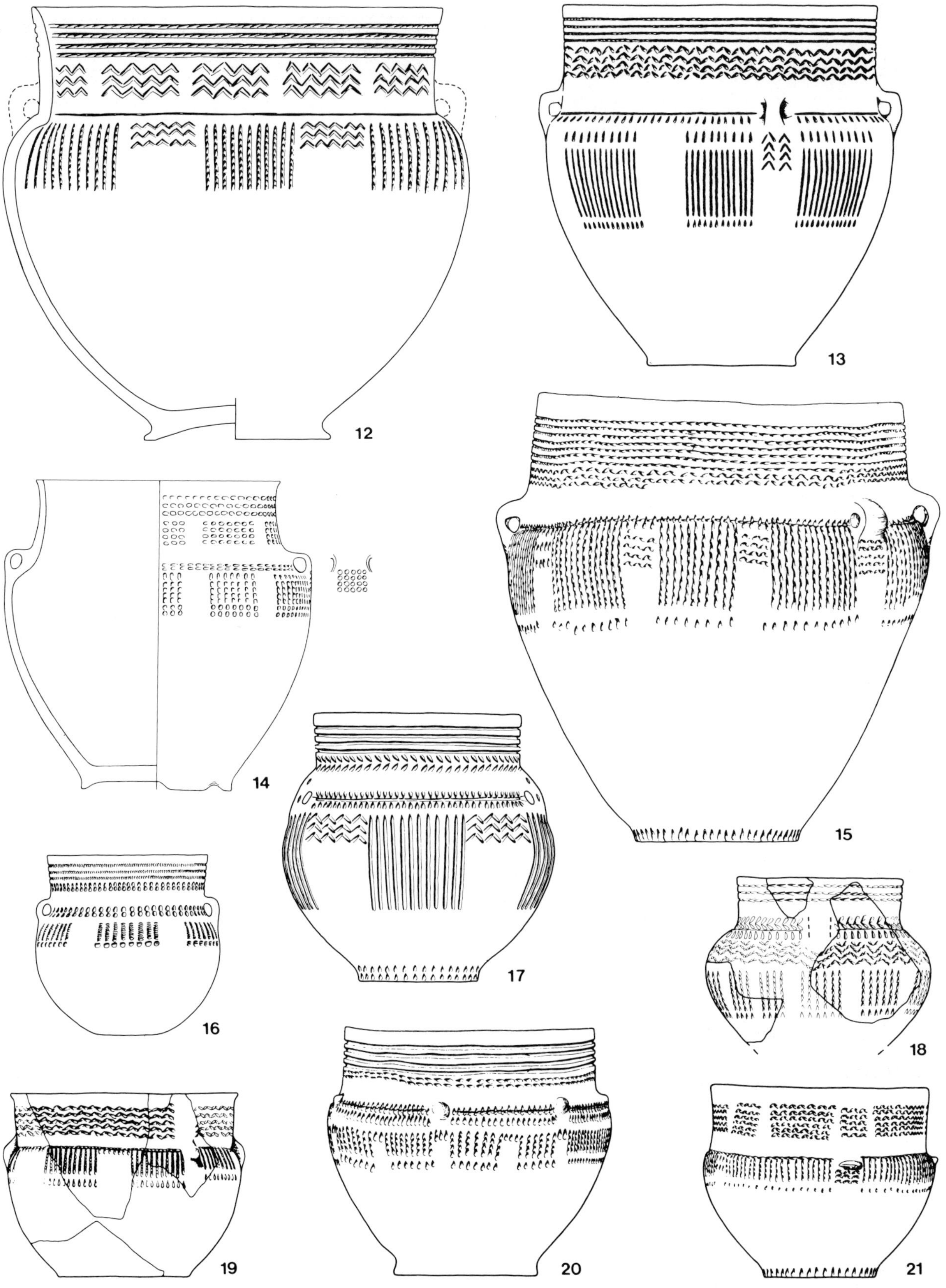

Fig. 7 (cont.).

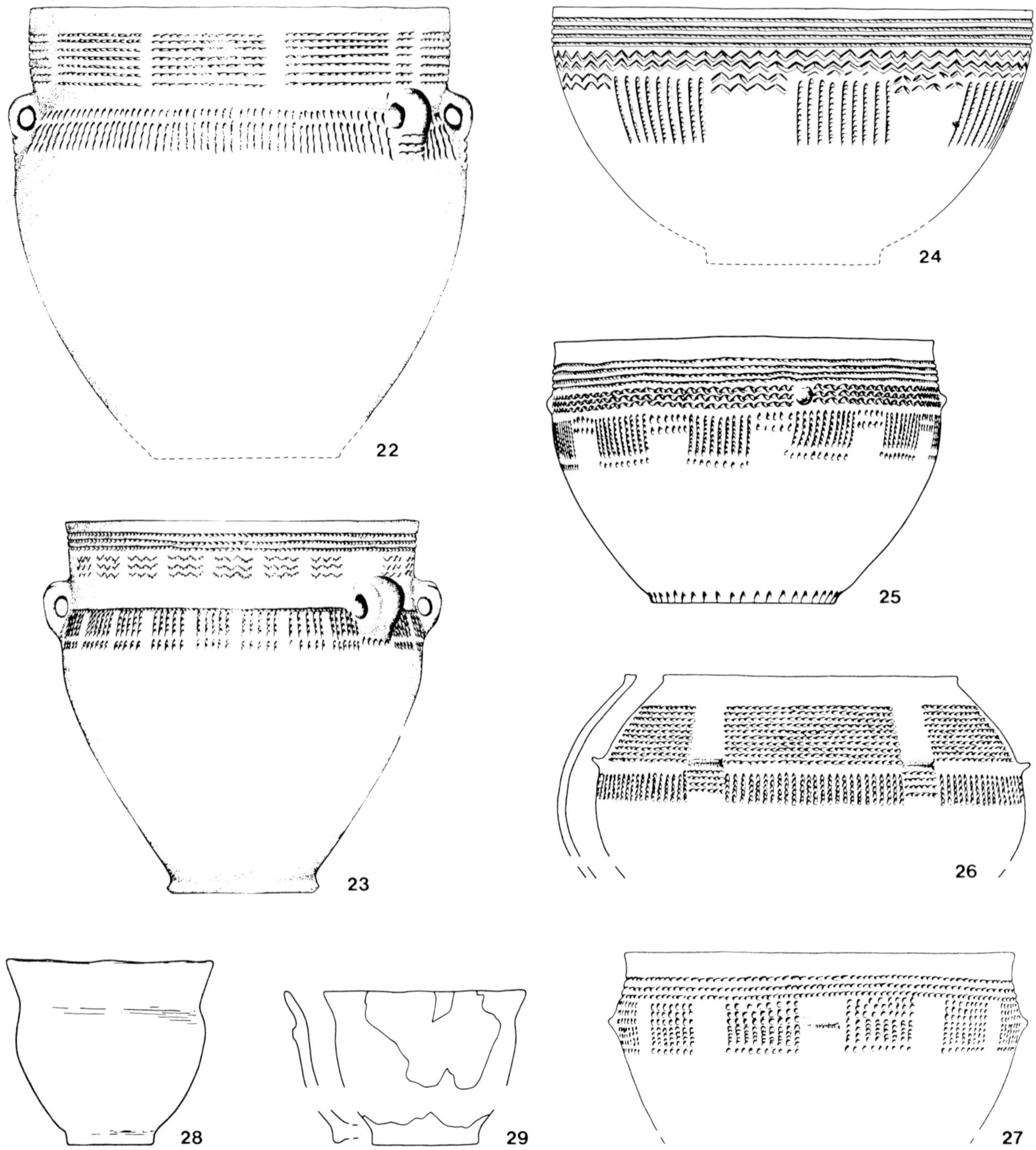

Fig. 7 (cont.).

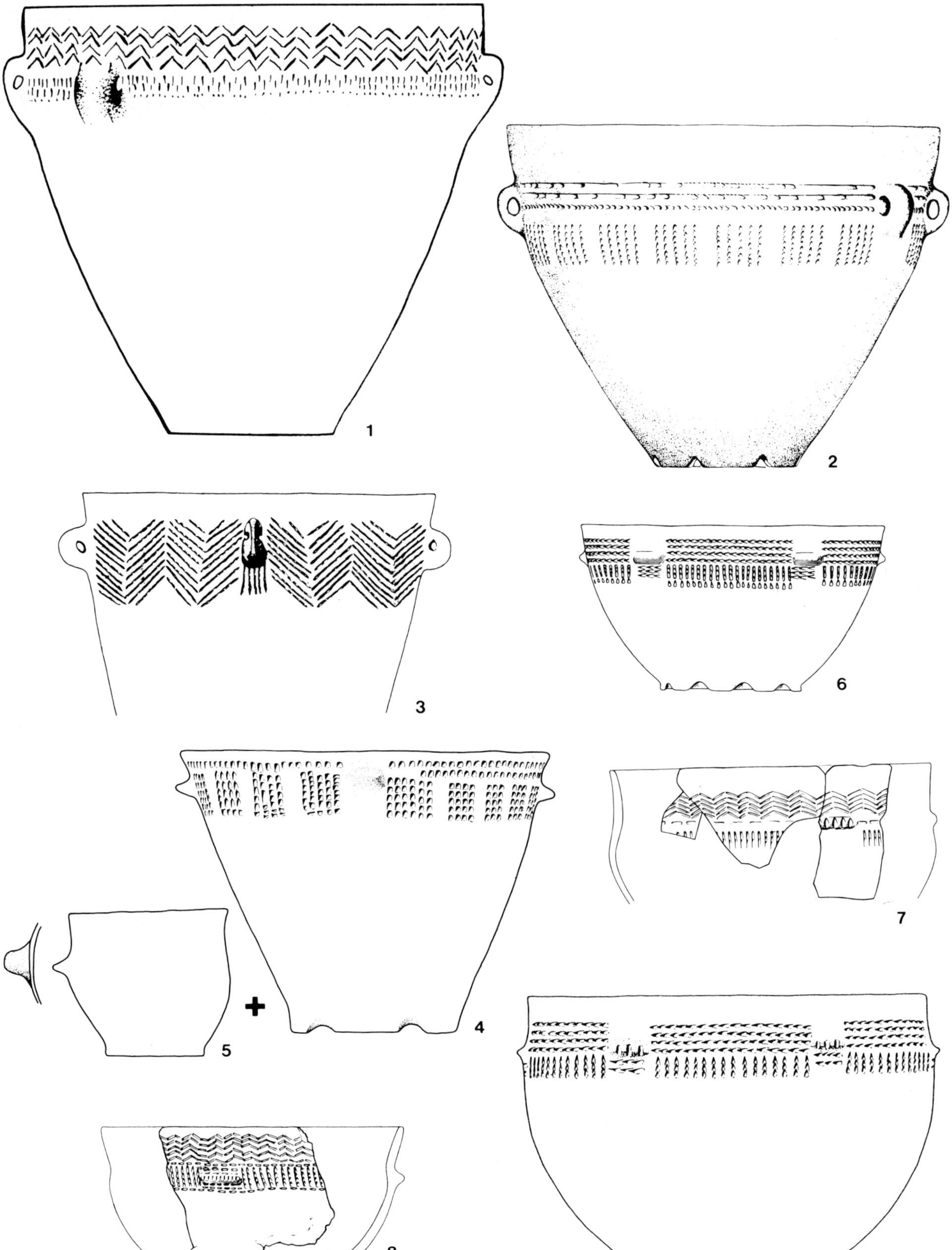

Fig. 8. Horizon 6. 1. Angelslo, flatgrave (after Bakker, 1979); 2. Harderwijk (after Manssen, 1980); 3. Emmeln (after Schlicht, 1968); 4-5. Glimmer Es, near G2; 6, 7, 9. Havelte, D53; 8. provenance unknown, collection B.A.I.

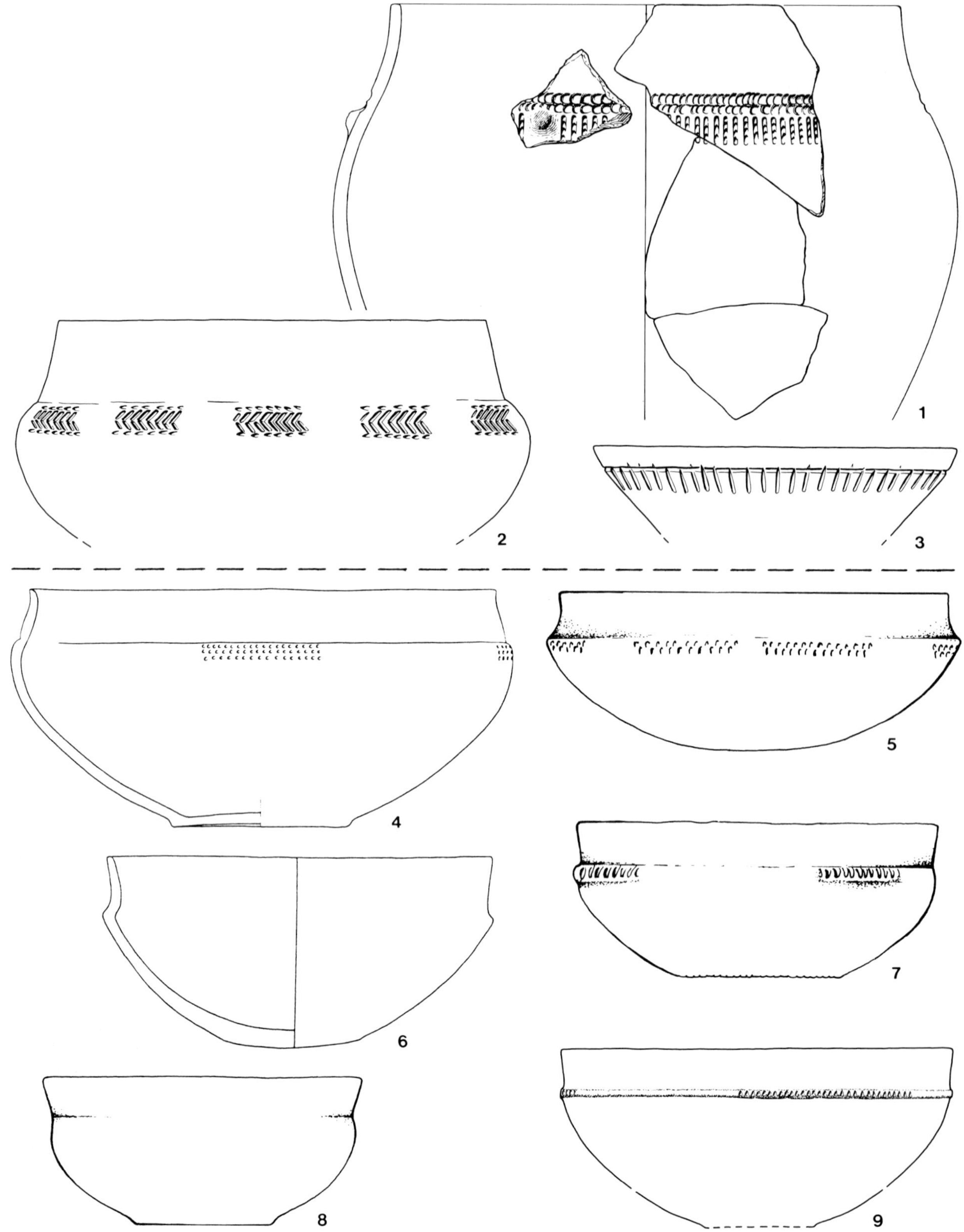

Fig. 9. Horizon 7 (1-3 typologically earlier). 1 and 12. Havelte, D53; 2, 3, 9, 15, 21. Glimmer Es, G2; 4. Leer, near megalithic tomb (after Bakker & van der Waals, 1973); 5. Hoge Loo, flatgrave (after Bakker, 1979); 6 and 20. Angelslo, flatgrave 1 (after Bakker, 1979); 7. Allardsoog, flatgrave? (after Bakker, 1979); 8. Schipborg, flatgrave? (after Bakker, 1979); 10. Angelslo, flatgrave 5 (after Bakker, 1979); 11. Nottuln (redrawn after Eckert, 1986); 13. Denekamp, flatgrave (after Bakker, 1979); 14. Angelslo, settlement pit 5 (after Bakker, 1979); 16. Losser (after Bakker & van der Waals, 1973); 17. Noordlaren, G1 (after Bakker, 1979); 18. Emmeln (redrawn after Schlicht, 1968); 19 and 22. Angelslo, flatgrave 3 (after Bakker, 1979).

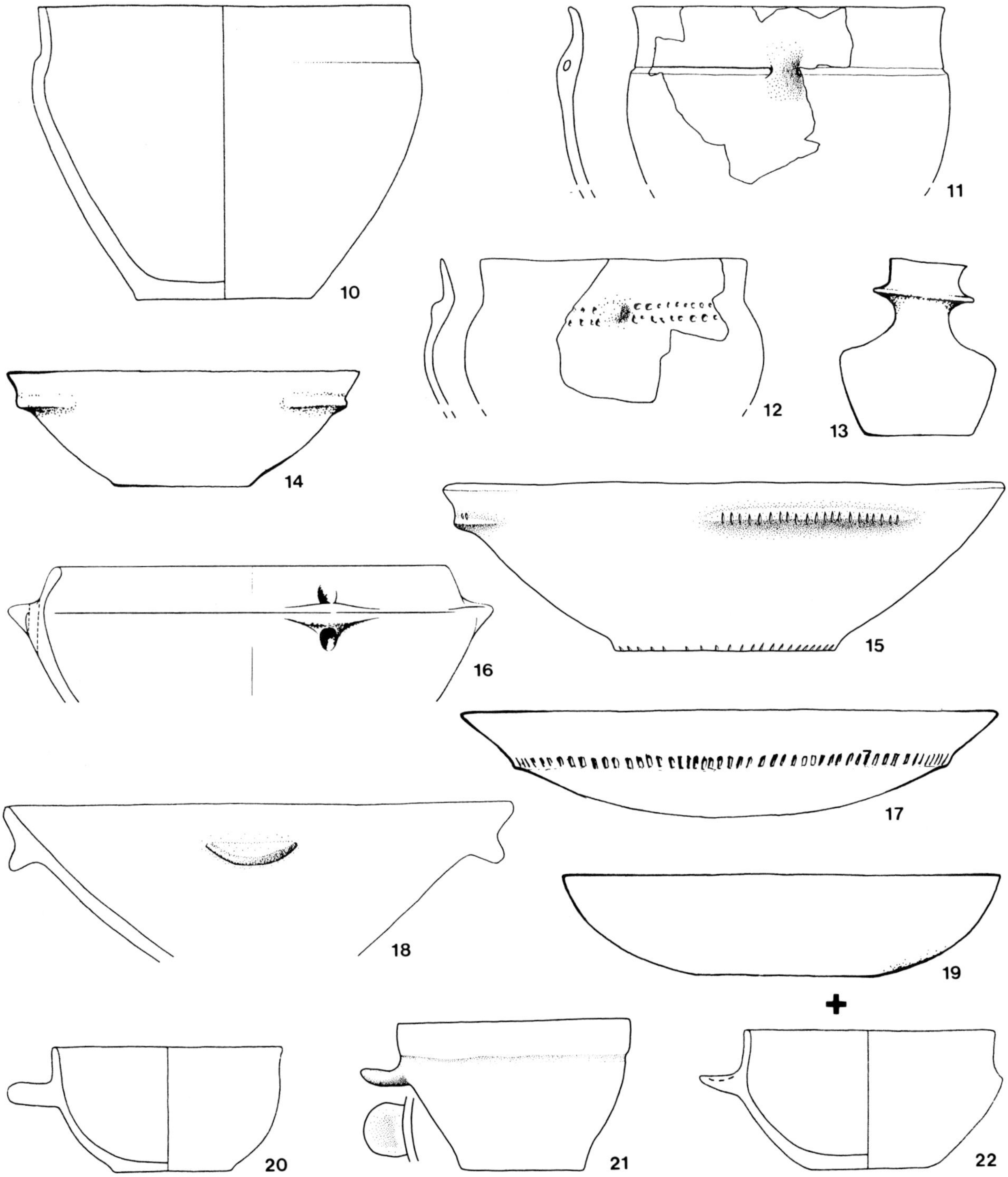

Fig. 9 (cont.).

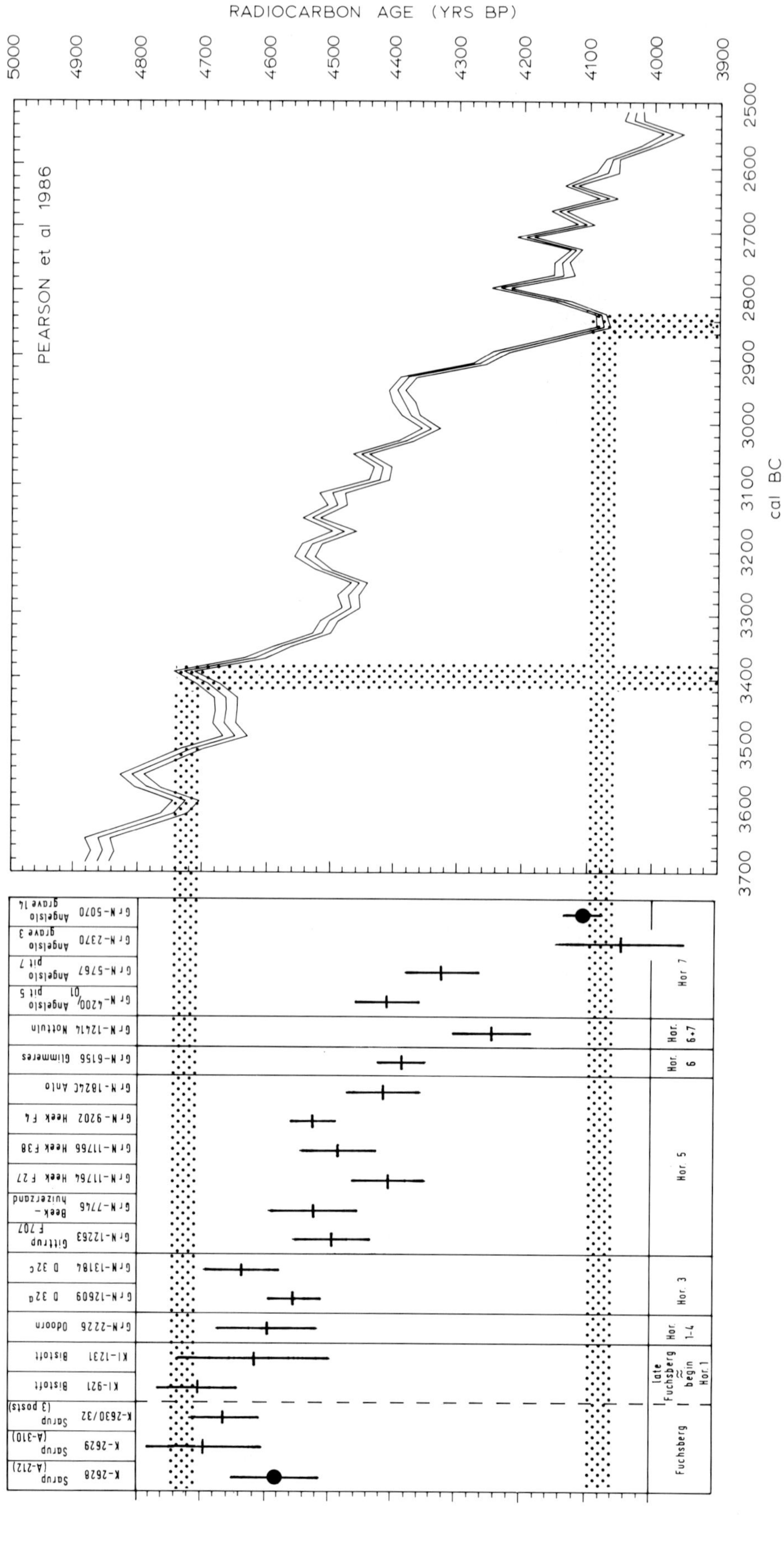

Fig. 10. Radiocarbon dates from Fuchsberg and TRB West Group contexts, in relation to the calibration curve, and arranged by horizon. The circular symbols indicate samples of short own life. The most likely beginning and end dates of the West Group are indicated by the shaded bars.

DROUWEN: END OF A 'NORDIC' RAINBOW?

J.J. Butler

Biologisch-Archaeologisch Instituut, Groningen, Netherlands

ABSTRACT: This article has three parts. Section 1 describes the Late Bronze Age hoard of bronzes in a pot, found near the village of Drouwen, Drenthe in December 1984.

Section 2 examines the character, origins and dating of this hoard. The pot and almost all of the bronzes are of types which were previously totally absent or very rare in this region. The hoard was apparently a collection of old objects imported as scrap metal, intended to be melted down for the manufacture of new bronzes. It seems to have originated in the Middle or Lower Elbe area, c. 200 to 300 km to the east.

Section 3 is a brief survey of the remarkable series of finds of Bronze Age material in the vicinity of Drouwen: the Early Bronze Age 'Sögel' chieftain's grave; the Late Bronze Age hoard of 1939, and other finds from the urnfield; a Scandinavian sword found nearby. These suggest some sort of extraordinary connection between Drouwen and the North European cultural area in the Bronze Age, especially in the Late Bronze Age.

KEYWORDS: Drouwen (Drenthe, the Netherlands), Late Bronze Age, hoard, bronze (scrap metal), social structure, Late Bronze Age.

1. THE DROUWENERVELD HOARD OF 1984[1]

A pot full of bronzes - about 70 Late Bronze Age objects and fragments of objects, almost all broken or incomplete (figs 2-8), weighing c. 1100 g - was recovered in December 1984 in the Drouwenerveld, approximately 1 km southwest of the present village centre of Drouwen, and of the site of the Late Bronze Age urnfield just west of the village (fig. 1). The new hoard is the largest Late Bronze Age hoard found in the north of the Netherlands. It came to light almost exactly 45 years after the discovery of the other rich hoard found in 1939 in the Drouwen urnfield (figs 20-22). The relationship of these two hoards with other finds in the neighbourhood, and with the South Scandinavian-North German culture area, will be discussed in section 3.

1.1. Find circumstances

The circumstances of the find have already been described by Kooi (1986), and need be reviewed only briefly here. Drouwen is a village lying along the eastern edge of the sand-covered old moraine ridge, the Hondsrug. It is roughly equidistant from the north end (Groningen) and the south end (Emmen) of the ridge. To the east was the Bourtanger Moor, certainly a formidable barrier to communication; and we do not know whether there were Bronze Age trackways here, as there certainly were in the Emmen area (Casparie, 1984). The site of the find is a featureless-looking field through which runs a slight sand ridge. In November 1984 the field was ploughed, to a depth greater than previously. As the field in question was known to have yielded flints, sherds and other archaeological material (not, however, of the Bronze Age), it was walked over hopefully by a number of amateur archaeologists. Among them was Mr. G. Holtrop of Rolde, Drenthe. On this occasion Mr. Holtrop found a number of fragmentary bronzes with unpatinated, recently broken edges, and potsherds, suggesting the presence of a hoard. Mr. Holtrop marked the spot and reported his finds to the Drents Museum in Assen, which in turn notified the Biologisch-Archaeologisch Instituut in Groningen. An investigation followed.

Then, a trial cutting of 2 x 2 metres was excavated. At the base of the plough soil, the top of the hoard - most of a pot, with bronzes inside it - became visible, and could be plotted in on the map.

The earth at the level of the hoard was naturally deposited sand, partially penetrated from above with ironpan. The hoard had presumably been deposited in a pit, but no trace thereof had survived the ploughing; nor were other archaeological soil traces observable.

Because of dark, threatening weather, it was decided to remove the pot in a block of earth and transport it to the B.A.I. in Groningen for detailed further examination. Some days afterward, the pot (about two-thirds of which had still been in situ,

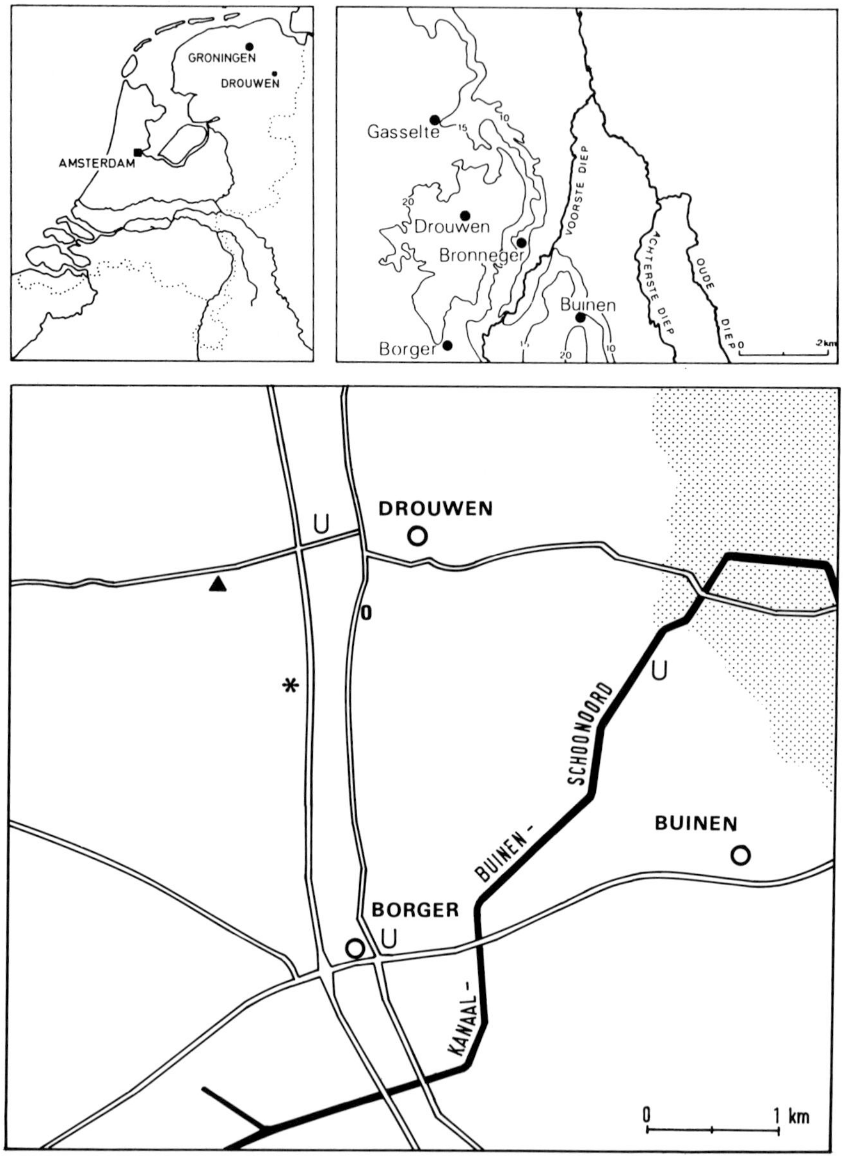

Fig. 1. Situation of Drouwen and its Bronze Age finds. **0** Early Bronze Age warrior's grave; ▲ Late Bronze Age sword; U Late Bronze Age urnfield; * Drouwenerveld, hoard of 1984; ⁘ former Bourtanger Moor.

lying on its side) and its contents (about 70 bronzes and fragments) were carefully 'excavated' at the B.A.I. by the technician K. Klaassens, in the presence of the original finder Mr. Holtrop, members of the B.A.I. and Drents Museum staff, students, photographers, press, and the present writer. Dr. P.B. Kooi made drawings of the position of the objects within the pot as the mini-excavation proceeded.

It appeared that the pot had simply been crammed full of bronze objects, mostly broken before deposition; many of the objects are represented only by one or more fragments. Only a socketed axe and a large number of buttons were quite intact. Some ploughed-out fragments found on the surface by Mr. Holtrop could be matched with pieces still in the pot. Other ploughed-out fragments are not likely to have escaped being found in the plough-soil, as the ground concerned was checked with a metal detector.

A subsequent excavation at the find-spot, conducted in March 1985 by J.N. Lanting (B.A.I.) in which an area 20 x 20 metres was opened, failed to yield any further objects or soil traces relevant to this find.

The inventory of the hoard includes tools, weapons, ornaments, and a number of fragments of bronze casting waste. It is evidently a collection of scrap metal, intended for melting down and the casting of new objects.

There is no evidence as to why the hoard was deposited where it was, or why it was left in the ground there. Opinions are divided as to whether such metal-worker's deposits are to be regarded as purely functional (e.g. Levy, 1982) or are to be considered as ritual deposits (e.g. Hundt, 1955). To import a load of scrap metal from a considerable distance, and then deposit it as a votive offering, would be rather self-defeating; unless, perhaps, it was part of a larger shipment, and only this part was used as an offering. Or it could have been deposited for safekeeping by a smith or a trader, and abandoned for whatever reason. This would imply some sort of workshop or storage building close by; but no traces of structures have so far been found.

In section 2 below we shall go into detail as to the origin of the material in the hoard. Here we can anticipate our findings by noticing the remarkable fact that practically all the objects in the hoard – the pot, the bronze types – are strangers to this area, being either totally unknown hitherto, or at least very scarce, in the Netherlands and Northwest Germany. The hoard seems to have been imported as a whole from the lower or middle Elbe area, but it also includes some South Scandinavian objects, and at least one of Central European origin or inspiration.

1.2. Dating

The dating evidence for the individual types contained in the hoard will be discussed in detail in chapter 2. Here we can say, in general, that the Scandinavian and 'Middle Elbe' types in the hoard are mostly types that originate in Northern Period IV, but continue to be used and deposited in Period V. The tanged knife (fig. 4:15) and the socketed axe (fig. 4:4) seem, indeed, to be Montelius V products.

The Urnfield socketed knife (fig. 4:30) is the only object in the hoard of a type which is well represented in other finds in this region (i.e. the Netherlands and Northwest Germany), where its associations are invariably with types with a North European dating of Montelius V. But the Drouwen socketed knife seems to be an early example of its kind, with Central European HaB1 connections. We must therefore conclude, on present evidence, that the Drouwenerveld hoard, despite the presence in it of possibly older objects (not in itself surprising in a scrap-metal hoard), was probably deposited within the North German Period V, though probably at an early date within that phase.

The Drouwenerveld hoard of 1984 was thus deposited within the same period as the Drouwen urnfield hoard of 1939, the period which includes most of the Late Bronze Age hoards found in the Netherlands.

Dendrochronological evidence from Late Bronze Age settlements in the Swiss lakes has shown that HaB3 settlements were being constructed there in the 9th century B.C.[2] The 8th-century date for our main 'Hunze-Ems' Late Bronze Age hoard horizon which we have previously used, on the basis of the chronology of Müller-Karpe (1959, etc.), can therefore be updated by a century or thereabouts. The Swiss tree-ring chronology can now, according to recent publications, be accepted as accurate and not subject to further corrections (Becker et al., 1985); how the archaeological materials are to be related to it will, of course, be subject to further refinement.

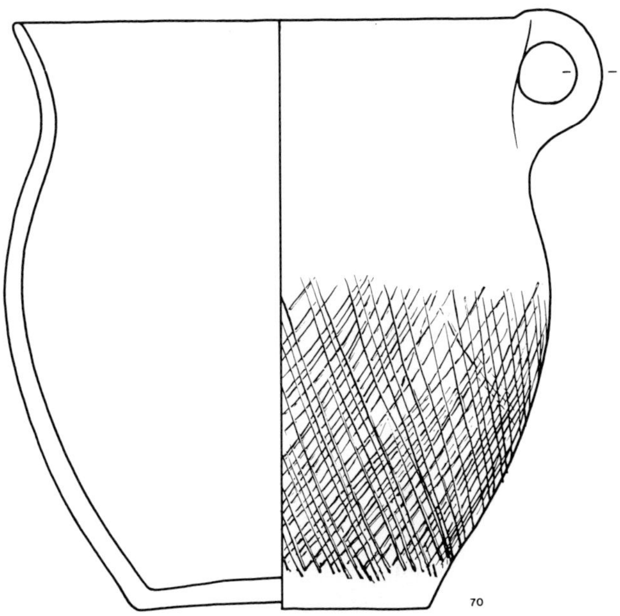

Fig. 2. Drouwenerveld 1984: the pot (70). Scale 1:2.

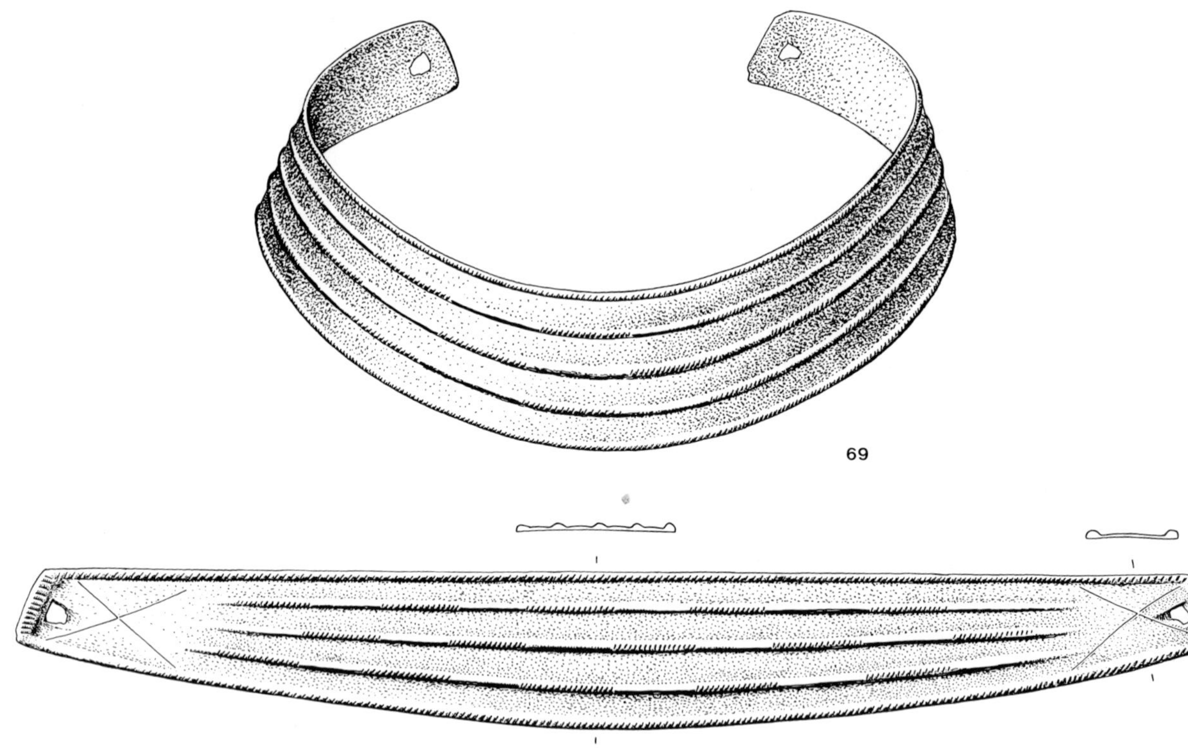

Fig. 3. Drouwenerveld 1984: the collar (69). Scale 1:2.

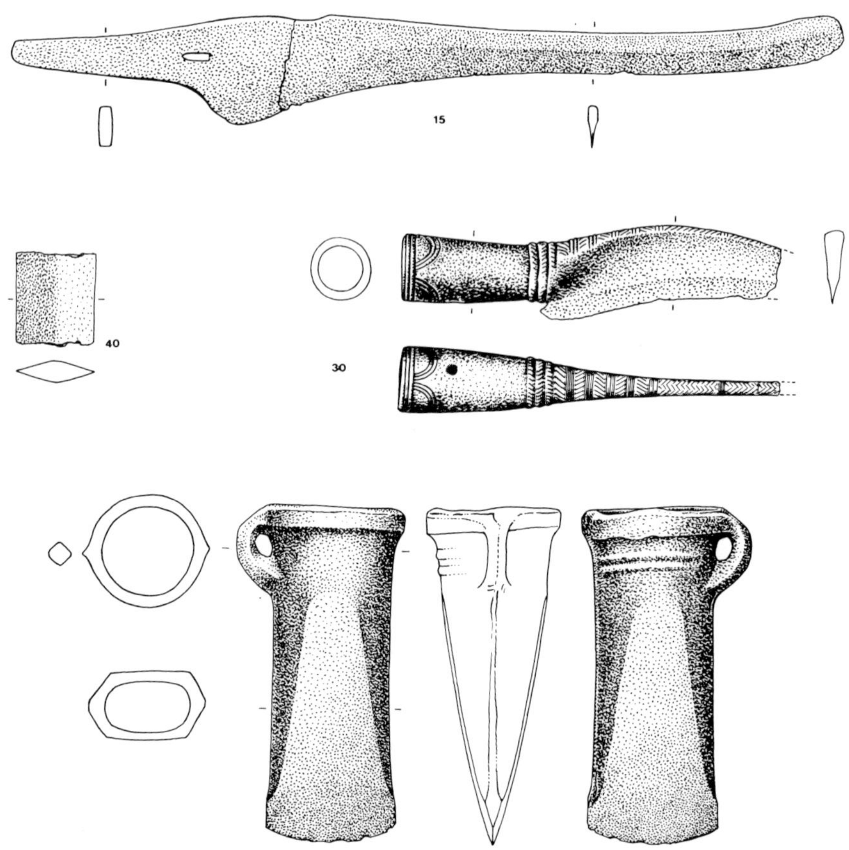

Fig. 4. Drouwenerveld 1984: tanged knife (15), socketed knife (30), knife fragment (40), socketed axe (4). Scale 1:2.

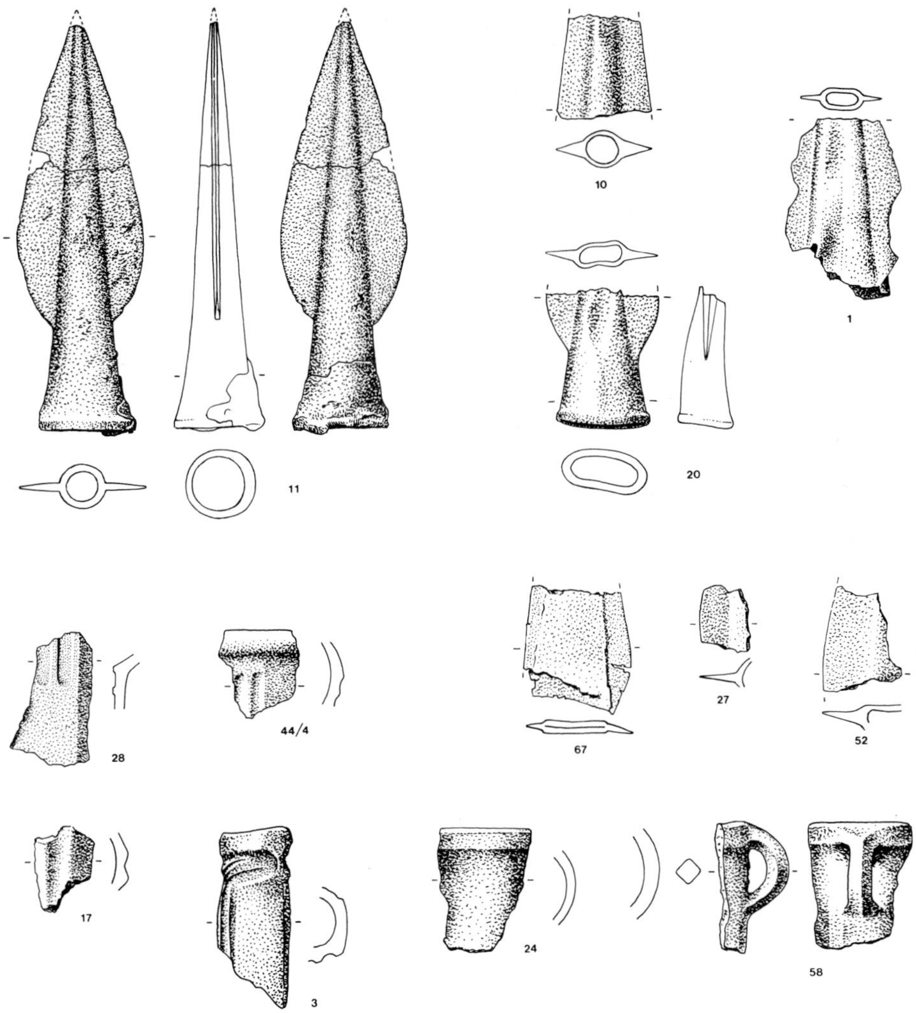

Fig. 5. Drouwenerveld 1984: spearheads (1, 10, 11, 20, 27, 52, 67), socked axe fragments (3, 17, 24, 28, 44/4, 58). Scale 1:2.

Period V in North Germany and Scandinavia has voluminous trade contacts with Central Europe in the Late Urnfield period (HaB3), and these two phases must be more or less contemporary; so that the absolute dates for HaB3 can also be regarded as applying, more or less, to Montelius V. The Drouwenerveld hoard could therefore have been deposited somewhere around 850 BC.

2. ORIGINS AND DATING OF THE HOARD OF 1984

2.1. General observations

The most striking facts about the Drouwenerveld hoard can be summarized as follows:

1. The hoard consists of Late Bronze Age objects, mostly of types previously unknown, or at least of very rare occurrence, in the Northwest German-Northern Netherlands region.

2. The pot in which the bronzes were found is also a stranger to this region.

3. The hoard is not characteristically 'masculine' or 'feminine'; it includes tools and weapons as well as ornaments, and also founders' scrap.

4. Almost all of the objects were deposited in fragmentary or damaged condition.

From this it must be clear that we have to do with an imported collection of old bronzes, not intended for use or exchange as finished objects, but rather as old metal which could be melted down for the manufacture of new objects. Such scrap metal finds (*Brucherzfunde* in German) are well-known in many parts of Europe, but no sizable example was previously known in the north of the Netherlands. Since there are no natural sources of metal in this region, all metal would necessarily have to be imported: either as ingots, new or second hand objects, or scrap. Ingot finds are unknown in this region, so finished objects and scrap must have been the most important elements in the supply.

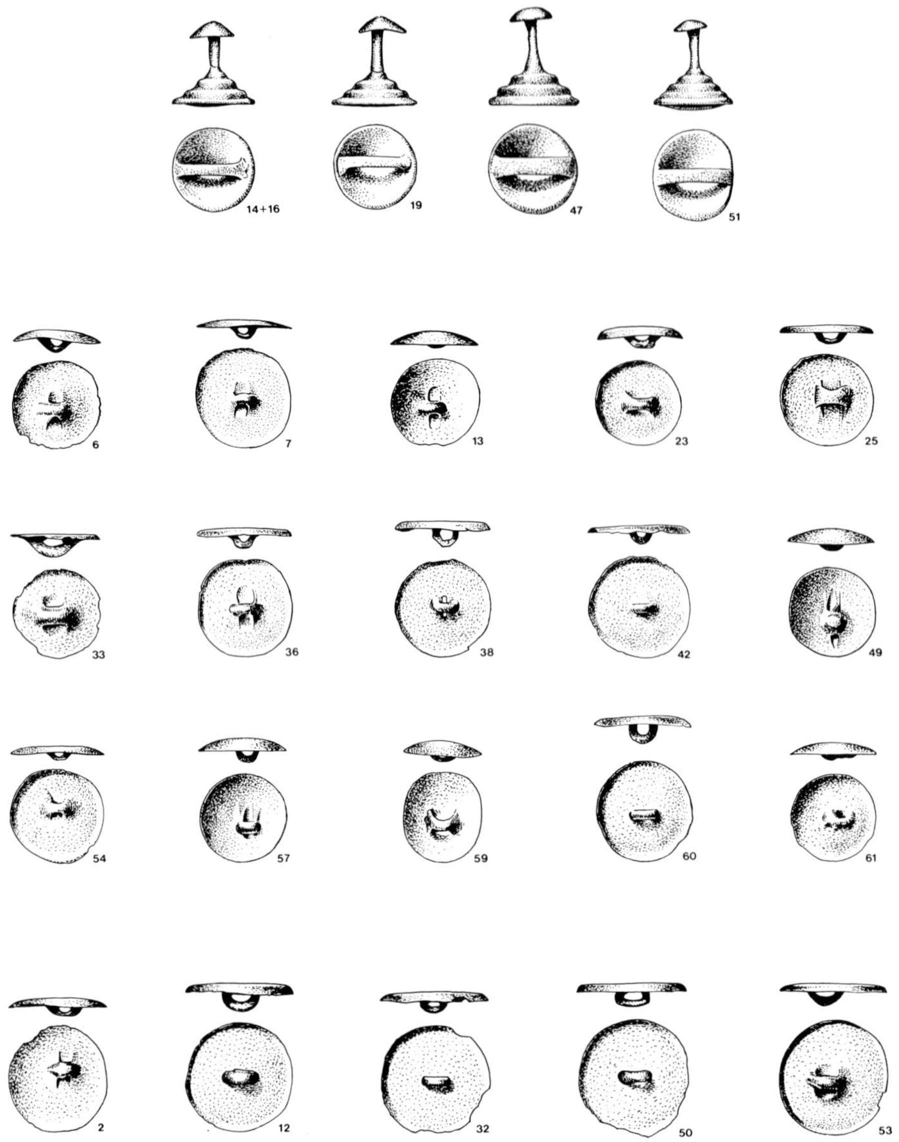

Fig. 6. Drouwenerveld 1984: tutuli (14+16, 19, 47, 51) and buttons. Scale 1:2.

The source of the Drouwenerveld scrap can be determined at least broadly on the basis of the types contained in the hoard. Some of the types are, of course, too general in their occurrence to be useful as source indicators, but others can be pinned down to a region of origin. Among these we can distinguish types of 'Middle Elbe' origin; types characteristic of the South Scandinavian cultural area; types generally 'North European' but not specifically pinned to a smaller sub-region; and (in only one case) a type of ultimate West Central European 'Urnfield' background. The types concerned can now be considered in detail.

2.2. Types of middle/lower Elbe origin

A Late Bronze Age '*Kulturprovinz* Mittelelbe' was defined by Sprockhoff (1937: pp. 60-61, Abb. 20, Karte 33), on the basis of the distribution of certain characteristic metal types; among these were the stepped tutuli with mushroom-shaped head. Actually, as later writers have pointed out (cf. Tackenberg, 1971; von Brunn, 1968: p. 227; Horst, 1981), the Mittelelbe concept does not hold together very well. The tutuli concerned (fig. 6) have, indeed, a 'Lower Elbe' rather than a 'Middle Elbe' centre of distribution (fig. 9; von Brunn, 1968: Karte 16). The horizontally ribbed collar occurs in much the same area (though it also has a secondary centre farther east: cf. fig. 10). In this lower Elbe region we also find many types of South Scandinavian origin, which commonly occur together with the Elbe types in hoards.

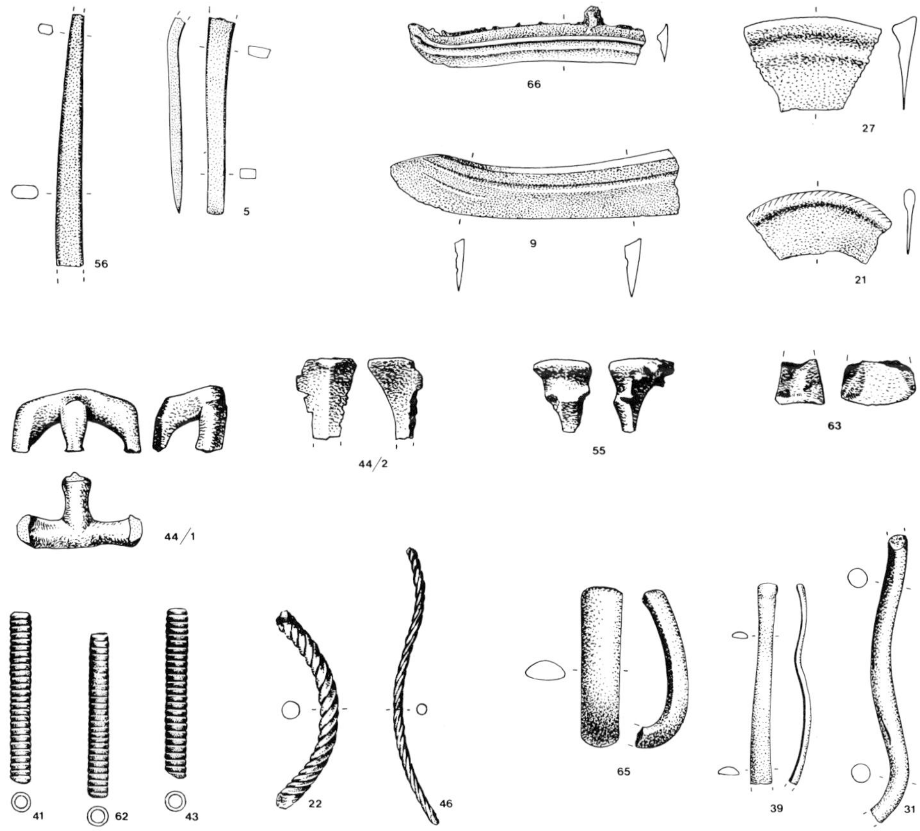

Fig. 7. Drouwenerveld 1984: punch fragments (5,56), sickle fragments (9, 21,27, 66), casting jets (44/1, 44/2, 55, 63), ribbed tubes (41, 43, 62), twisted wire fragments (22, 46), bracelet fragments (39, 65), plain wire (31). Scale 1:2.

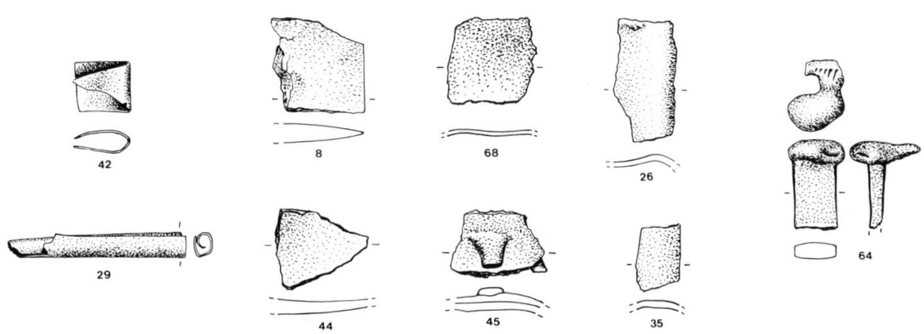

Fig. 8. Drouwenerveld 1984: miscellaneous fragments. Scale 1:2.

2.2.1. *The tutuli* (fig. 6)

These are a sort of decorative button, with a central pointed projection. Numerous varieties were in use in Eastern, Central and Northern Europe in many different periods. A special variety, with stepped sides and a lens-shaped (mushroom-shaped) cap, derived from Middle Bronze Age predecessors common in Northern Europe (e.g. Schubart, 1972: p. 34), begins toward the end of Period III in Mecklenburg. This variety (made by *cire perdue* casting, according to Drescher, 1956) is known from 27 finds in North Germany, and, rarely, Scandinavia. There are several remarkable things about their distribution (fig. 9; Struve, 1979: Tafel 62 has a few erroneous spots). There is one main concentration east of the Elbe – in Mecklenburg – and another south and west of the Elbe, in the Lüneburger Heide and Altmark areas. Mecklenburg has many examples; in half a dozen hoards (presumably trader's stocks) no less than 279 examples were found. The most extravagant single find, however, is the only major find east of the Oder: the Period V hoard from Hohenwalde, Kr.

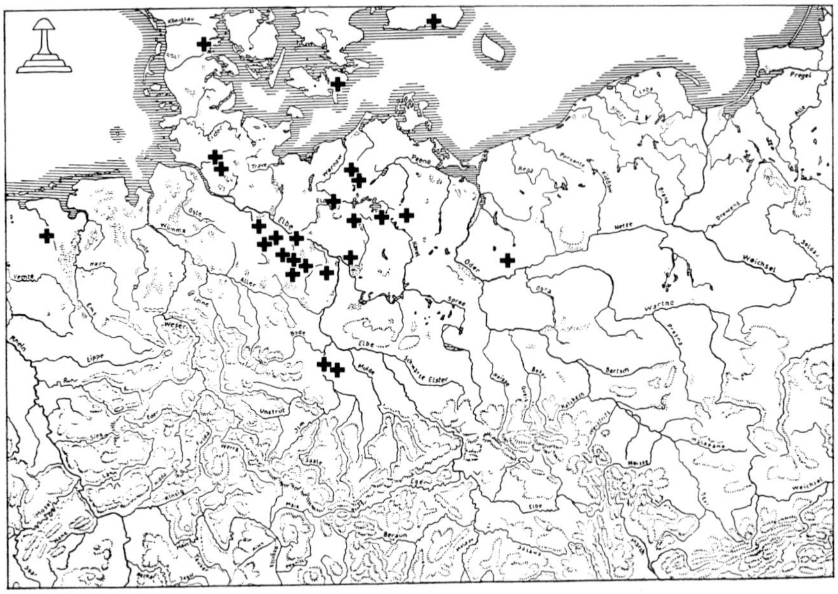

Fig. 9. Stepped tutuli with lens-shaped cap. After Sprockhoff, 1937, with additions from Von Brunn and Thrane.

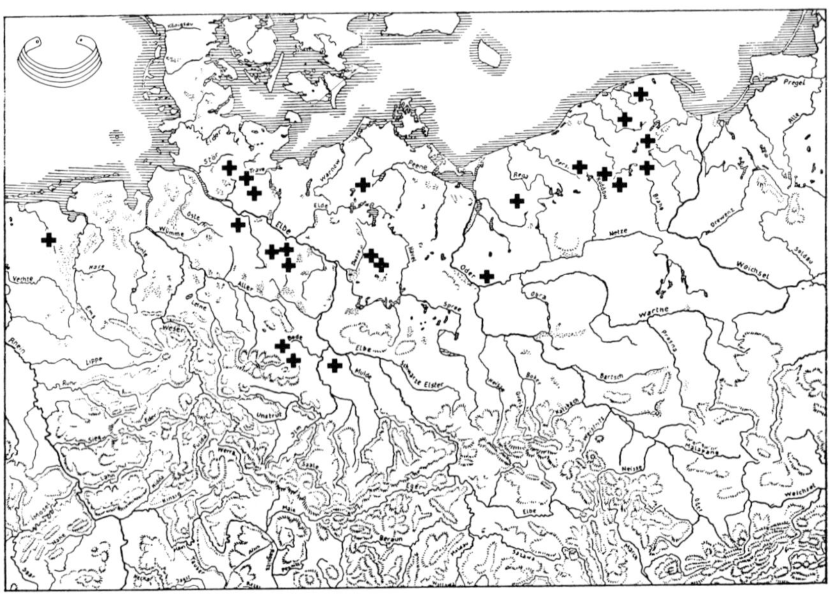

Fig. 10. Horizontally ribbed collars. After Sprockhoff, 1937.

Landsberg an der Warte (cf. Pescheck, 1949/50: pp. 21-25, Abb. 1; Landsberg is now Gorzów Wielkopolski), where no less than 326 tutuli occurred in the one find!

Remarkably, in view of its abundance in the 'Mittelelbe' (Unterelbe) province, there are no finds at all of such tutuli in Northwestern Germany west of the Lüneburger Heide, or in the Netherlands, until we come to Drouwenerveld, with its four examples.

Although we cannot now say whether our Drouwen specimens came from Mecklenburg or from the Lüneburger Heide, there can hardly be doubt that they are Lower Elbe products. Most of the datable finds (chiefly hoards, a few graves) are of Period IV, but some important hoards (the Hohenwalde find already mentioned, the Central German hoard of Bernburg with 172 examples, the North Jutland hoard of Saesing) are assigned to Period V.

2.2.2. *The ribbed collar* (fig. 3)

Various sorts of ribbed and unribbed bronze collars were common both in the Central European tumu-

Chronological table. Central and Northern Europe, Northwest Germany, the Netherlands.

D.D.		Central Europe	N.Eur.	NW Germany	NL	Hoards and Rich Graves NW Germany	Hoards and Rich Graves NL
	HaCl	HALLSTATT	VI	WESSEN- STEDT	LATE URN- FIELDS		
850	HaB3 (B.f.IIIb)	LATE (späte) URNFIELDS/ EARLIEST HALLSTATT	V	V		Bäk, Schinna Obernbeck Bargfeld Rethwisch	Drouwen 1939 Havelte Schoonebeek Drouwen 1984 Barger- oosterveld 1900
	HaB2(?)	YOUNGER (jüngere) URNFIELDS	V	V	MIDDLE URNFIELDS		
950 1000	HaB1 (B.f.IIIa)		IV	IV			
1050	HaA2 (B.f.IIb)	MIDDLE (mittlere) URNFIELDS	IV	IV	EARLY URNFIELDS	Warendorf 5	Gasteren
	HaA1 (B.f.IIa)	EARLIER (ältere) URNFIELDS	III	TANGEN- DORF	B		
	D (B.f.I.)	EARLIEST (frühe) URNFIELDS/ LATE TUMULUS	III / II	BERGEN- BLECKMAR	MIDDLE		Epe Elp Weerdinge
	C2	YOUNGER (jüngere) TUMULUS	II	BONS- TORF	BRONZE		
	C1	MIDDLE (mittlere) TUMULUS	II	WESTEN- DORF	AGE		
	B	EARLIER (ältere) TUMULUS	I	SÖGEL- WOHLDE	A		Overloon Drouwen 1927
1650	A2 (A3, A4)	(later part of) EARLY BRONZE AGE	I		EARLY		
	A1	(earlier part of) EARLY BRONZE AGE	I		BRONZE AGE		Wageningen

Notes to the chronological table:

D.D. = Dendrochronology: absolute dates (slightly rounded off) for the founding of settlements in Switzerland
- 850: Auvernier Nord (HaB3)
- 950: Cortaillod Ost (HaB1)
- 1050: Greifensee-Böschen (HaA2)
- 1650: Meilen-Schellen (last phase of the Early Bronze Age; variously cited as Langquaid phase, A2, A3, or *Stufe* 4)

Roman numbers = South Scandinavian periods, according to Montelius/Broholm/Baudou

Names in capitals = standard period names (NL after Lanting & Mook)

HaC1 ... A1 = former period designations (Müller-Karpe, etc.)

B.f.I ... B.f.IIIa = Bronze final I ... IIIA, East French period designations according to Hatt, adapted by Briard/Mohen/Blanchet

Lower case names = find-spots of hoards or graves with characteristic content

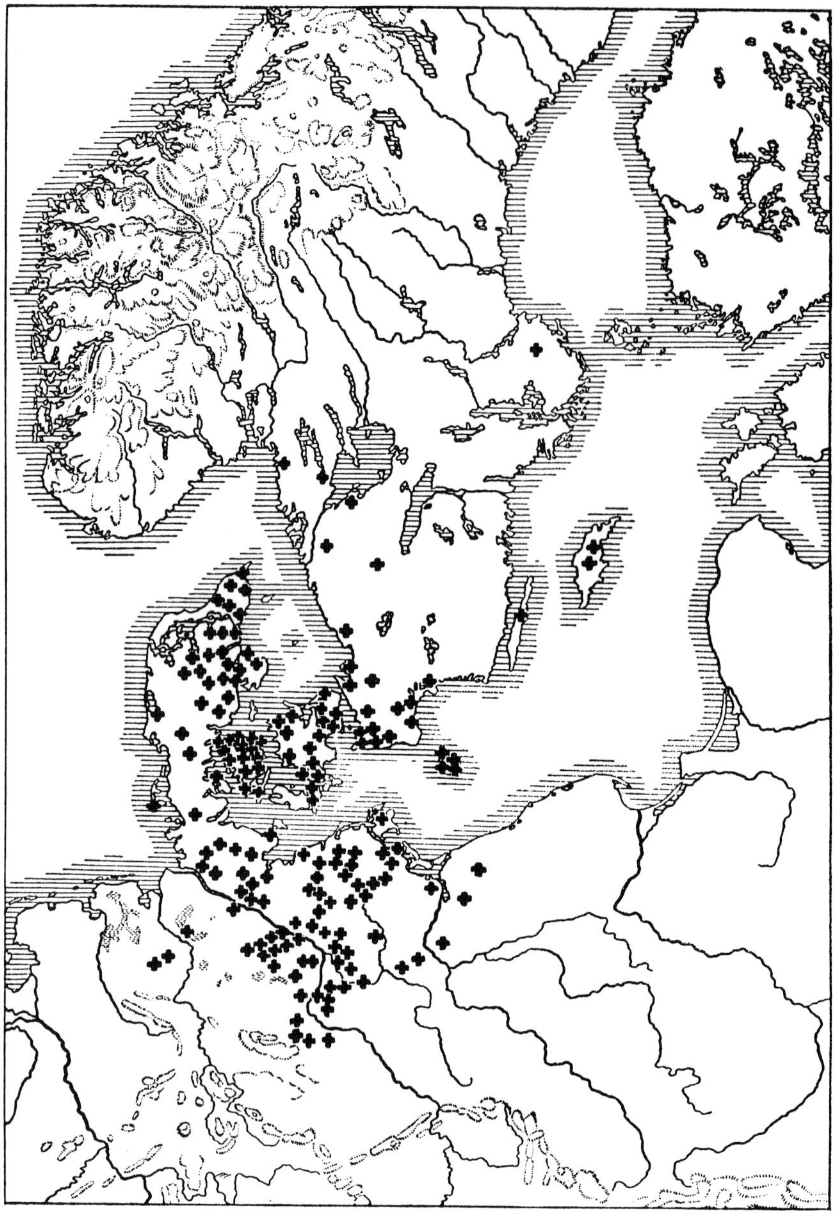

Fig. 11. The South Scandinavian culture area in the Late Bronze Age, as defined by hoards of Period IV (left) and Period V (right). After Sprockhoff, 1937.

lus culture and in North Europe during the Middle Bronze Age. The special variety of ribbed collar as it occurs in our Drouwenerveld hoard, usually with the triangularly perforated end-panels as we have at Drouwen, is known from 24 finds (cf. fig. 10) in the German literature. Sprockhoff discusses this type under the name *steile längsgerippte Halskragen* (1937: pp. 41-42, 89-90, Karte 20; 1956, I: pp. 19, 141-143; II: Taf. 24:1-2).

More than half of them occur (fig. 10) in an area centring on the river Elbe, extending north-south from Holstein to Central Germany. Mecklenburg has only one find (most surprisingly, since its Period III predecessors are common there); whereas there are three finds in Central Germany, and there is a considerable group in the east, roughly between the Oder and the Vistula, occurring there sometimes in hoards of Period V, though the dated finds of the 'Mittelelbe' area are of Period IV.

The ribbed collar, the tutuli, and the bronze buttons are perhaps part of a costume set; the combination is not unknown in the Elbe region.[3]

2.3. Types of 'South Scandinavian' origin

By 'South Scandinavian' we mean, in this context, the South Scandinavian Late Bronze Age cultural area, sometimes in the past described as the *Nor-*

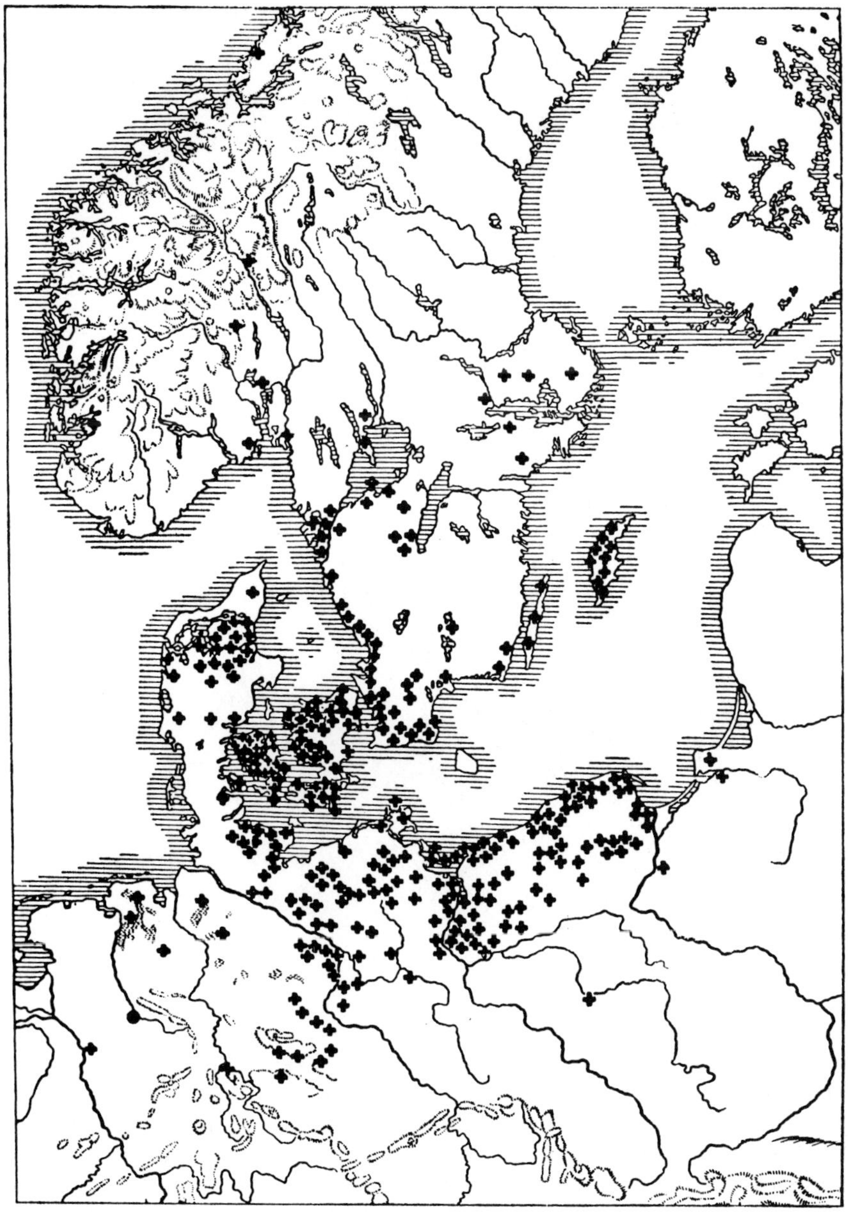

Fig. 11 (cont.).

dische Kreis, which is archaeologically defined by the distribution of its many characteristic types. Its extent can perhaps best be visualized with the help of the distribution of hoards with 'Nordic' material (fig. 11; Sprockhoff, 1937: Karte 1, 2; 1956, II: Karte 53; Baudou, 1960; Struve, 1979: Tafel 39). This area includes not only Denmark, Southern Sweden and Southern Norway, but also Schleswig-Holstein and some adjacent parts of Central North Germany. In Period IV it extends roughly to the Oder; in Period V it goes farther to the east, as far as the Vistula mouth area.

2.3.1. *The tanged knife* (fig. 4:15)

In the Scandinavian literature, tanged knives such as that from the Drouwenerveld hoard have been discussed under the title of *Griffangelmesser*; Baudou (1960: pp. 28-29, Taf. VII, XB) counted 130 examples, of which more than half have a perforation (usually round) in the tang. They are most common on Sjaelland and Fyn, relatively rare in South Sweden and Jutland. Thrane (1972: pp. 213-214, 219, fig. 22) has published a map of the *Griffangelmesser mit Nietloch*; he lists about 60 examples (his *Fundliste* 4, pp. 227-228). He divides them into three varieties, based on the outline of the

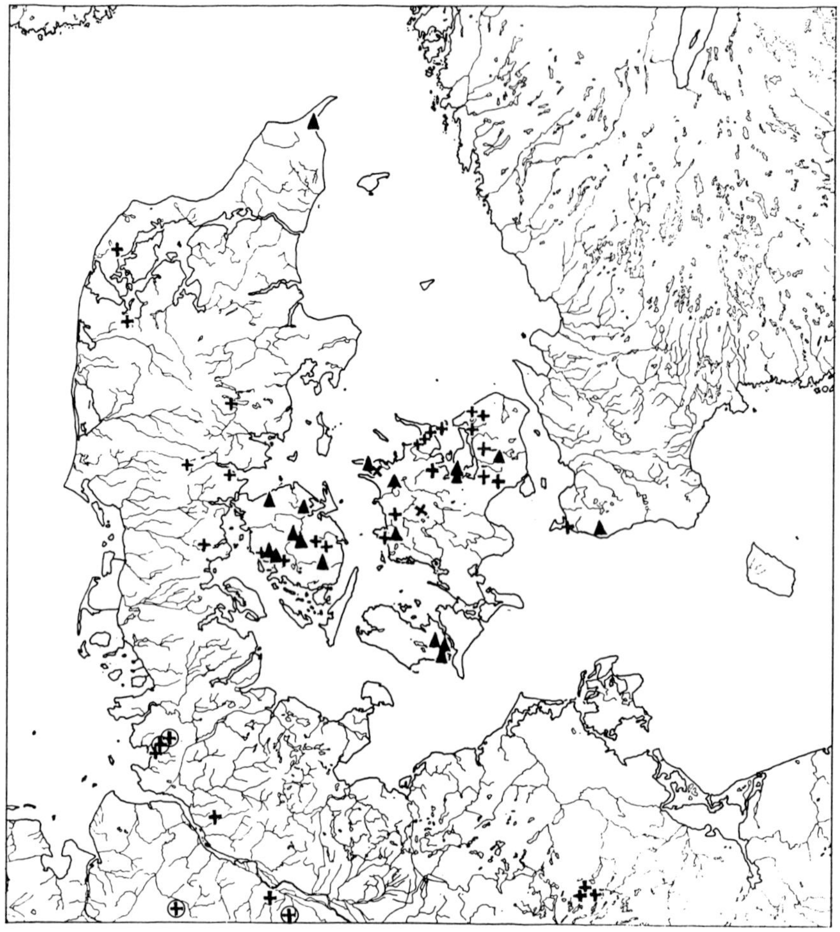

Fig. 12. Tanged knives with perforation in tang. After Thrane, 1972.

back the blade (1 convex, 2 straight, 3 S-curved). More recently Prüssing (1982, pp. 104-106), in dealing with the few such tanged knives in Schleswig-Holstein and Niedersachsen, has modified this typology somewhat. The Drouwenerveld knife most closely resembles the larger examples of his Type Bunsoh. The few examples of this type occur in Ditmarschen and on the Lüneburger Heide (Prüssing's No. 232-236). Characteristic for this small group of knives is the jog in the line of the back caused by the irregular junction of the blade and the tang. Interestingly, the find from Armstorf (Prüssing's Taf. 27H; his No. 236 and 238) contains both a Bunsoh knife (though a small example) and a small, atypical knife with an elongated perforation: the only published example we know of with a perforation resembling that of the Drouwenerveld knife. Whether examples with the 'Bunsoh back' and/or the 'Armstorf perforation' also occur in Denmark is not clear from the literature. If examples with the 'Bunsoh back' exist in Denmark, they would presumably be classified by Thrane in his Variety 3, with ogival back line, which he dates to Period V. An unperforated knife with a 'Bunsoh back' occurs in the Period V hoard from Hemmelsdorf, Kr. Ostholstein (Prüssing, 1982: No. 245; Struve, 1979: Taf. 59).

We can thus conclude that the Drouwenerveld tanged knife is an import from the South Scandinavian area, but perhaps is a representative of the small 'Bunsoh' group from the periphery of that area (fig. 12). It is, in any case, the only known example of a Nordic *Griffenangelmesser* this side of the Lower Elbe area.

As to dating, Baudou states that the *Griffangelmesser* occur for two-thirds in Period IV and one-third in Period V. Prüssing, however, specifically assigns his Type Bunsoh knives to Period V, on the basis of two examples with Period V pins and, in one case, a Period V neckring, as well as on a typological argument, namely that the shape of the back is a variant of the HaB *Krückenrücken*. A Northern Period V dating for the Drouwenerveld tanged knife would thus be probable.

2.3.2. *The lugged sickle* (fig. 7:66)

The lugged sickle, represented in the Drouwenerveld hoard by the fragment No. 66, is a purely South Scandinavian type (Baudou, 1960: p. 47, Typ XIV B, *Rückenzapfensichel*; list pp. 230-234; Karte 27). Baudou gives the number of known find-spots as totalling c. 240 in Denmark, c. 100 in Sweden, and only 7 in Schleswig-Holstein; there were none south of Schleswig-Holstein, nor in Norway or Finland.

On the basis of c. 70 associated finds, the larger number belong to Period IV, a smaller number to Period V; Baudou does not go into further detail.

2.3.3. *The fragmentary socketed axes*

The socketed axe fragment No. 28 (fig. 5:28) has flat-bottomed grooves on the face, which suggest, despite the smallness of the fragment, that it comes from one of the *Tüllenbeile mit geraden Blenden* (Baudou, 1960: Type VIIA1, pp. 17-18, Karte 8). The distribution is practically limited to the Danish Islands and Skåne.

Fragment No. 3 (fig. 5:3) apparently comes from one of Baudou's *Tüllenbeile mit Y-förmig verzweigten Blenden* (his type VIIA2, pp. 18-19, Karte 9), with a similar distribution.

Both of these are very common and characteristic Scandinavian types.

The other socketed axe fragments (fig 5:17, 24, 44:4) are not closely definable as to type.

2.4. Types of generally 'North European' origin

2.4.1. *The complete socketed axe* (fig. 4:4)

The complete socketed axe in the hoard (No. 4) has a large arch-shaped facet on each face, and thus belongs to the large family of socketed axes which Sprockhoff (1956: pp. 90-91) has grouped under the designation *keilförmige Tüllenbeile mit bogenförmigen Breitseiten*; or, as Tackenberg (1971) prefers, *mit glockenförmig abgesetzten Breitseiten*. Sprockhoff subdivides these according to the number and arrangement of horizontal ribs under the socket-mouth moulding. If there are no ribs, we have to do with Typ Schinna; if there are one to four ribs, it belongs to Typ Obernbeck; if there is a complicated rib pattern, it belongs to the type *Tüllenbeil mit profiliertem Tüllenmund* (cf. Butler, 1960a; 1960b; Kibbert, 1984: pp. 136-139 under Typ Bargeroosterveld). Tackenberg has added additional finds to Sprockhoff's lists, and has carried the subdivision slightly further by listing and mapping the examples with a single neck rib separately (his Liste 6 and Karte 12).

Our Drouwenerveld axe is remarkable in that its two sides differ: on one face there are two ribs below the socket-mouth moulding, but these ribs are absent on the other face. (A small number of axes with this kind of duality are mentioned by Tackenberg, 1971: pp. 255-256, Liste 6: Nos 1, 6, 18). In the typology of Sprockhoff the face without the ribs belongs to Typ Schinna, whereas the ribbed face is an example of Typ Obernbeck.

Axes of this family seem to have originated in Southeastern Europe (Petrescu-Dîmbovitța, 1978: Taf. 223 ff; Kemenczei, 1984: pp. 74-76, 169-170, Taf. CXCVII), and from there to have jumped somehow (largely skipping Central Europe) to North Germany and Scandinavia. Most common in the north is the Schinna type; the ribbed-necked Obernbeck variants have a similar distribution but are comparatively scarce (for the Schinna type, Sprockhoff and Tackenberg list nearly a hundred examples in North Germany; for all the variants of the Obernbeck type together they list only 22 examples, and only five of these are specifically described as being two-ribbed). Taking all these varieties together, it is noteworthy that they are

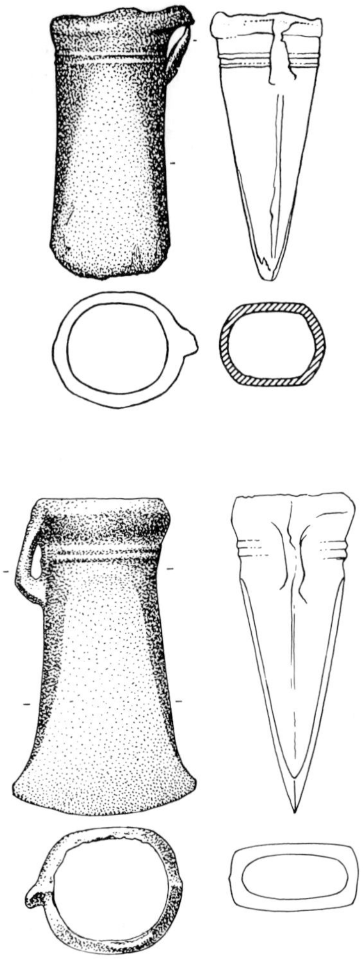

Fig. 13. 'Obernbeck' socketed axes from Frederiksoord, Drenthe, and Leeuwarden, Friesland. Scale 1:2.

quite common in Denmark, Schleswig-Holstein, and Northwest Germany, yet strangely uncommon on the Middle Elbe and in Central and East Germany.

Socketed axes of the Schinna and Obernbeck varieties are certainly not common in the Netherlands. Schinna axes are quite unknown this side of the border, and we know of only two Obernbeck axes: an example was recently found in Leeuwarden (fig. 13a; Kramer, 1985) and another (fig. 13b) is from Frederiksoord, Drenthe (private possession).[4]

It would seem, then, that our Drouwenerveld socketed axe must be an import from somewhere in Denmark, Schleswig-Holstein or Niedersachsen. At present we are unable to localize its origin more narrowly than that. A detailed special study could possibly narrow down the possibilities.

The Schinna/Obernbeck axes are typologically the forerunners of the more elaborate *Tüllenbeile mit profiliertem Tüllenmund* which in Period V are among the characteristic axes of the Ems-Hunze area (Butler, 1965).[5]

2.5. A type from West Central Europe (fig. 4:30)

2.5.1. *The socketed knife*

The socketed knife differs from the objects in the hoard previously discussed in regard to its origin and distribution.

The single-edged socketed knife has its ultimate origin in the Central European Urnfield culture. It is one of the end-products of a long development of Urnfield knives, extending over a period of something like 500 years. But the socketed knife does not occur everywhere within the Central European Urnfield area; the eastern Urnfield area used knives with other types of handle attachment. But there is a strong concentration of finds of socketed knives in the lakeside settlements of Western Switzerland and Eastern France: especially along the lakes of Biel, Neuchâtel, Leman and Bourget. In some of these settlements, too, casting moulds for socketed knives, made of clay or stone, have been found; which is why Sprockhoff (1950-51; 1956) discussed the socketed knives under the name *Pfahlbaumesser*; and this area is considered by most commentators as being the primary home of this type.

As to distribution, the main difference, at least with respect to our concerns in this article, is this: while finds of the 'Middle Elbe' and South Scandinavian types discussed above were all almost completely absent in Western Germany and the north of the Netherlands, socketed knives are rather uncommon in the former areas, and comparatively well represented in the latter. The distribution map (fig. 15) makes this abundantly clear.

This distribution (in so far as the finds were then known to him) led Sprockhoff to describe the

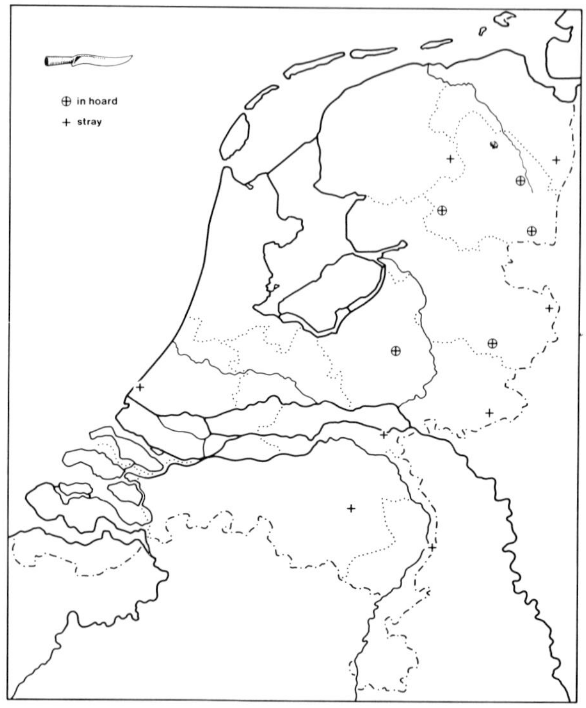

Fig. 14. Urnfield socketed knives in the Netherlands.

socketed knives as one of the *Leitformen der Ems-Weserkreis*. The comparatively large number of socketed knives in the Netherlands (with Drouwenerveld, now, 14 examples; listed in Appendix I, below; cf. figs 14-15) has been unrecognized in the international literature (though we published a list of them in Butler, 1968/69).

Sprockhoff and subsequent authors have given thought to the question as to whether the socketed knives found up north are to be thought of as actual imports from the West Swiss-East French lake area, or whether there might have been a manufacture of such knives in the northern part of Europe. There is general agreement that such local manufacture in the north is possible, and indeed probable. Unfortunately, there is no agreement as to which knives are import and which knives are northern imitations: when it comes to identifying criteria for distinguishing import and imitation, there is a somewhat embarrassing lack of definition.[6]

It might be interesting to survey briefly the range of possibilities. From the Swiss-East French lake area there was apparently a radial spread of the socketed knife in all directions except (with few exceptions) eastward. The numerous North Italian socketed knives, recently published by Bianco Peroni (1976), are obviously related to the West Central European ones, but are sufficiently different that one unhesitatingly has classified them as local variants. The socketed knives in the Rhône area and Southern France do not, insofar as we can

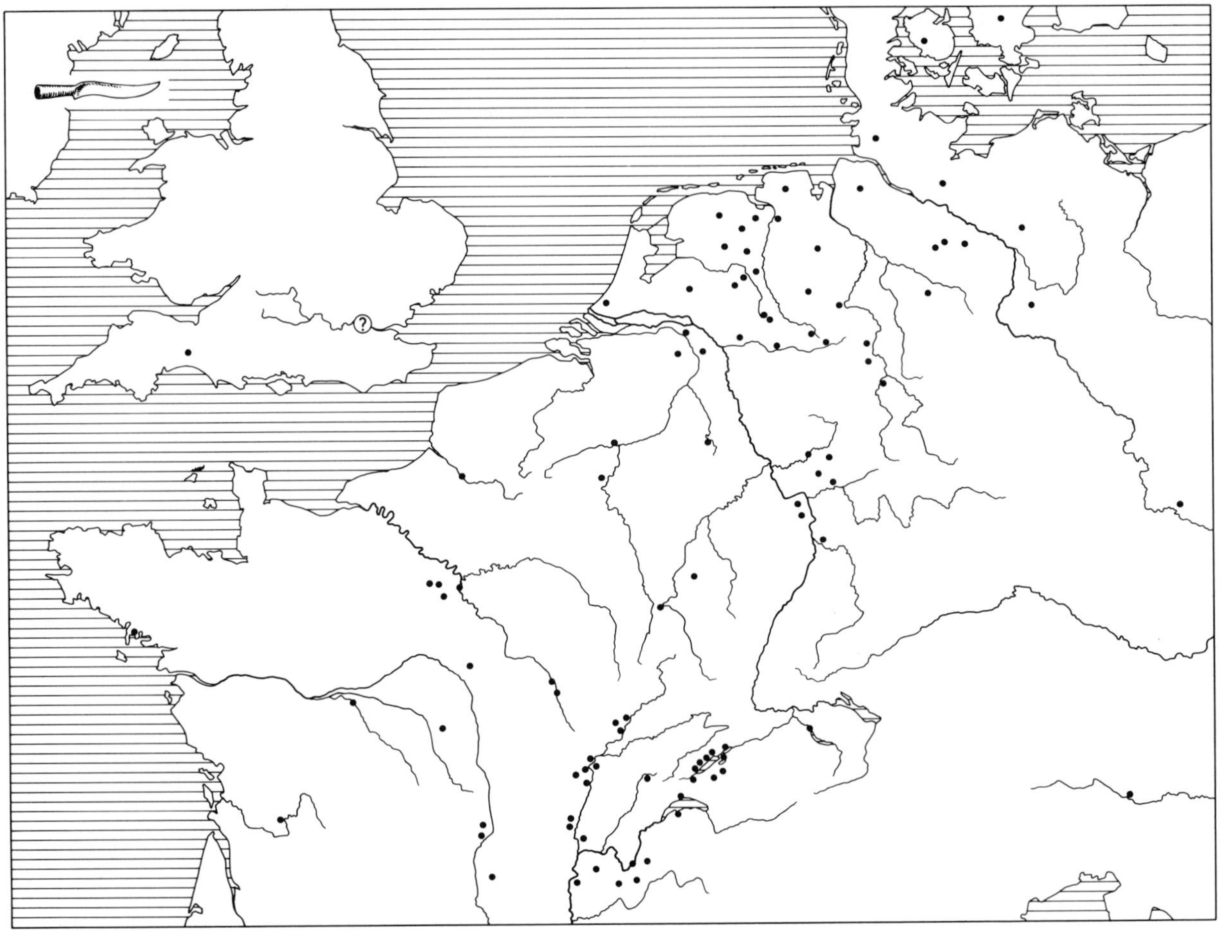

Fig. 15. Urnfield socketed knives in Europe. After Hundt, 1986, with additions and corrections.

oversee them (the state of publication is not entirely favourable), seem to differ significantly from those of the core area. The same seems to be true of the socketed knives that spread to the west, to the valleys of the Seine and the Loire, and even farther to the Vénat hoard in the southwest, the Locmariaquer hoard in Brittany, the Amiens-le Plainseau hoard on the Somme, and even, rarely, England (Jockenhövel & Smolla, 1975; Jockenhövel, 1980a).

To the north, there seem to be several tiers of penetration. In the first instance, there are a number of finds, chiefly in hoards, in the Rhine-Main area, with a few outliers in the Saarland and Lorraine. The following tier consists of an east-west belt of finds extending from the Rhine-Lippe junction on the west to Saxo-Thuringia. The finds in Belgium (including the remarkable group of ten examples from the Han-sur-Lesse cave exit) and those along the Maas and the Ems can reasonably be viewed as a further radiation from that belt, as can the finds in the east and north of the Netherlands on the left flank and those along the Elbe and on the Lüneburger Heide on the right. The final tier would be the few known examples in Denmark.

Tackenberg has rightly called attention to the lack of homogeneity among the socketed knife finds on the North European plain. Not only is there a considerable variation in form, but the various West Central European decorative styles, with incision, *pointillé*, and ribbing, all seem to occur with equal abandon in the north, though undecorated examples are the most numerous in both regions. There is no regional socketed knife style on the North European plain. We cannot, therefore, say whether any particular knife occurring in the north was made in the West Swiss-East French area, the Rhine-Main area, the Lippe-Saxo-Thuringian tier, the Lower Elbe, or the Hunze-Ems area. It does seem certain, however, that socketed knives were not manufactured east of the Elbe or in the Scandinavian cultural area.

And now the dating evidence for our socketed knife. Urnfield socketed knives have been found not only in the settlements along the West Swiss and East French lakes, but also in a large number of hoards in France, Western and Northern Germany, and the Netherlands. Practically all of these are dated to one chronological phase: Hallstatt B2/B3 in Central Europe, and its equivalents, B.f.IIIb in France and Montelius V in the north. There seems to be only one single example of an earlier dating: the very oddly located hoard from Velky Berezny in the Carpatho-Ukraine (Borkovsky, 1934/35: pp. 101-102, Tab. VI: 1-6; Müller-Karpe, 1961: Pl. 35C), a HaB1 hoard. And in Denmark a presumed Period V socketed knife found in a grave, at Hedwigslyst near Kalundborg, was accompanied by a Period IV razor; impelling Thrane (1972: pp. 184, 209-210) to propose, as a compromise solution, a date for this find of HaB2 in the sense of Müller-Karpe. A few socketed knives have been assigned to HaB1 on typological grounds; i.e. Chieming, Bavaria (Müller-Karpe, 1959, I: p. 308; II: Pl. 197J); the typologically 'early' example from Réallon, Hautes-Alpes, is in a HaB3/B.f.IIIb hoard (Audouze & Courtois, 1970: Pl. 27:51). It would therefore seem that, as the most recent commentators have made clear (Rychner, 1979; Prüssing, 1982: pp. 142-148), the spread of the socketed knife is almost exclusively an affair of the HaB2(3)/-B.f.IIIb/Montelius V stage.

But it is also clear (as V. Rychner has emphasized in a letter to the present writer) that the Drouwenerveld socketed knife has no features which were not already current in the HaB1 phase. The bend of its back is not close to the handle, as is customary in HaB3, but rather the earlier (HaA2/HaB1) gradual curve. The decoration on the back of the blade is typically HaB1; so also are the concentric semicircles on the socket. The cord-imitation ribs at the junction of the blade and socket do not occur on any other knife known to us, but similar ribbing is found, for example, at the socket mouth of 'Saxo-Thuringian' spearheads also decorated with incised motifs of the concentric-semicircle family. Rychner suggests that the Drouwenerveld knife belongs to the very rare group of HaB1 socketed knives in the style of the Réallon knife, which is typologically assignable to HaB1 (although the Réallon hoard itself is considered, despite some possibly earlier components such as the knife, to belong to the French Bronze final IIIb (= HaB3) phase).

But, alas, the typologically HaB1 features all seem to persist into HaB3/Montelius V. This can be shown typologically (i.e. the HaB1 patterns occur also on knives with the HaB3 hump: cf. the socketed knives from Aurich, Sterley, Havelte, the tanged knife from Forstort Fortwinkel: Prüssing Nos 286, 287, 266; Butler, 1961: fig. 14; Havelte and Forstort Fortwinkel are Montelius V hoards) and, as already mentioned, by the associations in the many other hoards. A certain ambiguity thus remains in the dating of the Drouwenerveld knife, as indeed for the hoard as a whole.

Incidentally, the features cited also provide links between the socketed knives and the knives with double-T hilt (Butler, 1973a; Tackenberg, 1971: p. 125; Prüssing, 1982: pp. 87-89; Thrane, 1972: pp. 167, 186-189, fig. 1), all eight known examples of which have the 'HaA2/B1' back profile: the pattern on the back of the Drouwenerveld socketed knife is at least partially matched on the double-T knife from Valthe, and the concentric-semicircle complex is found on the blade of the Appelscha knife. The double T-knives also hover uneasily on the chronological border between Periods IV and V.

In short, we seem to be confronted with so many borderline cases that we may seriously wonder whether a chronological borderline actually exists, at least as far as the Northwest German-Northern Netherlands area is concerned. Still, hoards like Rethwisch, Bargeroosterveld 1900, and Berg-en-Terblijt do seem to have an earlier date than Havelte, Schoonebeek, Drouwen 1939, etc.; and the hoard from Drouwenerveld may fall chronologically in between these two hoard groups.

2.6. Types of uncertain origin

2.6.1. *The spearheads* (fig. 5)

The complete spearhead (No. 11) does not belong to any of the distinct types recognized by Jacob-Friesen (1967) in Northern Europe. Two rather similar spearheads are known from a rich urn grave in a stone packing in Central Germany, at Düsedau, Kr. Osterburg (Magdeburg district), and were published recently by Horst (*Inventaria Archaeologica* DDR 25). Horst assigns this grave to his *Stufe Zedau-Stölln* of his Elbe-Havel Gruppe, Period (IV)/V. A possibly related example is the spearhead from Schnega, Kr. Lüchow-Dannenberg (Niedersachsen), in a Period IV hoard (Sprockhoff, 1963: p. 81, Abb. 28).

2.6.2. *The buttons* (fig. 6)

Buttons with D-loop, of the two varieties found in the Drouwenerveld hoard, seem to be quite common throughout the Late Bronze Age of the Central European and North European areas. They are, in fact, so common that it has not seemed worth any one's trouble to list and map them, so that we do not have concrete information available as to their possible origin. It is, however, noteworthy that they have often been found together with the tutuli, ribbed collars, and others of the types we have discussed above (see Note 3), and are presumably part of a costume set involving such components. There is no reason to suppose that they could not also have been manufactured in the 'Middle Elbe' province.

A rich grave found in Western Poland, at Banie, pow. Gryfino (formerly Bahn, Kr. Greifenhagen in Pomerania) included, along with an elaborate collection of women's ornaments, 24 looped buttons (Hommerberg, 1946: pp. 74-78, Bild 75-76).

We have no information concerning the occurrence of these buttons in Northwest Germany. In the Netherlands they are, apart from the 20 examples from Drouwenerveld, certainly very scarce. A single example is present in the Drouwen hoard of 1939; another (but with a bar rather than D-shaped loop) is in the hoard from Hijken (Butler, 1965: fig. 10). An example of that type is also present in the richest Belgian Late Bronze Age hoard of women's goods, that from Port Arthur, Gent (Mariën, 1952, with previous references).

2.7. The pot (fig. 2)

It is not possible to say much at present on the origin of the pot which contained the Drouwenerveld bronzes. It does not seem to be related to local Late Bronze Age pottery. Nor, for that matter, do we at present know of close parallels anywhere else.

It is, however, noteworthy that many pots in the Lausitz and sub-Lausitz tradition in the central part of the North German plain, of quite varied forms, are decorated, on the lower half only, by incision, sometimes in fairly neat lattice patterns, more often as vertical incisions or rather at random. This is rather rare farther west, but is known occasionally (e.g. a settlement pot in Westphalia, from Telgte-Raestrup; Wilhelmi, 1983: p. 59, Abb. 50), or the urn in the Harpstedt tradition, with fingernail-impressed rim, from the Drouwen urnfield itself (excavation 1951, burial No. 1, in a deep pit, with sherds of another pot and 'some bronze': Kooi, 1979, pp. 91-94, figs 91, 92:1, 1a).

More detailed study will be needed before we can put the Drouwenerveld pot in its place; but we would not be surprised were it to appear that the pot as well as the bronzes were imported all together.

Professor W. van Zeist (B.A.I.) has pointed out that the grain of bread wheat in the side of the pot might point in this direction, as bread wheat grows best in a richer soil than would be found in Drenthe, but might be more appropriate to the North German morainic soils.

3. DROUWEN AND NORTHERN EUROPE

Can it be pure coincidence? Or is there a deeper underlying significance in the fact that both in the Early Bronze Age, around 1600-1500 BC, and again in the Late Bronze Age, perhaps 600 years later, the neighbourhood of Drouwen has produced striking evidence for prestigeous presences? And that in both of these cases a strong North European connection is involved?

The question we thereby pose is at present unanswerable; in this article we can only attempt briefly to summarize the existing archaeological evidence, and invite the reader to judge for himself.

3.1. Drouwen and Sögel: the Early Bronze Age

In the year 1927, A.E. van Giffen (1930, I: pp. 84-93; II: Abb. 78; cf. Butler, 1971, with further references) excavated the battered fragment of a prehistoric burial mound at Drouwen, and uncovered one of the richest Early Bronze Age graves ever found on the North European plain (fig. 16a-c). For richer Early Bronze Age burials we must go as far as the *Fürstengräber* of the Saale valley in Saxo-Thuringia, or the equally pretentious tumuli on the western end of the Armorican peninsula, or the richest of the chiefly graves of Wessex.

By luck, the central inhumation burial under the Drouwen tumulus was still almost entirely undisturbed when van Giffen got there. He found, in a rectangular pit under a four-post mortuary house, a warrior's grave, presumably that of a chiefly person. None of his grave goods – the sword with decorated blade; the flanged axe (*geknickte Randbeil*); the set of finely worked flint arrowheads; the polished

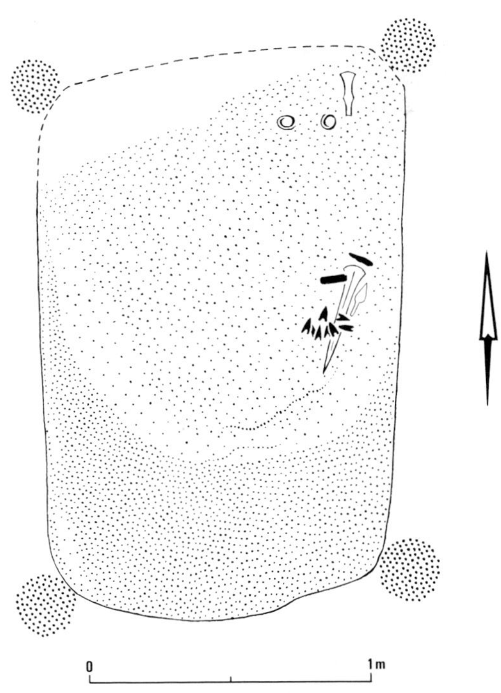

Fig. 16a. The Early Bronze Age 'Sögel' grave from Drouwen. After Butler 1969.

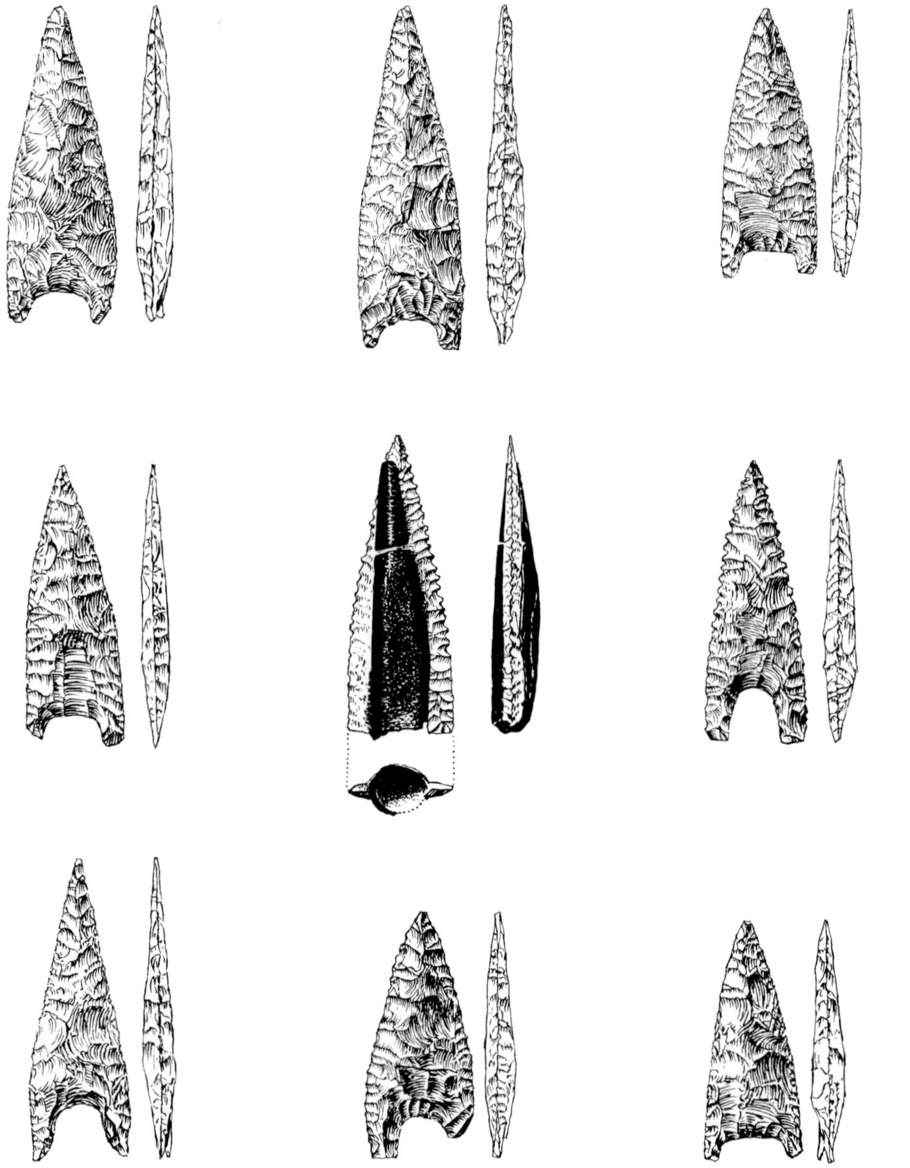

Fig. 16b-c. The finds from the Early Bronze Age 'Sögel' grave from Drouwen. Drawings P.C. van der Kamp, B. Kuitert. Scale: 16b 1:1, 16c 1:2.

whetstone; the flint strike-a-light; the coiled-wire gold earrings – are at all likely to be of local manufacture; they are all rare objects in the Netherlands. Probably the warrior himself came from a distance; though it is of course possible that he was a local figure who had acquired exotic accoutrements. Almost all the items have parallels in the 'Sögel' (or 'Sögel-Wohlde') group of Early Bronze Age male burials, extending across Northwest Germany to Jutland and Mecklenburg and southward to Hessen, though none of them contain so much of them all together. But, if the Drouwen warrior's grave goods are exotic, the fact that he was buried there under a monumental tumulus (a recent excavation by J.N. Lanting, in October 1985, has shown that the tumulus was surrounded by a ring-ditch some 30 metres in diameter) argues that in life he must have had local authority.

3.2. The gap: the Middle Bronze Age

We do not have evidence of anything in or near Drouwen which might suggest a dynastic continuity from the Sögel warrior through the Middle Bronze Age to the period to be discussed below. This may simply be a matter of chance, or it might represent a real gap in the history of Drouwen as a seat of authority. Interestingly, however, a Scandinavian Middle Bronze Age connection has only just come to light near Drouwen:[7] the half of a typically

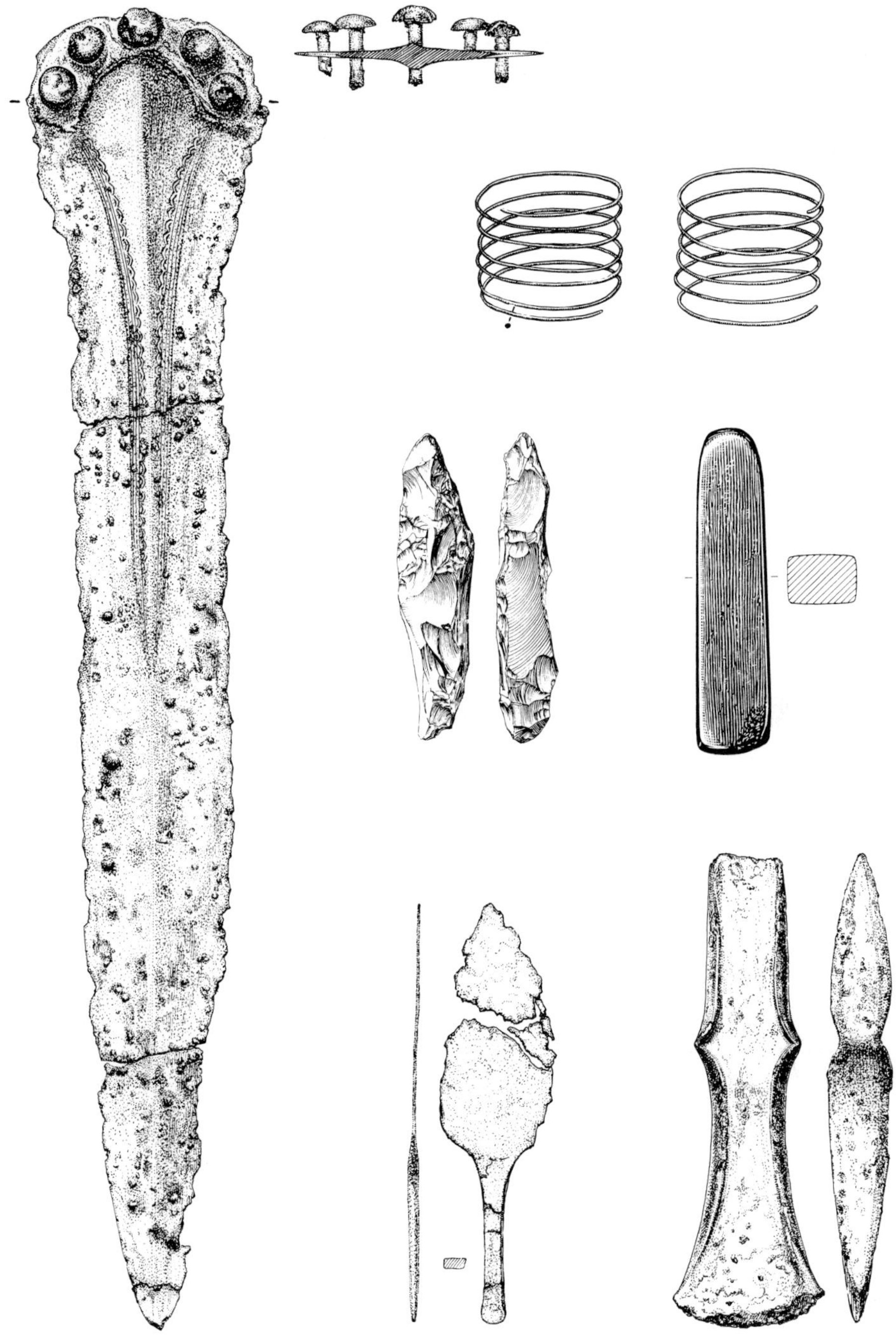

Fig. 16c.

Scandinavian palstave (fig. 17), a type very rare indeed in the Netherlands and Northwest Germany. Unfortunately, the find is without context, and the exact significance of its presence in the area is obscure; it does, however, indicate some kind of contact (direct or indirect?) with Northern Europe in Period II.

3.3. Drouwen and Northern Europe in the Late Bronze Age

After the gap of the preceding period, there is a remarkable series of finds connecting Drouwen with the North European cultural area in the Late Bronze Age. These include several urnfield burials, the two bronze hoards, and one probably ritual deposit of a bronze sword.

The Drouwen urnfield is a link between a number

of these finds. It is unfortunate that the urnfield at Drouwen is not one of those that was excavated completely or even in large part, though a series of small excavations, occasioned by accidental discoveries during heathland reclamation activities, have given some sort of picture (summarized by Kooi, 1979: pp. 90-96). In his reconstruction, the Drouwen urnfield must rank as one of the largest known in the north of the Netherlands. This need

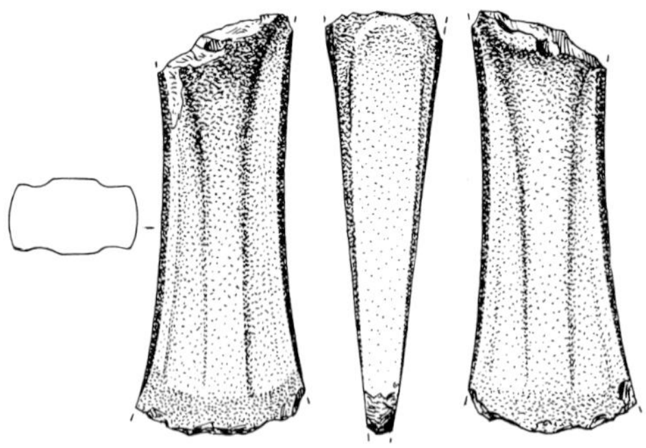

Fig. 17. Palstave fragment. Drouwen, Drenthe. Scale 1:2.

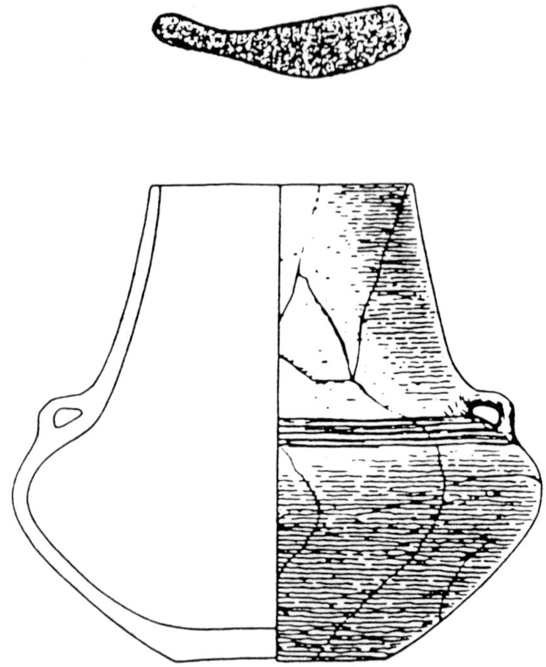

Fig. 19. Razor from Grave 1941:57, Drouwen urnfield. Scale 1:2.

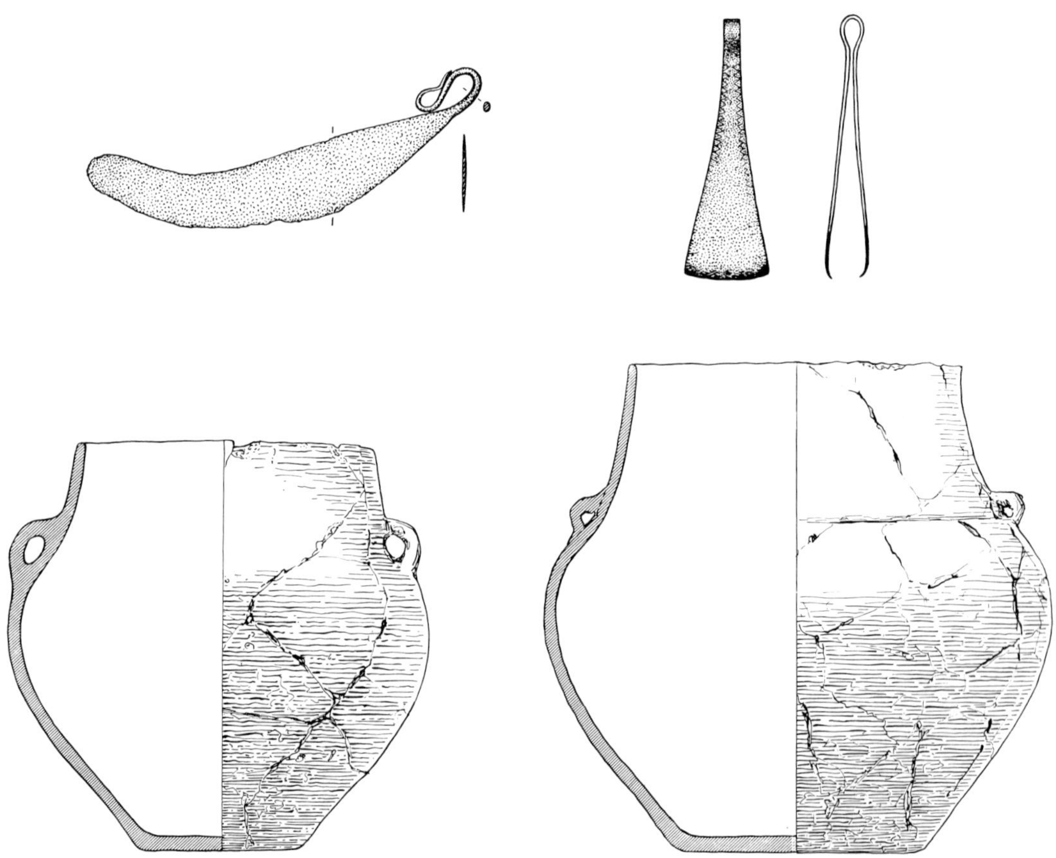

Fig. 18. Urns, razor and tweezers from the stone-packed grave 1939:8 of the Drouwen urnfield. Scale 1:2.

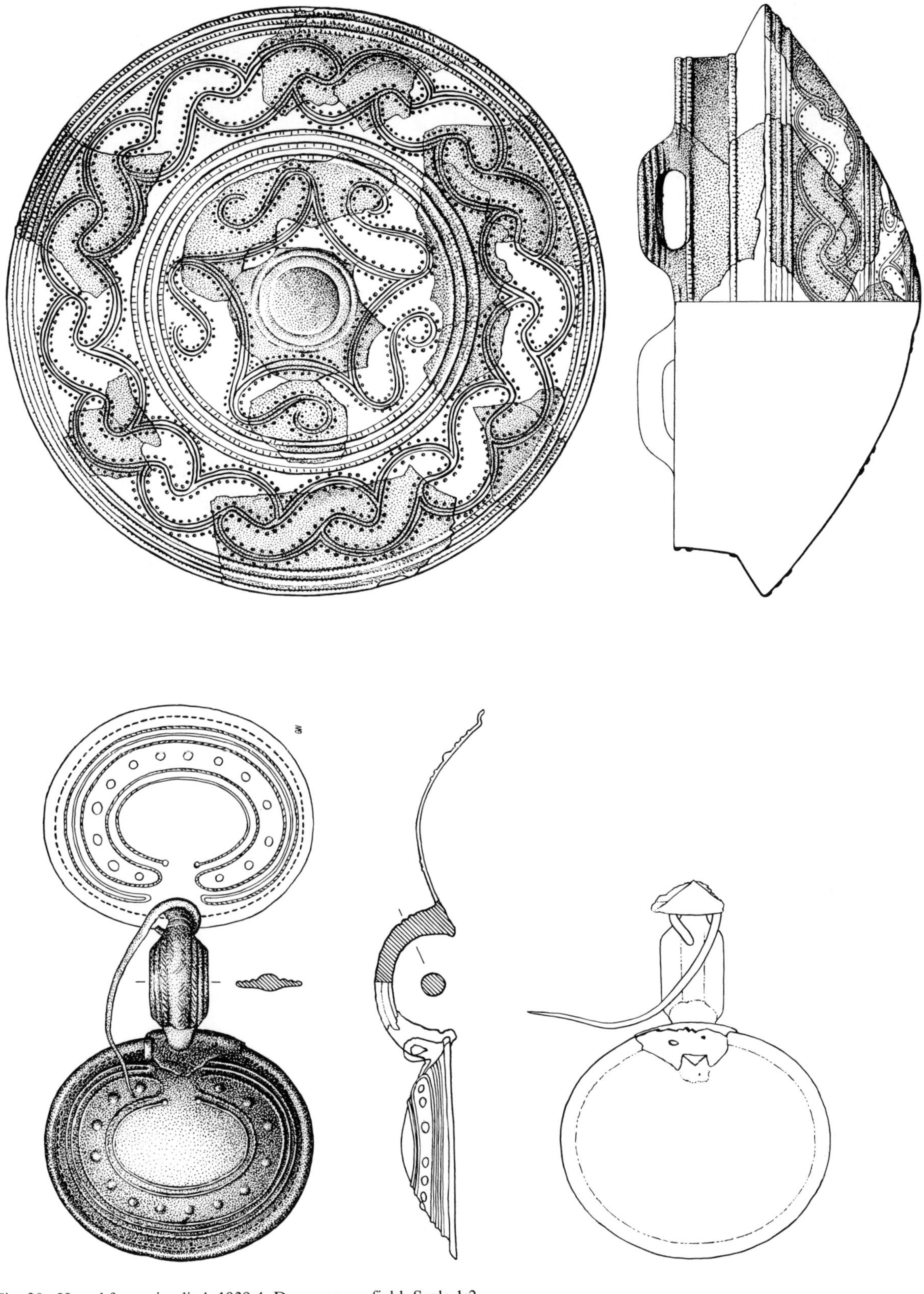

Fig. 20. Hoard from ringditch 1939:4, Drouwen urnfield. Scale 1:2.

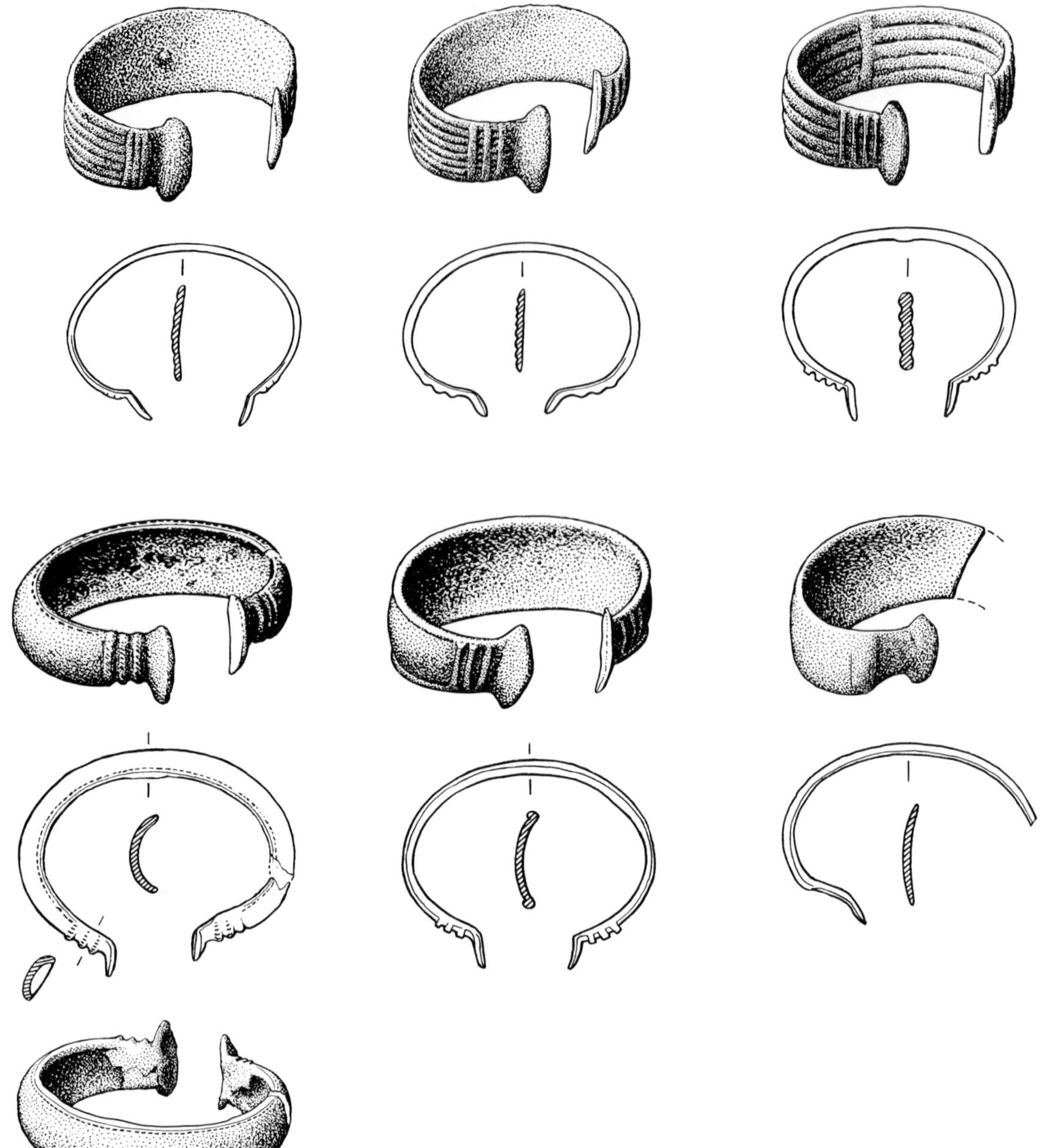

Fig. 21. Hoard from ringditch 1939:4, Drouwen urnfield. Scale 1:2.

not necessarily imply a very large population (Kooi, 1979: pp. 167-174), but there must at any rate have been an as yet undiscovered (and, possibly, already destroyed) settlement of some consequence close by.

Of special interest in this context are the burials 4 and 8 from the excavation of 1939 and grave 57 from the excavation of 1941.

Grave 8 (fig. 18) was in a pit with a stone packing (in itself a most unusual grave form in the Netherlands) and contained two urns of the *zweihenklige Terrine* form. In one of the urns was a 'Nordic' single-edged razor and a decorated pincette: the former rare in the Netherlands, the other unique in this area, and both very probably imports, though their exact source cannot be localized. The razor is assigned, in the recent study by Jockenhövel (1980b: p. 157, No. 572, Taf. 82D) to his type '*einschneidige Rasiermesser mit S-förmigem Griff, Var. II (mit nach oben gebogenes Klingenende)*'. It is dated by him to Period V, with reference to German finds at Wittenhusen, with a Period V socketed knife, and Albersloh, with a Period V *kleine Kugelkopfnadel*. The association of *zweihenklige Terrine* and razor

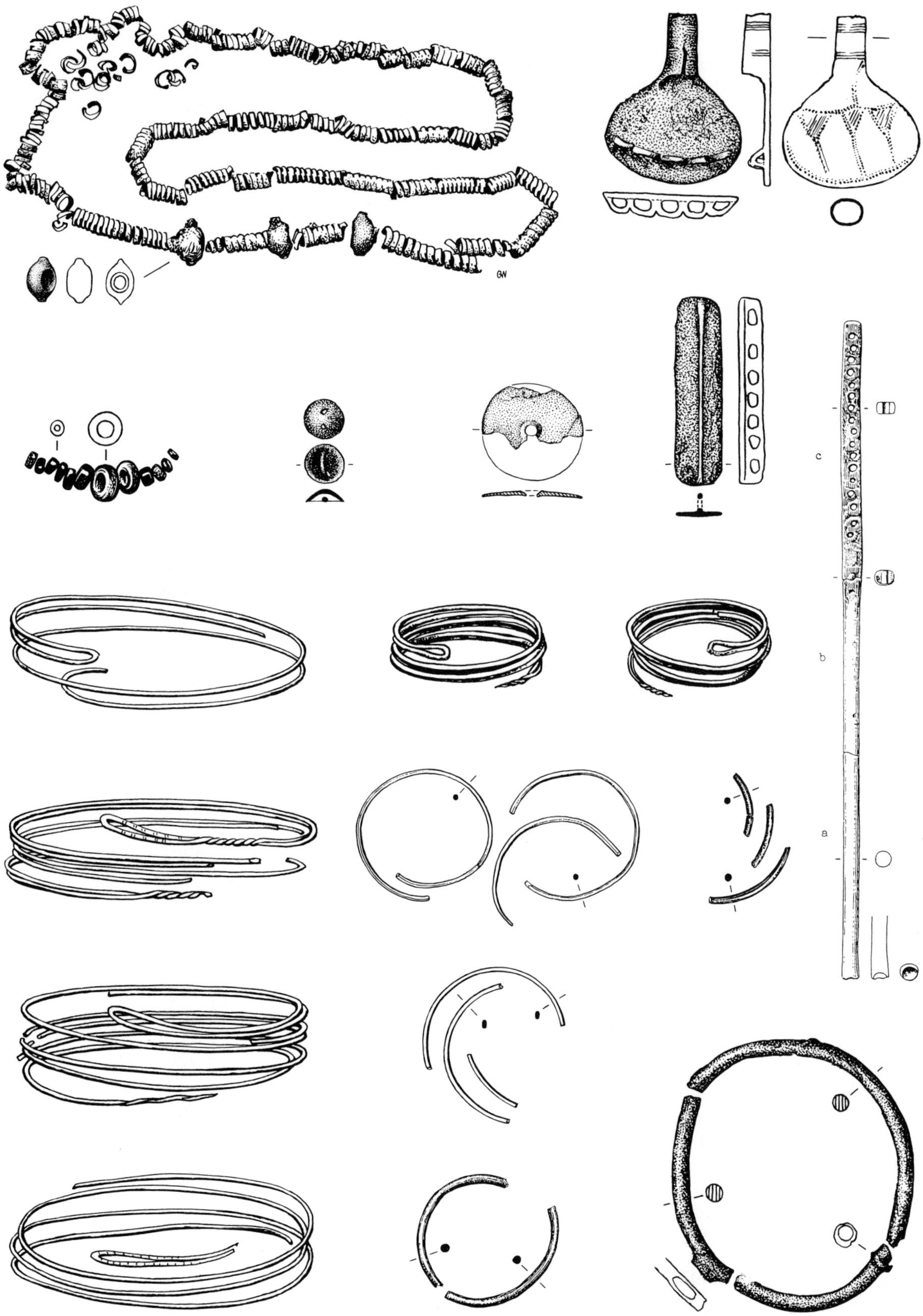

Fig. 22. Hoard from ringditch 1939:4, Drouwen urnfield. Scale 1:2.

Var. II occurs also in a secondary grave in a tumulus at Harenermolen, Groningen (Jockenhövel, 1980b: No. 568). Altogether there are about 10 'Nordic' single-edged razors in the Netherlands.

Grave 57 contained another 'Nordic' single-edged razor, which Jockenhövel (1980b: pp. 168-169, No. 630) has classified under the heading *Rasiermesser mit breitem, rechteckigem bis abgerundet-dreieckigem Griff*. This razor is also dated by him to Period V, with reference to two Westphalian graves: one at Schale with a *kleine Vasenkopfnadel*, the other at Haltern with a *schmale Pinzette*. This Drouwen razor accompanied a high-conical-necked urn related to the *zweihenklige Terrine* family, with groove lines on the shoulders (and not *mit Buckeln auf der Schulter* as stated by Jockenhövel).

Only five metres away from the stone-packed grave No. 8 was found, in an upper level of the ring-ditch that surrounded grave 4 (in which no central urn burial was recovered), the bronze hoard of 1939 (figs 20-22). This hoard has been several times illustrated (van Giffen, 1930; 1943; Jacob-Friesen, 1956; Butler, 1965; 1969; 1979; Kooi, 1979: p. 93, fig. 88), and some of its contents have been discussed in the literature (Butler, 1979) though there has never yet been a full published description. A student paper concerning this hoard (Bouman, 1985) is also unpublished. The hoard consists mostly of ornaments, which were apparently deposited in intact condition, although many suffered severe plough damage before discovery. The deposit seems therefore to have been the personal property of a woman; and, to judge by the rarity and quality of the contents, the woman concerned must have been a person of wealth and authority. In a popular publication (Butler, 1969) we described her, by way of speaking, as 'the princess of Drouwen'; but of course we do not really know what sort of ranking system was then in use or her place within it.

The most spectacular pieces in the hoard – the richly decorated 'hanging bowl' and the spectacle fibula – are imports from the Scandinavian area, and extremely rare in this region. Apparently of local manufacture is the set of seven bracelets, *cire perdue* castings, each an individualized piece (Butler, 1965). Some other objects in the hoard are either unique to this find or are known in only a very few finds in the region.

A find of this scope and character would be considered to be a rich find even in Denmark, where it would be one of many such hoards. Levy (1982) has made an interesting analysis of the ranking system which might be deduced from the large number of votive hoards of female goods found in that land: there are around 100 such finds in Denmark in Period V alone. She notes that the 'female sumptuary sets' contained in these hoards often have up to five main components: hanging vessel, belt plate, neckring(s), bracelet(s) and fibula.

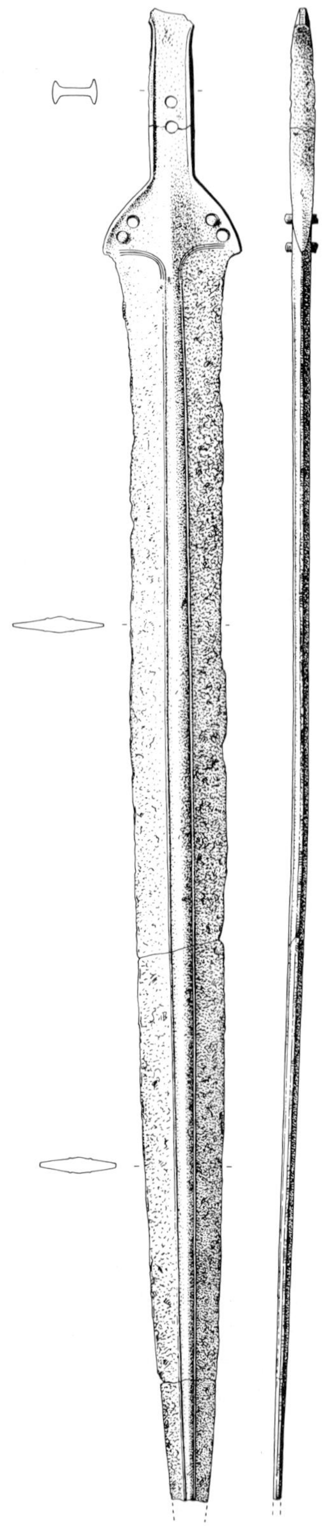

Fig. 23. Sword with narrow hilt-plate: stray find from Drouwen. Scale 1:3.

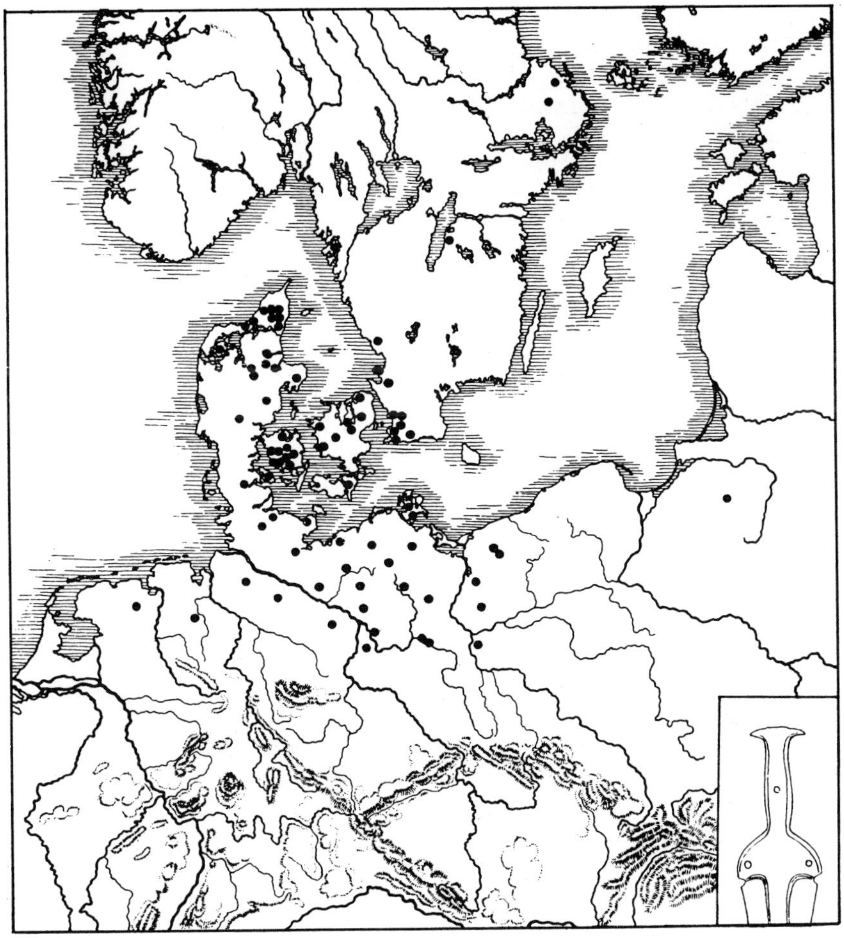

Fig. 24. Swords with narrow hilt-plate. After Sprockhoff, 1931 (Drouwenerveld 1984 added).

Of the hundred hoards, only four have all five of these components, and 12 have four components. Our Drouwen hoard has at least three of these components – the hanging bowl, the fibula, the bracelets; perhaps four of them if we may consider the string of beads as being equivalent to a neck-ring. On this basis, the Drouwen lady's find could be considered to rank by Danish standards with the top 16%. In our region, where it is a find of quite exceptional character, its relative importance is obviously very much greater.

Roughly 2 km southwest of the urnfield, apparently in a low boggy spot, was ploughed up, somewhere around or before 1941, a bronze sword with narrow flanged hilt (*Grebtungesvaerd met smal tunge, Schwert mit schmaler Zunge*; fig. 23). This sword has been in private possession since its finding, and only became known in 1985; it has not previously been published. We are grateful to J.N. Lanting for the details, and for his permission to illustrate it here.

Once again we have to do with a type characteristic of the Scandinavian Late Bronze Age culture area (fig. 24; Sprockhoff, 1931: pp. 26-35, Taf. 29; Broholm, 1946/49: pp. 32-34, Pl. 4; Baudou, 1960, pp. 9-10, Typ Ia1, list p. 153, supplementary to Sprockhoff's list, Karte I). The Danish islands seem to be its primary centre of distribution, with radiations to North Jutland, Scania and North Germany. But its rarity in Northwest Germany is striking; until the Drouwen find there was only one example known west of the Weser (Sprockhoff, 1931: p. 101, No. 7; from Huntlosen, Amt Wildeshausen, Oldenburg, found in a *Hügelgrab* with other, unknown objects).

Swords of this type belong traditionally to Period IV. Sprockhoff (1931: pp. 28-33) found that a number of the closed finds in which these swords occurred were of Period V date, but Broholm (1946/49) and Baudou (1960) did not follow him in this, at least with respect to the Danish and Swedish finds, which they attribute uniformly to Period IV. It seems, however, that at least the North German hoard from Stölln, Kr. Westhavelland (Sprockhoff, 1956: pp. 63-64; cf. Jacob-Friesen, 1967: p. 276, No. 420) must be continued to be regarded as a HaB3/M

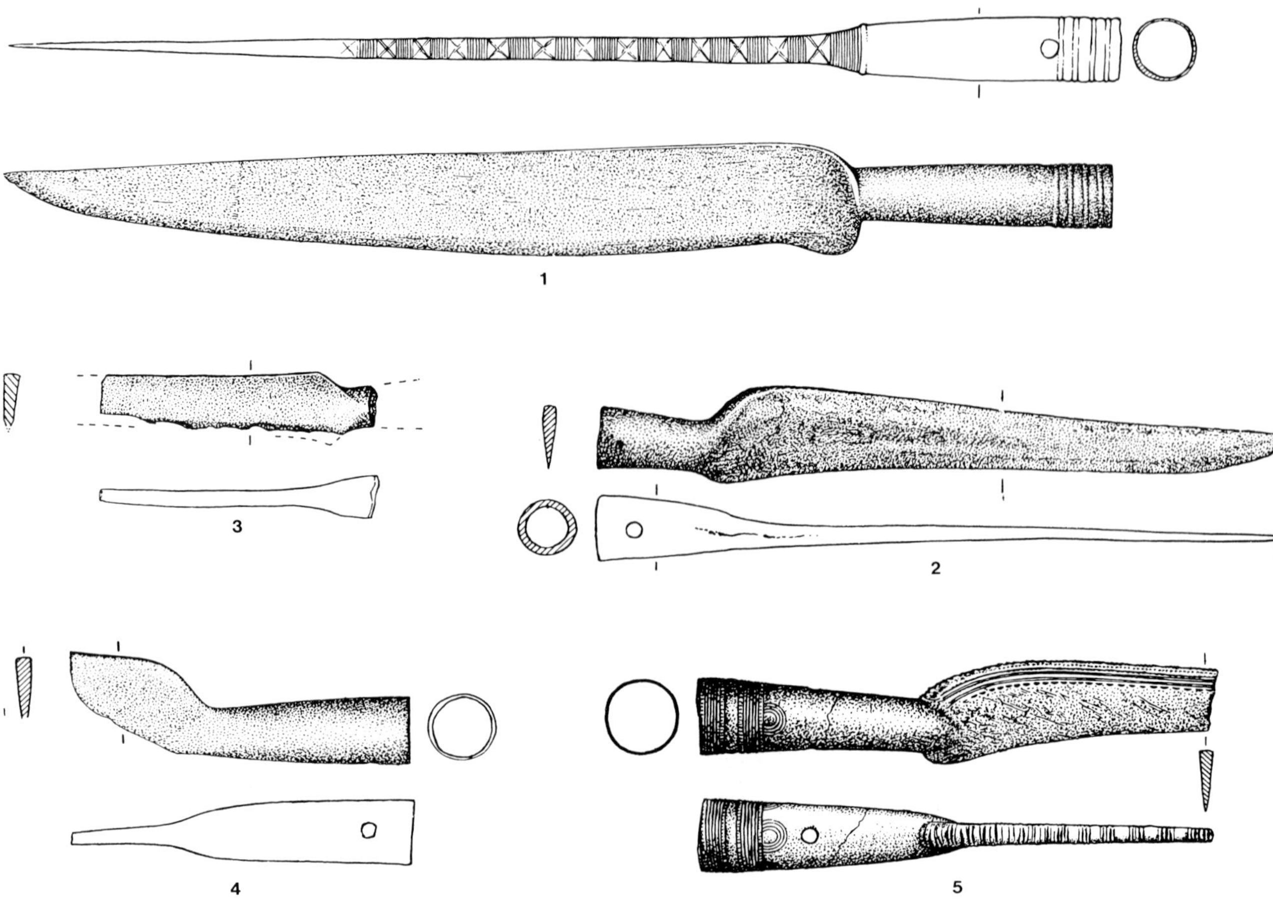

Fig. 25. Socketed knives from hoards in the Netherlands. 1, 2 Schoonebeek, Drenthe; 3 Bruggelen, Gelderland; 4 Elzen, Overijssel; 5 Havelte, Drenthe.

V find; it contains for example an Auvernier sword).

The sword found at Drouwen must have been deposited in a wet environment; the present owner told J.N. Lanting that when originally found the sword still had part of a wooden haft preserved. Part of the parcel in which the sword was found was indeed formerly a boggy depression, according to Lanting. The sword was thus presumably a one-piece votive deposit in a bog or pool.

Swords of this type were evidently objects of considerable value in their homeland. Particular evidence of this is the fact that five examples have hilts decorated with gold leaf, and one with gold wire. In Central Sweden (thus at the extreme opposite end of their distribution) an example was placed in a rich grave in Kung Björns Hög near Uppsala, a classic example of a Bronze Age *Fürstengrab*. The Drouwen sword is the only Scandinavian sword known in the Netherlands, and one can presume that its owner was a person of no small importance here.

Prehistoric chance finds and systematic research at Drouwen have thus managed to assemble – uniquely for the Late Bronze Age in this region – a varied assortment of personalities: the very prestigeous female figure represented by the hoard of 1939; the authoritative male figure symbolized by the sacrificed sword; the presumable male buried with the razor, pincette and urns in the grave with stone packing; the other presumable male with the razor and grooved urn; the bronze smith implied by the hoard of scrap metal imported from along the Elbe – all in addition to the rather undifferentiated mass of ordinary burials in the large urnfield. And we have surely emphasized sufficiently the fact that all the leading characters have some connection with the South Scandinavian culture area and/or the 'Mittelelbe' province, although such connections are otherwise rare in this region.

Our grasp of what was really going on at Drouwen is limited by the fragmentary nature of the available evidence. For one thing, we cannot say exactly what length of time was involved with these

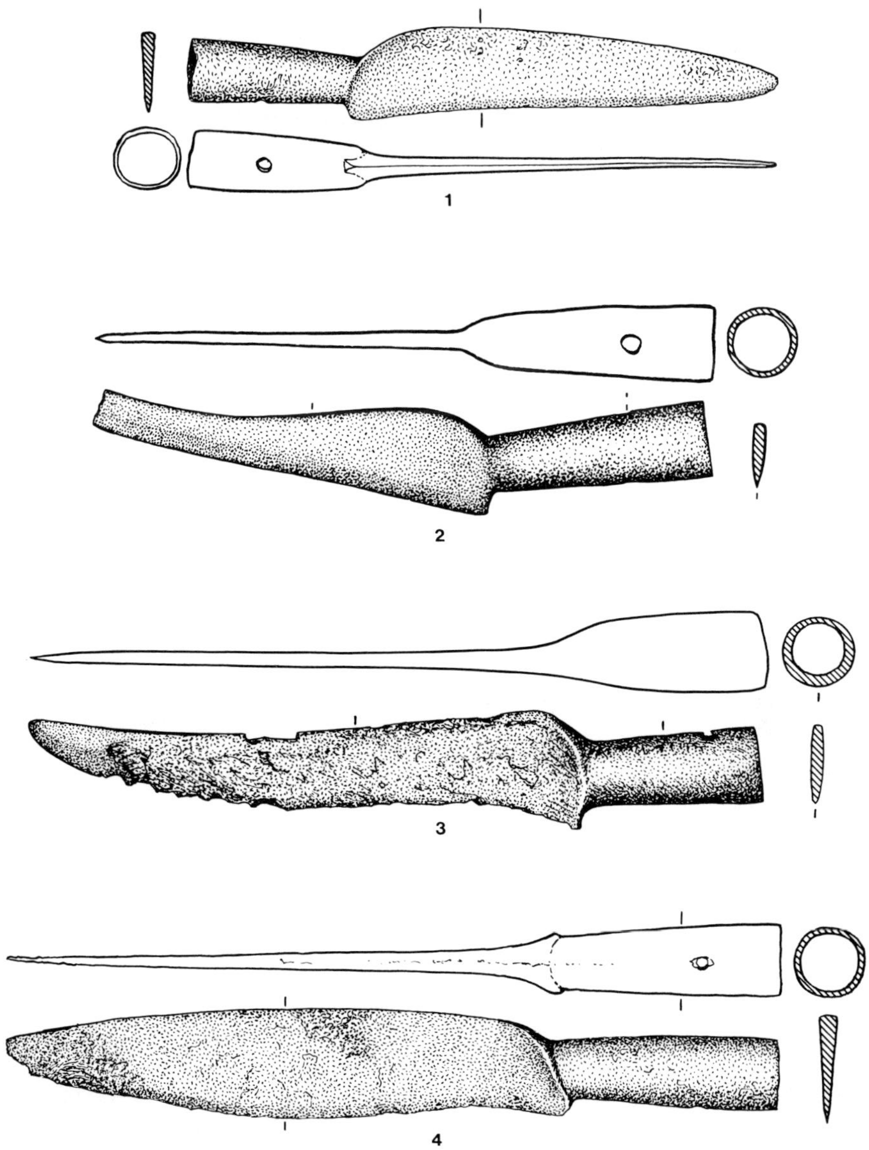

Fig. 26. Socketed knives from the Netherlands. 1 Nijmegen, Gelderland; 2 between Weper and Haule, Friesland; 3 *gem.* St. Oedenrode, North Brabant; 4 near Denekamp, Overijssel.

events. The archaeological evidence tells us that types of Northern Period IV, Period IV or V, and Period V are involved, but that their deposition was very probably all in Period V. But are we to imagine that these deposits were all very close together in time – say, all in the same generation, at one extreme – or whether they were spread out over something like a century? Another uncertain point: does our cast of characters at Drouwen consist of actual Northerners, settlers from afar: did we have an influx of Two-eared Terrineans from the Elbe Mouth region, or Keyhole Ditchers from Westphalia, or such-like? Or do we have to do with local folk somehow acculturized to 'Urnfield' rituals, who have, thanks to some sort of special contact situation, been able to acquire an unusual share of North European metal prestige goods? Or are we dealing with an intrusive overlord (m/f) and his/her attendants, who have, by invitation or imposition, established a hegemony over the local population?

There is little doubt that chiefdoms existed in the rich South Scandinavian-North German cultural area in the Late Bronze Age (cf. for example Levy, 1982). Whether this was true also of the poorer, fragmented areas west of the Weser is another question. Van der Waals (1987: pp. 13-16, 46) has lately asserted a presupposition that the north of the Netherlands remained in a tribal, pre-chiefdom state at least up to the Middle (pre-Roman) Iron Age, i.e. around 500 BC. Are, then, our chiefly Bronze Age figures at Drouwen chiefs without a chiefdom?

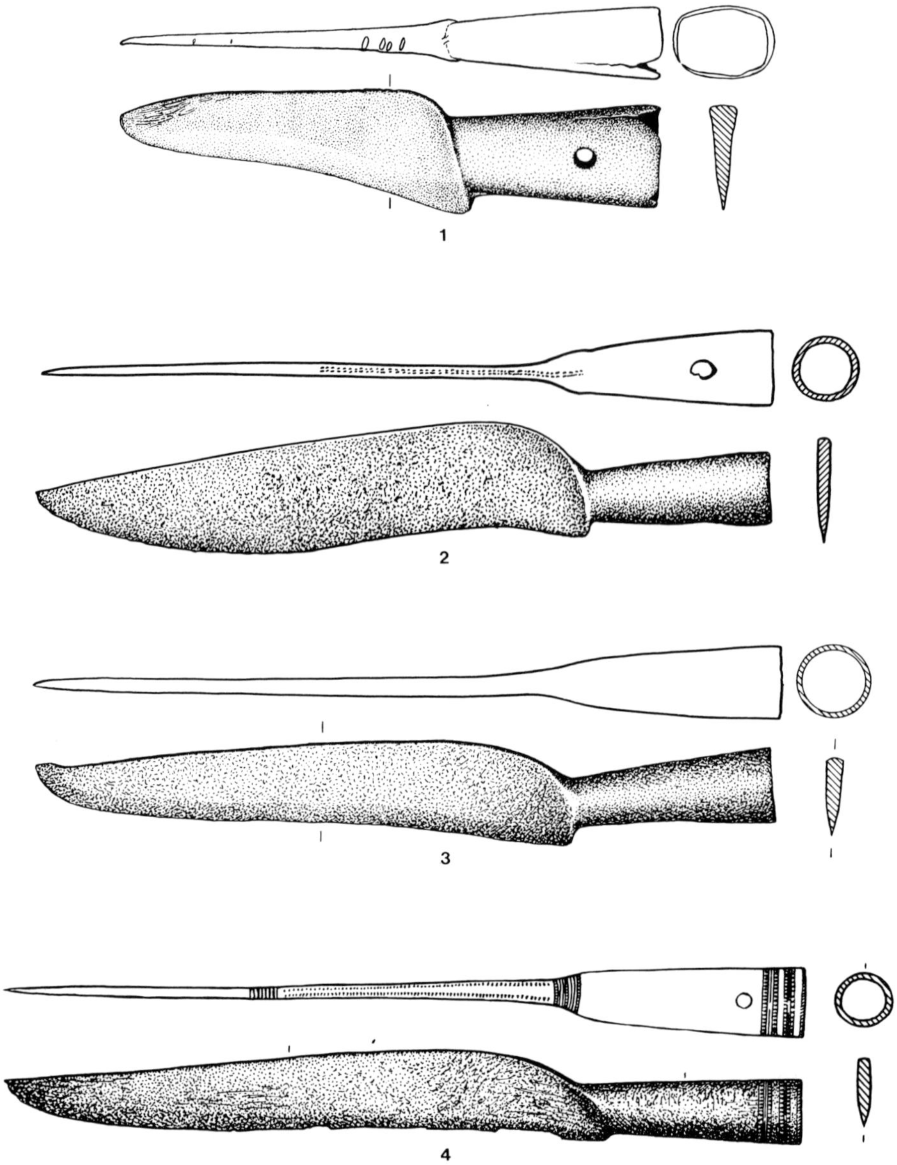

Fig. 27. Socketed knives from the Netherlands. 1 near Venlo, Limburg; 2 near Aalten, Gelderland; 3 Voorschoten, South Holland; 4 Oud-Schoonebeek, Drenthe.

We do not believe that we can at present decide whether Late Bronze Age Urnfield Society in Drenthe was at the 'tribal' or the 'chiefdom' level of social organization, or at some intermediate level between these ideal types. There is, admittedly, little evidence for 'chiefdom' complexity in this area. But chiefs can and do exist at the tribal level, even if their status is less institutionalized than it would be in a chiefdom society. We can, therefore, have our chiefs at Drouwen in either case.

Personally, we like to imagine a marriage to a local person of a chiefly figure from the Elbe area – perhaps from the Lüneburger Heide, or from Holstein, or from Mecklenburg – coming with rich exotic gifts, bringing along a subservient smith with loads of scrap metal, and other retainers. Drouwen might then have constituted a territorial power centre. Was it the chiefly capital of the Hondsrug, or of the whole of the north of the Netherlands, or the whole of the Hunze-Ems region?

In the present state of knowledge, it would seem that the Emsland, the Münster basin, and the north of the Netherlands had a common culture in the Late Bronze Age. They had in common, at least, pottery types such as those of the *Doppelkonus* family (most recent maps Tackenberg, 1985: Karten 26-27), burial monuments such as the keyhole-shaped ditches (latest map Wilhelmi, 1983: Abb. 13). In the same areas the 'Hunze-Ems' socketed axes circulated. If there was an ethnic and economic

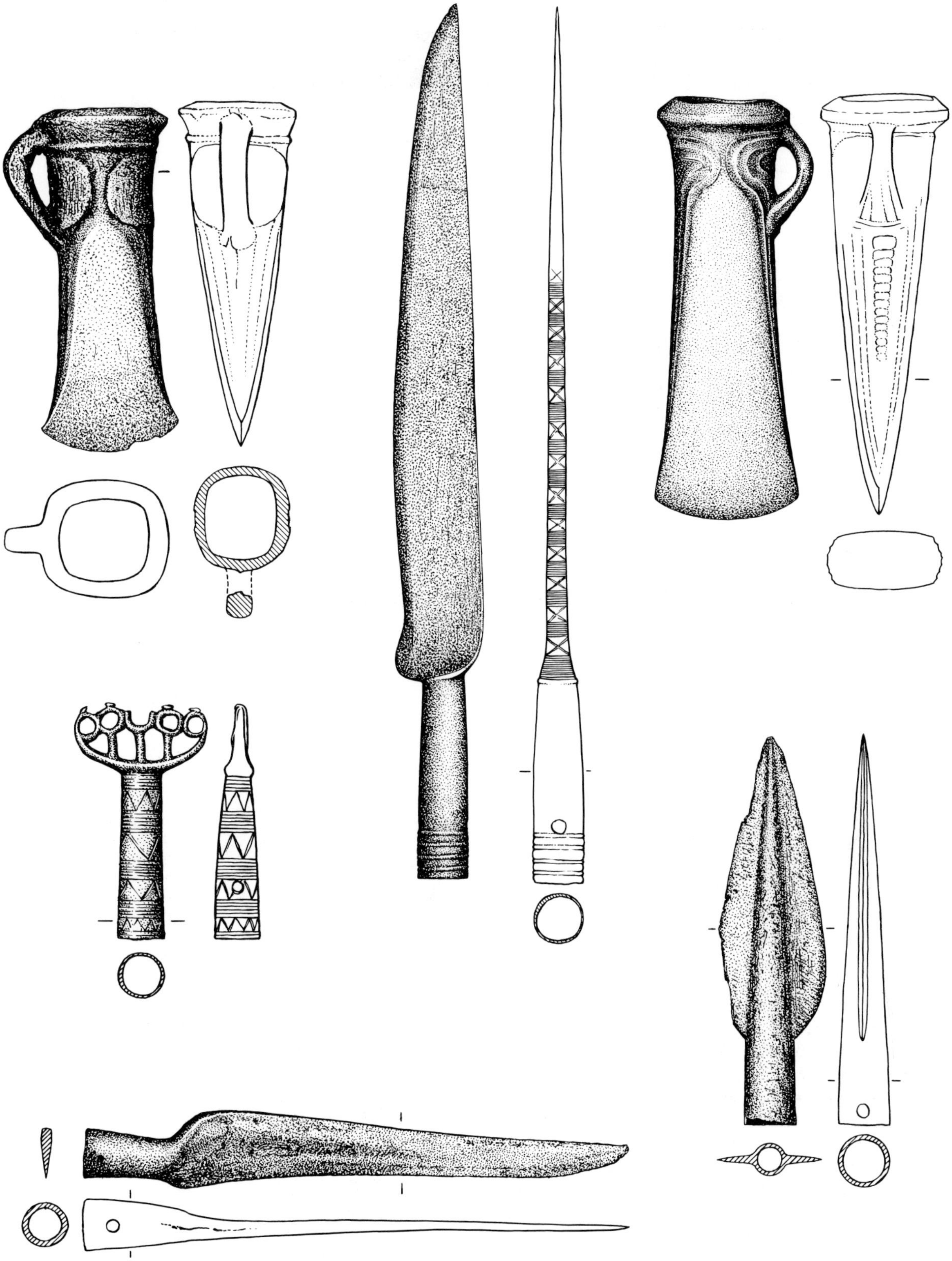

Fig. 28. Hoard from Schoonebeek, Drenthe.

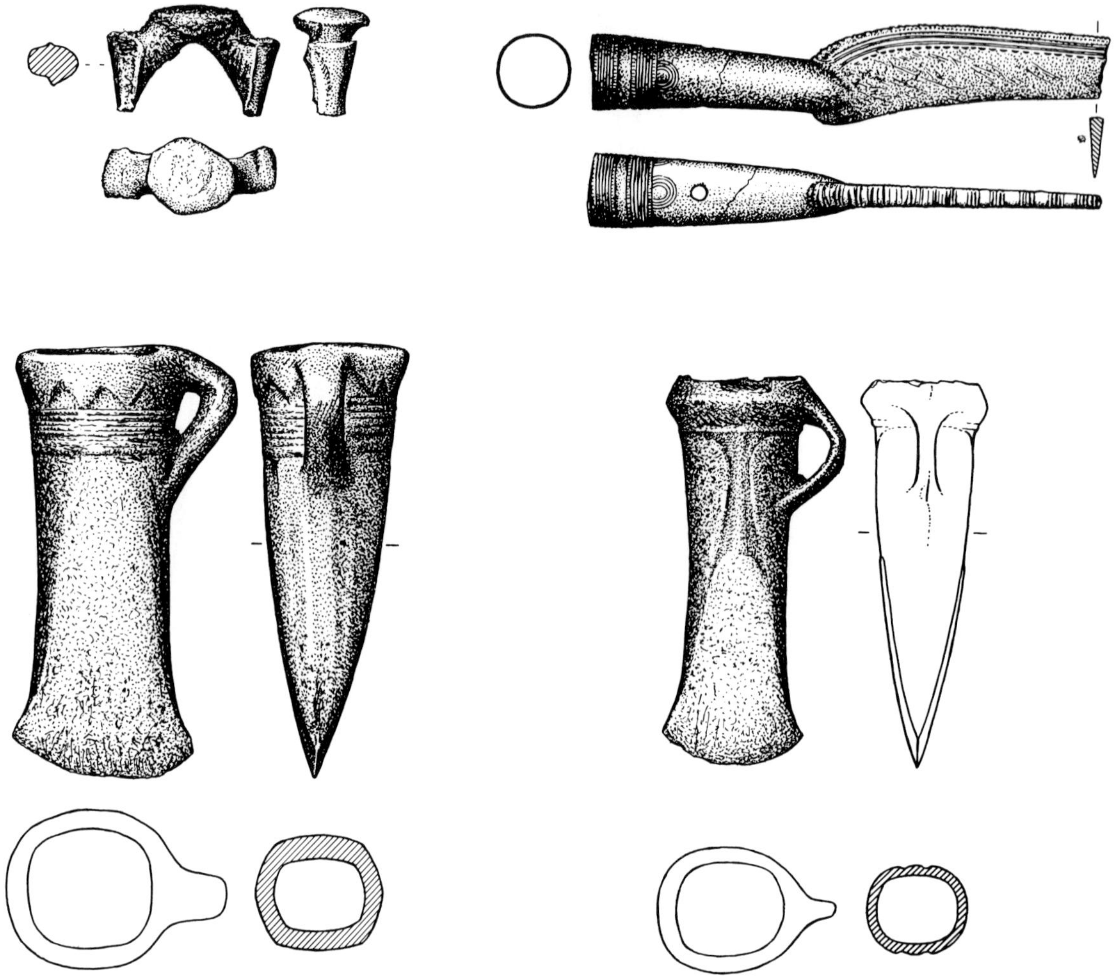

Fig. 29. Hoard from Havelte, Drenthe.

community between these areas, it is not entirely inconceivable that they might, for a shorter or longer time, have lived under a common political authority. And as Drouwen is now the only chiefly centre that we can identify, it could be that Drouwen served as the centre of authority for the entire region. This may be pure fantasy; but future research and discovery may add to or subtract from the picture we have suggested.

It was, perhaps, also then a redistribution centre for scarce luxury imports from farther east. In this respect we can think of finds such as the Scandinavian spectacle brooch from the Bonnerveen, *gemeente* Gasselte, only 6 km to the north of Drouwen (fig. 30), or the gold bracelets from Hijkersmilde and the bronze hoard from Hijken (including a socketed knife, bracelets related to those of the 'Princess', and a looped button), c. 25 km farther west (Butler & van der Waals, 1960; Butler, 1965: figs 9-10).

Obviously, any number of scenarios different from this can be imagined to fit the same facts. What in any case appears certain is that Drouwen had some sort of special relationship with the North European area, seemingly persistent over a long period, but reaching a remarkable climax in the Late Bronze Age.

4. NOTES

1. A Dutch version of this article (Butler, 1987; kindly and laboriously translated by O.H. Harsema) appeared in the *Nieuwe Drentse Volksalmanak* for 1987. The drawings are by H. Roelink. We hereby express our gratitude to them, and to J.N. Lanting and P.B. Kooi (all B.A.I. colleagues), for their generous collaboration.
2. This is not to say, however, that some part of HaB3 may not eventually be shown to continue on into the 8th century, as suggested tentatively by Gross (1984).
   Müller-Karpe (1959, etc.) divided the Central European Later Urnfield period into three parts, then referred to as

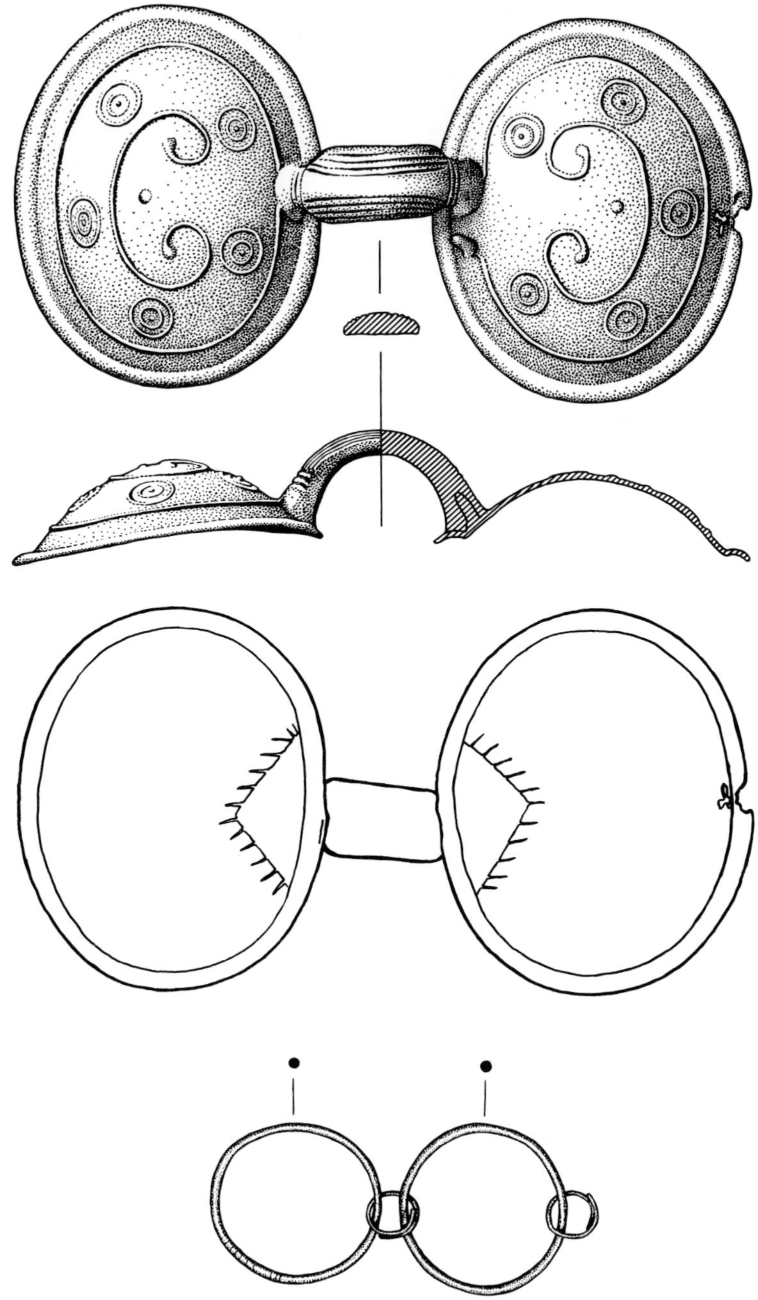

Fig. 30. Hoard from Bonnerveen, *gem.* Gasselte, Drenthe.

HaB1, B2 and B3 respectively; later (1974) he substituted the designations *jüngere Urnenfelder* (B1 and B2) and *späte Urnenfelder* (B3). This usage has been followed by many recent writers.

While HaB1 and HaB3 have long established and generally accepted contents, the Müller-Karpe HaB2 phase has not been clearly identifiable in many areas, and its existence is doubted or denied by various specialists. The concept is, however, still used by some (e.g. Kibbert, 1984: p. 151, who regards such hoards as Obernbeck, Rethwisch, Berg-en-Terblijt, Niederolm, Ostrhauderfehn, and Onstwedder Holte as having been deposited during the MK HaB2 phase, if also containing older types). In Switzerland it has recently been suggested by Gross (1984: pp. 69-70) that there is room, on the basis both of pottery typology and absolute chronology, for a HaB2 phase at around 950-875 BC. This phase has not yet been found in the West Swiss lakeside settlements, but he names some grave finds and hill settlements which he would assign to HaB2 on the basis of pottery typology (graves at Elgg and Rafz; settlements Bischofsstein bei Sissach; Wittnauer Horn, Courroux *unterer Hüttenplatz).*

Where, however, there is no HaB2, it is not generally agreed whether the B2 metallic material claimed by Müller-Karpe and his followers should be properly assigned to HaB1 or to HaB3.

It is interesting, however, to notice that in Switzerland Grave

1 at Elgg (Ruoff, 1974: Taf. 12:9-13, 13:1-10) assigned by Gross to his B2 phase, has a *kleine Vasenknopfnadel mit gebogenem Schaft*. The pin with small vase-head and bent shaft is in North Germany regarded as a typically Montelius V object, and as one of the *Leitformen des Ems-Weser Kreises* in that period (Sprockhoff, 1950/51: pp. 125-126; 1956: pp. 229-230; new map Tackenberg, 1985: Karte 7).
A related terminological complication is that the designation 'HaB2' is used by some authors in Switzerland to refer to the Müller-Karpe HaB2, while with others it is more or less identical in content to Müller-karpe HaB3.
3. *Hoards with similar buttons, mushroom tutuli, and ribbed collar*: Lentföhrden, Kr. Segeberg (S.-H.), Struve, 1979: Taf. 47-49; Hinzenhagen, Kr. Güstrow (Meckl.), Hundt, 1955: pp. 127-128, Abb. 4:1-21.
Hoards with buttons and mushroom tutuli: Bäk, Kr. Herzogtum Lauenburg (S.-H.) (also with sickle fragments, socketed axe fragments, bracelets, twisted wire, casting jets), Hundt, 1951; Parchim, Kr. Parchim (Meckl.), Hundt, 1955: pp. 126-127, with further refs., Abb. 2:1-4; Hagen, Kr. Lüneberg, Drescher, 1956: pp. 83-92; Bargfeld, Kr. Uelzen, Bath, 1953/55: pp. 79-82 (also with Rethwisch-Bargeroosterveld *Nierenringe*, sickles, twisted wire, founders' waste); Schnega, Kr. Lüchow-Dannenberg (also with similar spearhead, frag. Obernbeck socketed axe), Sprockhoff, 1963: pp. 29-31, Abb. 28. Bahrendorf, Kr. Lüchow-Dannenberg, Sprockhoff, 1932: Taf. 1.
Hoard with buttons and ribbed collar: Tüschau, Kr. Lüchow-Dannenberg, Sprockhoff, 1932: Taf. 14-15; 1937: p. 22.
4. The Leeuwarden socketed axe was found by the gardener J. Koning in 1985 in transported earth (probably of local origin, dredged out of the Van Harinxmakanaal) in a flower bed on the Zwettestraat in the industrial estate Nijlân, on the south side of Leeuwarden. It contains the fragment of a wooden shaft, identified as of oak by W.A. Casparie (B.A.I.). The axe is now in the museum at Leeuwarden (FM 1985-V-1).
Tackenberg lists as a Schinna axe an example from Exloërmond, Drenthe (No. 1 in his Liste 7), after the illustration by Pleyte (1882, Drenthe: Pl. 33:7). This axe has plastic pendant triangles on the neck (Butler, 1965) and is evidently a local product in the north of the Netherlands.
5. *Type Schinna*: Sprockhoff, 1956, II: pp. 17-19 (list), Abb. 17:5, Karte 6; Tackenberg, 1971: p. 256, Liste 7, Karte 7; Sprockhoff, 1956, II: Karte 6.
Type Obernbeck: Sprockhoff, 1956, I: p. 90; II: pp. 19-20 (list), Abb. 17:6, Karte 7; Tackenberg, 1971, Liste 6, Karte 6; Baudou, 1960: Karte 12 (single-rib); Tackenberg, 1971: Liste 5, Karte 5; Baudou, 1960: pp. 178-188 (2 to 4 ribs);
mit profiliertem Tüllenmund: Sprockhoff, 1941: pp. 86 ff; 1956, II: pp. 20-21 (list), Karte 7; Butler, 1960a; 1960b; Tackenberg, 1971: Liste 8, Karte 8.
Baudou (1960) provides extensive lists for Scandinavia, but unfortunately for present purposes he does not differentiate between examples with and those without arch-facets on the face.
Kibbert (1984: pp. 136-137), who had only a few examples to deal with in his West German area, combines all the ribbed variants under the term Typ Løvskal (after a Danish hoard); he says that there are 36 examples in Denmark and 17 in South Sweden.
6. While it is difficult to cite features on socketed knives that occur in the north and not in the West Swiss-East French area, it is perhaps worth calling attention to a converse point that has practically been overlooked in the literature. The junction between blade and socket can take two different forms. The first we can call Form V: the socket is externally projected into the back of the blade. The other can perhaps be called Form T: the cone is truncated where it joins the blade. Both forms occur in the West Swiss-East French knives; whereas Form V seems to be absent on the North European plain (only the Havelte knife has a hybrid form). But seven of the ten specimens from the cave of Han are of the V type (Mariën, 1984).
7. Due to a misunderstanding, we reported the find-spot of this previously unpublished palstave fragment as 'Borger' (the next village south of Drouwen) in Butler, 1987. The correct find-spot was, according to the finder, about 1 km east of the centre of the village of Drouwen, along the Stobbenweg.

5. BIBLIOGRAPHY

ANON., 1986. Kroniek van opgravingen en vondsten in Drenthe in 1984. *Nieuwe Drentse Volksalmanak* 103, pp. 146-150.

AUDOUZE, F. & J.-C. COURTOIS, 1970. *Les épingles du Sud-Est de la France* (= Prähistorische Bronzefunde XIII:1). München.

BATH, F.C., 1953/55. Der bronzezeitliche Hortfund von Bargfeld, Kr. Uelzen. *Hammaburg* 4 (Heft IX/X), pp. 79-82.

BAUDOU, E., 1960. *Die regionale und chronologische Einteilung der jüngeren Bronzezeit im Nordischen Kreis.* Stockholm.

BECKER, B. et al., 1985. *Dendrochronologie in der Ur- und Frühgeschichte. Die absolute Datierung von Pfahlbausiedlungen nördlich der Alpen im Jahrring-Kalender Mitteleuropas* (= Antiqua 11). Basel.

BIANCO PERONI, V., 1976. *Die Messer in Italien* (= Prähistorische Bronzefunde VII,2). München.

BORKOVSKY, J., 1934/35. Tři hromadné nálezy bronzů Podkarpatské Rusi. *Památky Archaeologické* N.R. 4-5, pp. 99-102.

BOUMAN, H., 1985. Het depot van het urnenveld te Drouwen. Unpubl. typescript, B.A.I., Groningen.

BROHOLM, H.C., 1946/49. *Danmarks Bronzealder.* III, IV. Copenhagen.

BRUNN, W.A. VON, 1968. *Mitteldeutsche Hortfunde der jüngeren Bronzezeit* (= Römisch-Germanische Forschungen 29). I-II. Berlin.

BUTLER, J.J., 1960a. Drie bronsdepots van Bargeroosterveld. *Nieuwe Drentse Volksalmanak* 78, pp. 205-231.

BUTLER, J.J., 1960b. A Bronze Age concentration at Bargeroosterveld. With some notes on the axe trade across Northern Europe. *Palaeohistoria* 8, pp. 101-126.

BUTLER, J.J., 1960c. Het bronsdepot van Onstwedder Holte. *Groningse Volksalmanak*, pp. 117-122.

BUTLER, J.J., 1961. De Noordnederlandse fabrikanten van bijlen in de late bronstijd en hun produkten. *Nieuwe Drentse Volksalmanak* 79, pp. 199-233.

BUTLER, J.J., 1965. Ook eens iets voor de dames. Een bijzonder facet van de Noordnederlandse bronsbewerking in de 8e eeuw v.Chr. *Nieuwe Drentse Volksalmanak* 83, pp. 163-198.

BUTLER, J.J., 1968/69. Een urnenvelden-kokermes uit de Hidsmeden bij Onstwedde. *Groninger Volksalmanak*, pp. 206-223.

BUTLER, J.J., 1969. *Nederland in de bronstijd* (= Fibulareeks 31). Bussum.

BUTLER, J.J., 1971. *Bronze Age grave groups and hoards of the Netherlands* I. (= Inventaria Archaeologica NL11-20). Bonn.

BUTLER, J.J., 1973a. The big bronze knife from Hardenberg. In: W.A. van Es et al. (eds.), *Archeologie en Historie.* Bussum, pp. 15-27.

BUTLER, J.J., 1973b. Einheimische Bronzebeilproduktion im Niederrhein-Maasgebiet. *Palaeohistoria* 15, pp. 319-343.

BUTLER, J.J., 1979. A Late Bronze Age drawing instrument? *Palaeohistoria* 21, pp. 195-203.

BUTLER, J.J., 1987. Drouwen: het einde van de Noordse regenboog? *Nieuwe Drentse Volksalmanak* 104, pp. 103-150.

BUTLER, J.J. & J.D. VAN DER WAALS, 1960. Three Late Bronze Age gold bracelets from the Netherlands. *Palaeohistoria* 8, pp. 91-99.

CASPARIE, W.A., 1984. The three Bronze Age footpaths

XVI(Bou), XVII(Bou) and XVIII(Bou) in the raised bog of Southeast Drenthe (the Netherlands). *Palaeohistoria* 26, pp. 41-99.

CHANTRE, E., 1875. *Age du Bronze: recherches sur l'origine de la métallurgie en France*. Paris.

DRESCHER, H., 1953/55. Werkzeug- und Gebrauchsspuren an den Fundstücken des Hortfundes von Bargfeld, Kr. Uelzen. *Hammaburg* 4 (Heft IX/X), pp. 131-143.

DRESCHER, H., 1956. Die Knöpfe des Hortfundes aus Hagen, Kreis Lüneberg. Ein Beitrag zum bronzezeitlichen Metallguss nach Wachsmodellen. *Offa* 15, pp. 83-92.

GIFFEN, A.E. VAN, 1930. *Die Bauart der Einzelgräber* (= Mannus-Bibliothek 44-45). Leipzig.

GIFFEN, A.E. VAN, 1943. Opgravingen in Drente. In: J. Poortman (ed.), *Drente, een handboek voor het kennen van het Drentsche leven in voorbije eeuwen*. I. Meppel, pp. 393-564.

GROSS, E., 1984. Die Stratigraphie von Vinelz und ihre Ergebnisse für die Chronologie der westschweizerischen Spätbronzezeit. *Jahrbuch der Schweizerischen Gesellschaft für Urgeschichte* 67, pp. 61-72.

HARCK, O., 1972/73. *Nordostniedersachsen vom Beginn der jüngeren Bronzezeit bis zum frühen Mittelalter* I-II (= Materialhefte zur Ur- und Frühgeschichte Niedersachsens 7). Hildesheim.

HARCK, O., 1978. *Das Gräberfeld auf dem Heidberg bei Billerbeck, Kr. Lüchow-Dannenberg* (= Materialhefte zur Ur- und Frühgeschichte Niedersachsens 13). Hildesheim.

HEERINGEN, R.M. VAN, 1986. De late bronstijd en vroege ijzertijd in de Nederlandse delta. In: *Rotterdam Papers 5*. Rotterdam, pp. 27-48.

HÖCKMANN, O., 1976. Zu einem Brüchstück eines nordischen gegossenen Bronzebeckens aus Corcelettes in der Schweiz. *Archäologisches Korrespondenzblatt* 6, pp. 131-136.

HOMMERBERG, S., 1946. *Hallstattidens relativa kronologi. I. Hallstatt A och B*. Lund/Copenhagen.

HORST, F., 1981. Die jungbronzezeitlichen Stämme im nördlichen Teil der DDR. In: W. Coblenz & F. Horst (eds.), *Mitteleuropäische Bronzezeit. Beiträge zur Archäologie und Geschichte*. Berlin, pp. 137-187.

HUNDT, H.-J., 1951. Ein Hortfund der Periode IV von Bäk, Kreis Herzogtum Lauenberg. *Offa* 9, pp. 40-46.

HUNDT, H.-J., 1955. Versuch zur Deutung der Depotfunde der nordischen jüngeren Bronzezeit, unter besonderer Berücksichtigung Mecklenburgs. *Jahrbuch des Römisch-Germanischen Zentralmuseums Mainz* 2, pp. 95-140.

HUNDT, H.-J., 1978. Die Rohstoffquellen des europäischen Nordens und ihr Einfluss auf die Eintwicklung des nordischen Stils. *Bonner Jahrbücher* 178, pp. 125-162.

JACOB-FRIESEN, K.-H., 1956. Zwei jungbronzezeitliche Hortfunde aus Niedersachsen. In: P. Zylmann (ed.), *Zur Ur- und Frühgeschichte Nordwestdeutschlands* (Festschrift zum 70. Geburtstage von K.H. Jacob-Friesen). Hildesheim, pp. 123-143.

JACOB-FRIESEN, K.-H. (bearb. von G. Jacob-Friesen), 1963. *Einführung in Niedersachsens Urgeschichte. II. Bronzezeit*. Hildesheim.

JACOB-FRIESEN, G., 1967. *Bronzezeitliche Lanzenspitzen Norddeutschlands und Skandinaviens*. I-II. Hildesheim.

JOCKENHÖVEL, A., 1971. *Die Rasiermesser in Mitteleuropa* (= Prähistorische Bronzefunde VIII:1). München.

JOCKENHÖVEL, A., 1980a. Tüllenmesser von den Britischen Inseln. *Archäologisches Korrespondenzblatt* 10, pp. 233-237.

JOCKENHÖVEL, A., 1980b. *Die Rasiermesser in Westeuropa* (= Prähistorische Bronzefunde VIII:3). München.

JOCKENHÖVEL, A. & G. SMOLLA, 1975. Le dépôt de Juvincourt-Damary (Aisne). *Gallia Préhistoire* 18, pp. 289-313.

KALTOFEN, A., 1985. *Die ur- und frühgeschichtliche Sammlung des Kreisheimatsmuseums Lingen/Ems: eine Quellenedition*. Hildesheim.

KEMENCZEI, T., 1984. *Die Spätbronzezeit Nordostungarns*. Budapest.

KIBBERT, K., 1984. *Die Äxte und Beile im mittleren Westdeutschland*. II (= Prähistorische Bronzefunde IX:13). München.

KOOI, P.B., 1979. *Pre-Roman urnfields in the North of the Netherlands*. Groningen.

KOOI, P.B., 1986. Het bronsdepot van het Drouwenerveld. In: W.A. Casparie et al. (eds.), *Vondsten uit het verleden: oudheidkundig bodemonderzoek* (= SNA Archeologisch jaarboek). Maastricht/Brussel, pp. 62-63.

KÖRNER, G. & F. LAUX, 1971. *Vorgeschichte im Landkreis Lüneburg*. Lüneburg.

KRAMER, E., 1985. Vroegste bodemvondst uit Leeuwarden. *De Tweebakstrommel* 16 (9), pp. 15-17.

LEVY, J.E., 1982. *Social and religious organization in Bronze Age Denmark: an analysis of ritual hoard finds* (= B.A.R. Intern. series 124). Oxford.

MARIËN, M.-E., 1952. *Oud-België. Van de eerste landbouwers tot de komst van Caesar*. Antwerpen.

MARIËN, M.-E., 1984. Couteaux de l'Age du Bronze final découverts dans la grotte de Han-sur-Lesse. In: *Éléments de pré- et protohistoire européenne* (Hommage à Jacques-Pierre Millotte). Paris.

MENKE, M., 1972. *Die jüngere Bronzezeit in Holstein: topographisch-chronologische Studien. Typographisch-chronologische Studien* (= Offa-Bücher 25). Neumünster.

MÜLLER-KARPE, H., 1959. *Beiträge zur Chronologie der Urnenfelderzeit nördlich und südlich der Alpen* (= Römisch-Germanische Forschungen 22). I-II. Berlin.

MÜLLER-KARPE, H., 1961. *Die Vollgriffschwerter der Urnenfelderzeit aus Bayern* (=Münchner Beiträge zur Vor- und Frühgeschichte 6). München.

MÜLLER-KARPE, H., 1974. Zur Definition und Benennung chronologischer Stufen der Kupferzeit, Bronzezeit und älteren Eisenzeit. *Jahresbericht des Instituts für Vorgeschichte der Universität Frankfurt a.M.*, pp. 7-18.

PESCHECK, CHR., 1949/50. Originale in der Sammlung des Seminars für Ur- und Frühgeschichte an der Universität Göttingen. *Germania* 28, pp. 21-25.

PETRESCU-DÎMBOVIŢA, M., 1978. *Die Sicheln in Rumänien, mit Corpus der jung- und spätbronzezeitlichen Horte Rumäniens* (= Prähistorische Bronzefunde XVIII, 1). München.

PLEYTE, W., 1882. *Nederlandsche Oudheden van de vroegste tijden tot op Karel de Groote: Drenthe*. I-II. Leiden.

PRÜSSING, P., 1982. *Die Messer im nördlichen Westdeutschland* (= Prähistorische Bronzefunde Europas VII:3). München.

RUOFF, U., 1974. *Zur Frage der Kontinuität zwischen Bronze- und Eisenzeit in der Schweiz*. Bern.

RYCHNER, V., 1979. *L'âge du bronze final à Auvernier* (= Cahiers d'archéologie Romande 15-16; Auvernier 1-2). Lausanne.

RYCHNER, V., 1986. Les eaux-vives à Genève. Aspects de la métallurgie lémanique à l'âge du bronze final. *Genava* N.S. 34, pp. 69-80.

SCHUBART, H., 1972. *Die Funde der älteren Bronzezeit in Mecklenburg*. Neumünster.

SPROCKHOFF, E., 1931. *Die germanischen Griffzungenschwerter* (= Römisch-Germanische Forschungen 5). Berlin/Leipzig.

SPROCKHOFF, E., 1932. *Niedersächsische Depotfunde der jüngeren Bronzezeit* (= Veröffentlichungen der Urgeschichtlichen Sammlung des Provinzial-Museums zu Hannover 2). Hildesheim/Leizpig.

SPROCKHOFF, E., 1937. *Jungbronzezeitliche Hortfunde Norddeutschlands (Periode IV)* (= Kataloge des Römisch-Germanischen Zentralmuseums zu Mainz 12). Mainz.

SPROCKHOFF, E., 1941. Niedersachsens Bedeutung für die Bronzezeit Westeuropas. *Bericht der Römisch-Germanischen Kommission* 31, Teil II, pp. 1-138.

SPROCKHOFF, E., 1950-51. Pfahlbaubronzen in der Südzone des nordischen Kreises während der jüngeren Bronzezeit. *Archaeologia Geographica* 1, pp. 120-128.

SPROCKHOFF, E., 1952. Zur jüngeren Bronzezeit von Hol-

stein. *Offa* 11, pp. 118-133.
SPROCKHOFF, E., 1956. *Jungbronzezeitliche Hortfunde der Südzone des Nordischen Kreises (Periode V)*. I-II. Mainz.
SPROCKHOFF, E., 1963. Das Hügelgräberfeld von Schnega. *Prähistorische Zeitschrift* 41, pp. 1-50.
SPROCKHOFF, E. & O. HÖCKMANN, 1979. *Die gegossenen Bronzebecken der jüngeren nordischen Bronzezeit*. Mainz.
STRUVE, K.W., 1979. *Geschichte Schleswig-Holsteins. II:2. Die jüngere Bronzezeit*. Neumünster.
TACKENBERG, K., 1971. *Die jüngere Bronzezeit in Nordwestdeutschland. I. Die Bronzen* (= Veröffentlichungen der Urgeschichtlichen Sammlung des Landesmuseums zu Hannover 19). Hildesheim.
TACKENBERG, K., 1985. *Westfalen in der Urgeschichte Nordwestdeutschlands. 45 Fundkarten vom Paläolithikum bis in der Zeit um Christi Geburt*. Münster.
THRANE, H., 1960. Drei dänische Hortfunde der jüngeren Bronzezeit mit einheimischen und fremden Formen. *Acta Archaeologica* 31, pp. 1-62.
THRANE, H., 1963. De første broncebidsler i Mellem- og Nordeuropa. *Aarbøger*, pp. 50-99.
THRANE, H., 1972. Urnenfeldermesser aus Dänemarks jüngerer Bronzezeit (Periode IV-V). *Acta Archaeologia* 43, pp. 165-228.
WAALS, J.D. VAN DER, 1987. *De kolonisatie van het terpengebied. Een ethno-archeologische benadering* (= Kroon-voordracht 10). Amsterdam.
WILHELMI, K., 1983. *Die jüngere Bronzezeit zwischen Niederrhein und Mittelweser* (= Kleine Schriften aus dem Vorgeschichtlichen Seminar Marburg 15). Marburg.
WIT, C. DE, 1964. De prehistorie van onze kuststreek. *Westerheem* 13, pp. 2-7.

APPENDIX I. Drouwenerveld 1984: Description of the objects

All of the objects (pot and bronzes) were acquired by and are inventorized in the Drents Museum, Assen, under the numbers 1984.XII.1-70. A summary list, in numerical order, has previously been published (*Nieuwe Drentse Volksalmanak* 103, 1986: pp. 147-150); in the following the objects are grouped by type.

The *pot* (No. 30), c. 2/3 preserved and restored. Of thin-walled, burnished brown fabric, with fine sandy and crushed-granite grits. S-shaped outline. Single strap-handle preserved, 4 x 1.8 cm, descending from rim. The circular perforation has a diameter of 1.5 cm. The lower half is decorated with a dense lattice pattern, incised with a round-ended instrument. There is a very slight base protrusion. Height 15.5 cm; diameter at rim 14 cm; body width 14.5 cm; base 8 cm.

A grain impression in the pot wall has been identified by van Zeist (verbal comm.) as bread wheat (*Triticum aestivum*).

The *bronzes* (all with dull green patina, somewhat encrusted with sand cemented by iron pan):

Socketed axe (No. 4): complete and well preserved. Mouth opening oval, from which springs a moderate-sized D-shaped loop. Axe body of oval cross-section. Each face has an arch-shaped flat facet, the lower edge of which is the cutting edge. The tips of the blade edge have been hammered flat so as to continue the straight line of the sides, which are parallel. On one side of the axe there are two thin horizontal ribs on the neck; these ribs are absent on the other side. There are prominent casting seams, partially smoothed by hammering. There is a horizontal rib inside the socket, at its base (= Baudou's *Stützleiste* Type A). The cutting edge is still sharp in part. There is a modern plough nick on one side. Length 8.9 cm; width 4.5 cm (with loop); blade width 3.4 cm; weight 197.5 g.

Socketed axe fragments:

(No. 3) Fragment of upper part of an axe. Single socket-mouth moulding; curved ribs and grooves on face. All breaks patinated. Length 4.8 cm; width 2.0 cm;

(No. 17) Small fragment, with groove and rib. Length 2.25 cm; width 1.7 cm;

(No. 24) Rim fragment, with single flattish moulding. Length 3.4 cm; width 2.45 cm; thickness 0.4 cm;

(No. 28) Body fragment, with part of face-side angle. The axe was apparently of hexagonal cross-section. On the face are parts of flat-bottomed grooves. Length 3.6 cm; width 2.2 cm;

(No. 44/4) Rim fragment. Single biconical socket-mouth moulding. Length 2.4 cm; width 2.15 cm;

(No. 58) Rim fragment, with D loop springing from single, flattish mouth moulding.

Knives

Tanged single-edged knife (No. 15): the more or less triangular tang is of rectangular section, and has a slit-like perforation (0.7 x 0.2 cm) by which handle-plates could have been fastened to the tang. The slightly concave back of the blade is rounded in section. The concave cutting edge suggests that there has been considerable re-sharpening. The knife was, presumably, bent so that it would go into the pot; the break is, however, recent plough damage. Original length c. 22 cm; maximum width 2.8 cm; thickness 0.4 cm.

Socketed single-edged knife (No. 30): truncated-conical socket; peg-hole top and bottom. Blade with curved, blunt rounded back. Incised decoration at base of socket, at junction of blade and socket, and on blade back. The cutting edge is barbed at its base. The tip of the blade was anciently broken off (break patinated). The lower part of the blade is slightly hollowed from sharpening. Length now 10 cm; width 1.4/1.9 cm.

Two-edged knife (blade fragment; No. 40): lozenge section; edges parallel. Length 2.5 cm; width 2.1 cm; thickness 0.6 cm.

Sickle fragments

(No. 9) Blade fragment. Flat on one face, a groove on the other. The back is concave and sloping; the convex edge is sharp. Length of fragment 7.6 cm; width 1.7 cm; thickness 0.5 cm. (A similar fragment is in the hoard from Bäk).

(No. 66) Blade fragment. Curved blade with slightly recurved tip. Lug on the back; groove on the blade face. Break ancient (patinated). Present length 6.3 cm; width 1.0 to 1.5 cm; thickness 0.2 cm.

(No. 21) Fragment from middle of blade; flat on one face, the other with a single, rounded backing rib. Breaks ancient (patinated). Present length 3.85 cm; width 1.6 cm; thickness 0.3 cm.

(No. 37) Fragment from middle of blade; flat on one face, two ribs on the other. Breaks ancient (patinated). Present length 3.65 cm; width 2.4 cm; thickness 0.6 cm.

Chisel fragments

(No. 5) One end straight and sharp; body of rectangular section, widening gradually; slightly bent. break ancient (patinated and corroded). Present length 5.15 cm; width 0.7 x 0.4 cm; length of cutting edge 0.45 cm.

(No. 56) Tapering, slightly curved shaft with flat faces, rounded sides. Preserved end is blunt. Break ancient (patinated and corroded). Present length 7.0 cm; width 0.8 cm; thickness 0.4 cm.

Spearheads

(No. 11) Complete but broken; the break is recent (unpatinated). Small example without peg-holes; leaf-shaped blade; a slight rib strengthens the socket-mouth opening. Part of the socket was anciently repaired by casting on (*Überfangguss*). Length 10.8 cm; width 3.4 cm. Inside the socket are remains of a brownish substance, perhaps pitch, apparently intended to hold the spearhead fast to its shaft.

(No. 1) Middle section (now 4.8 x 3,0 cm). Socket has been hammered flat. Breaks ancient (patinated).

(No. 10) Middle section (now 2.7 x 2.5 cm). Breaks ancient (patinated).

(No. 20). Basal section (now 3.7 x 3.0). Slight rib strengthens the socket-mouth. Short socket part (only 1.9 cm long). The socket has been hammered flat. Breaks ancient (patinated). Greyish matter inside socket.

(No. 27) Middle section; very small fragment. Length 1.9 cm.

(No. 52) Middle section (now 2.8 x 2.1 cm).

(No. 67) Middle section (now 3.2 x 2.9 cm). Breaks ancient (patinated). Socket tube hammered flat.

Collar

(No. 69) Almost complete, but recently broken into three pieces and deformed by the plough. Only a small fragment, about one cm square, is missing. The collar was originally cast with one side flat and the other side longitudinally ribbed. The reverse side still shows traces of casting roughness; the ribbed side was finely finished. The two outer ribs run round the ends, and thus form one continuous edge rib, given an imitation-cord pattern by diagonal nicking with a punch. The three horizontal ribs inside stop short of the ends, thus leaving end-panels, each of which bears a large incised X and a small, irregularly triangular perforation. The three inner ribs are also nicked, but not continuously, rather with an alternation of nicked and plain zones.

It is not easy see what damage might pre-date the modern plough damage, but it obviously could not have fit into the pot unless it had previously been folded. Flattened length 32.1 cm; width 4.2 cm.

Tutuli

Four virtually identical examples. Hollow conical body with stepped external surface, surmounted by thin shaft with a mushroom-shaped cap. Inside, a horizontal flat bar. Cast in one piece, apparently by *cire-perdue*.

(No. 14+19) Width 2.3 cm; height 2.4 cm; (No. 16+34) Width 2.2 cm; height 2.25 cm; (No. 47) Width 2.45 cm; height 2.55 cm; (No. 51) Width 2.4 cm; height 2.35 cm.

Buttons

Twenty examples; some dome-shaped, other flattish with turned-down edges; all with small cast D-shaped loop on the underside. They vary from c. 22 mm to 31 mm in diameter.

Dome-shaped: (No. 2) diameter 2.8 cm; (No. 13) diameter 2.3/2.4 cm; (No. 49) diameter 2.4 cm; (No. 57) diameter 2.35 cm; (No. 59) diameter 2.1/2.4 cm; (No. 61) diameter 2.3 cm.

Flattish, with turned-down edges: (No. 6) diameter 2.4 cm; (No. 7) diameter 2.65 cm; (No. 12) diameter 2.9 cm; (No. 23) diameter 2.2/2.35 cm; (No. 25) diameter 2.5 cm; (No. 32) diameter 2.95 cm; (No. 33) diameter 2.35 cm; (No. 36) diameter 2.65 cm; (No. 38) diameter 2.65 cm; (No. 42a) diameter 2.8 cm; (No. 50) diameter 3.1 cm; (No. 53) diameter 3.0 cm; (No. 54) diameter 2.4/2.6 cm; (No. 60) diameter 2.55 cm.

Bracelet fragments

(No. 39) Thin, of D section; slightly thickened end. Distorted. Break ancient (patinated). Present length 5.3 cm; width 0.6 cm at the break; thickness 0.3 cm.

(No. 65) Thick, of lenticular section; flattened end slightly widened end. Break ancient (patinated). Present length 4.2 cm; width 1.1 cm; thickness 0.65 cm at terminal.

Ribbed tubes

(No. 41) Cast cylinder, with 25 incised grooves, giving the effect of 26 ribs. One end is irregular, the other diagonal. Length 4.6 cm; outer diameter 0.6 cm; perforation diameter 0.35 cm.

(No. 43) Virtually identical with No. 41. More heavily worn than No. 41 and 62. 24 ribs; length 4.5 cm; width 0.55 cm; diameter perforation 0.4 cm.

(No. 62) Ditto; but 26 ribs. Length 4.4 cm; width 0.55 cm; tube diameter 0.3 cm.

Twisted wire fragments

(No. 22) Z-twist; worn. Ends patinated. Length 5.2; thickness 0.45 cm.

(No. 46) Thin, with Z-twist; worn smooth on one side; now with S-curve. Ends patinated. Length 7.6; thickness 0.25-0.35 cm.

Casting jets

(No. 44/1) Three runners with seams. Ends patinated. Length 3.4 cm. (No. 44/2) Single runner, with seams. Length 2.15; width 1.4 cm. (No. 55) Single runner, with seams. Length 1.95; width 1.7 cm. (No. 63) Single runner. Length 1.0; width 2.0 cm.

Miscellaneous fragments

(No. 31) Wavy bar; irregularly rounded section. Ends patinated. Length 7.6; thickness 0.75 cm.

(No. 42b) Sheet fragment.

(No. 45) Irregular cast fragment. Breaks patinated. 2.5 x 1.8 x 0.7 cm.

(No. 26) Irregular cast fragment. Breaks patinated. 3.2 x 1.6 x 0.3 cm.

(No. 35) Irregular cast fragment. Breaks patinated. 1.6 x 1.2 x 0.35 cm.

(Nos 48a, 48b) Two tiny fragments.

(No. 64) Fragment of bar of rectangular section; one end has an irregular swelling. Break patinated. Length 2.4 cm.

(No. 68) Irregular fragment. Breaks patinated. 2.2 x 2.2 x 0.15 cm.

(No. 8) Small fragment (of knife blade?). One edge is sharp. Breaks patinated. 2.5 x 2.5 x 0.5 cm.

(No. 44/3) Irregular triangular fragment. Breaks patinated. Length 2.4; thickness 0.4 cm.

APPENDIX II. List of urnfield socketed knives in the Netherlands and Belgium

Netherlands

Limburg

1. near Venlo. Mus. Maastricht 2434 (formerly Venlo 23). Subrectangular socket.

North Brabant

2. *Gem.* St. Oedenrode. Mus. 's-Hertogenbosch 23.

Gelderland

3. Nijmegen. Mus. Nijmegen 4.1947.4. Purchased from dealer Esser.

4. Bruggelen. Fragment in hoard. Mus. Arnhem, GAS 1958.2.11.

5. Aalten. Mus. RMO Leiden, e.1925.11.1. *Pointillé* on back.

Overijssel

6. Denekamp. Mus. Enschede. Ex coll. Eshuis.

7. Elzenerveen. Fragment in hoard. Mus. RMO Leiden d.1949.12.5.

Drenthe

8. *Gem.* Borger. Drouwenerveld. Hoard. Mus. Assen, 1984 (present paper).

9. Havelte. 'Het Lok'. Hoard. Mus. Assen, 1872.1.16. Incised decoration on blade, blade back, and socket.

10. *Gem.* Schoonebeek. Oud-Schoonebeek. Mus. Assen, 1907.-XI.3. *Pointillé* and incised decoration on back; incised at socket mouth.

11. *Gem.* Schoonebeek. Schoonebeek. Hoard. Mus. RMO Leiden, c.94.11.1. Decoration.

12. *Gem.* Schoonebeek. Schoonebeek. Hoard. Museum, Assen, 1894.XII.1b. Undecorated.

Groningen

13. *Gem.* Onstwedde. Hidsmeden. Mus. Groningen, 1964.-VIII.1. (Butler, 1968/69: pp. 206-223, fig. 2, Pl. 4).

Friesland

14. *Gem.* Ooststellingwerf. Between Weper and Haule. River Kuinder. Mus. Leeuwarden, 218-44.

South Holland

15. *Gem.* the Hague. Voorschoten. 1st Dune. Mus. the Hague HH 26-55 (loan ex coll. Eggink). De Wit, 1964.

Belgium

Prov. Namur

16. Thy-le-Chateau. Mus. Namur, 7536. Unpublished.

17-26. Han-sur-Lesse. 10 examples, from the river Lesse at its exit from the Han cave. One specimen (No. 4708) with two horizontal lines on face; nine (4701, 4702, 4704, 4706, 4713, 4714, 4718, 4720, 4721) undecorated. Mariën, 1984, No. III: pp. 13-22. Mus. Han-sur-Lesse.

THE TWO IRON AGE WOODEN TRACKWAYS XIV(Bou) AND XV(Bou) IN THE RAISED BOG OF SOUTHEAST DRENTHE (THE NETHERLANDS)*

W.A. Casparie

Biologisch-Archaeologisch Instituut, Groningen, Netherlands

ABSTRACT: In the raised bog of Southeast Drenthe, a part of the Bourtanger Moor, six prehistoric wooden trackways have been found, of which two can be dated to the Iron Age. The trackways were built on an ombrogenous peat-bog surface with different types of hummock-hollow systems. XV(Bou), the northern plank footpath near Bargeroosterveld, made of oak, is dated to 530±40 BC. It probably crossed the entire raised bog (c. 12 km), and shortly after its construction it was intentionally destroyed. It is supposed that after a bog burst, dated to c. 500 BC, the peat-bog surface was too dangerous to pass across. XIV(Bou), the hurdle trackway near Emmerschans, dated to 170±50 BC, is constructed of wattlework of willow on a substructure of roundwood of different wood species. The hurdles were made in exactly the same way as modern hurdles. The track was probably used for the transport of bog iron-ore, present in the peat at a distance of about 3 km from the bog margin, to Iron Age dwelling places on the higher sandy soils. This area was intensively occupied in the Iron Age, as is also demonstrated by the large Celtic Field complexes.

KEYWORDS: The Netherlands, Southeast Drenthe, Bourtanger Moor, Iron Age, wooden trackways, prehistoric and modern hurdles.

1. INTRODUCTION

1.1. Prehistoric wooden trackways of Southeast Drenthe

This is the third article about the wooden trackways in the raised bog area of Southeast Drenthe, in the region around Emmen. The first article was concerned with the Neolithic trackway of Nieuw-Dordrecht (Casparie, 1982), the second article dealt with the three Middle Bronze Age footpaths (Casparie, 1984). The present article concerns the two Iron Age bog trackways: a footpath and a trackway made of wattlework (fig. 1). In the previously published articles attention was devoted to various aspects of the geology, the vegetation development, the formation of peat, and the history of the occupation of this region. In the present study this previously published information will be referred to; here we shall confine ourselves to providing supplementary information. No discussion will be presented here of the history of occupation in the Iron Age, apart from a few general remarks. This subject is too broad to be dealt with adequately within the scope of this study, that is concerned with the archaeological information provided by the bog finds. With this third article an integral study of the peat-bog trackways of Southeast Drenthe has been completed.

1.2. The aim of the study and the framework of this paper

The two trackways have been discussed in various publications (i.a. Wesselink, 1924; Landweer, 1912; van Zeist, 1955a, 1955b; van Zeist & Casparie, 1966; Casparie, 1972, 1978; Casparie & Smith, 1978), none of which involved an integral approach. This article is intended to be the final report of the various excavations and of the other, palynological and peat-stratigraphical research, with the aim of explaining the presence of the bog trackways and their (possible) functions. Section 2 gives a broad description of the vegetation development from c. 800 BC until AD 100, and of the development of the bog surface in this period. Section 3 is concerned with the actual trackways; the main topics here include the method of construction, the manufacture of the constructional elements, the course of the trackways, and their destination and function. In section 4 an attempt is made to give a comprehensive picture of all the information and the relationship between the various aspects of these two bog trackways and the rest of the research.

* Figs 10-14, 27 and 28 can be found in the fold at the back of this volume.

1.3. The situation of the trackways: a broad overview

The 'northern plank footpath' XV(Bou) near Bargeroosterveld (fig. 2), that has been dated to 530±40 BC (GrN-4622), was presumably intended

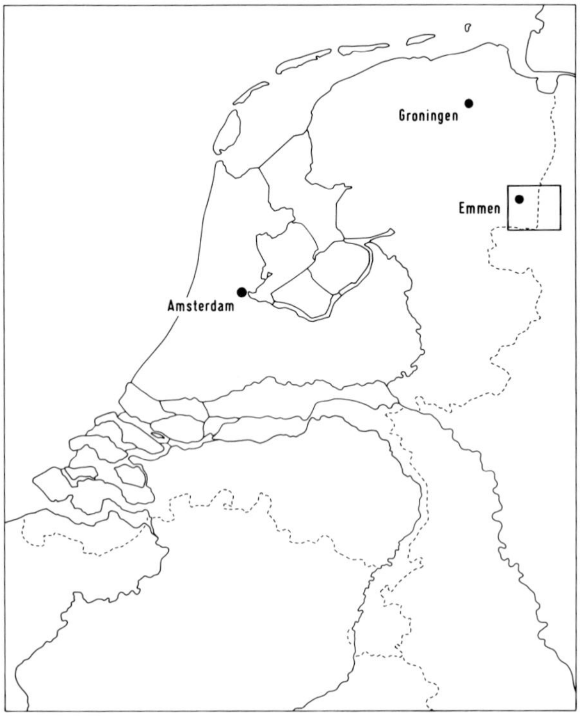

to be a passageway right across the whole Bourtanger Moor (fig. 1), over a distance of some 12 km. Part of it was excavated as early as in 1910. Those parts of it that were excavated between 1952 and 1967 lie 90 m north of the MBA footpath XVI(Bou) (Casparie, 1984). It was built from the Hondsrug, the higher sandy soils west of the raised bog (fig. 4). Traces of occupation of Iron Age date have been found there (B.A.I. excavations, pers. comm. P.B. Kooi).

The hurdle trackway XIV(Bou) near Emmerschans (fig. 3), that has been dated to 170±50 BC (GrN-4147), was intended to provide access to something in this part of the Bourtanger Moor, probably bog iron-ore (fig. 4). The trackway is situated ±750 m north of XV(Bou). It is probably the only prehistoric bog trackway made of hurdles in the Netherlands. The way in which these hurdles were made is very well known, since hurdles produced up until this century were made in exactly the same way as the Iron Age hurdles.

In one of the regions of bog iron-ore a worked oaken plank was found, dated to 575±35 BC (GrN-7899), that is indicative of the exploitation of this ore already in the early Iron Age (Casparie & Smith, 1978) (fig. 5; 3.2.9). Both the construction of the hurdle trackway and this exploitation took place from the Hondsrug.

The (presumed) starting-points of both bog trackways can be situated on the eastern flank of the Hondsrug, the eastern, partly ice-pushed border of the Drents Plateau (fig. 1). This plateau had already

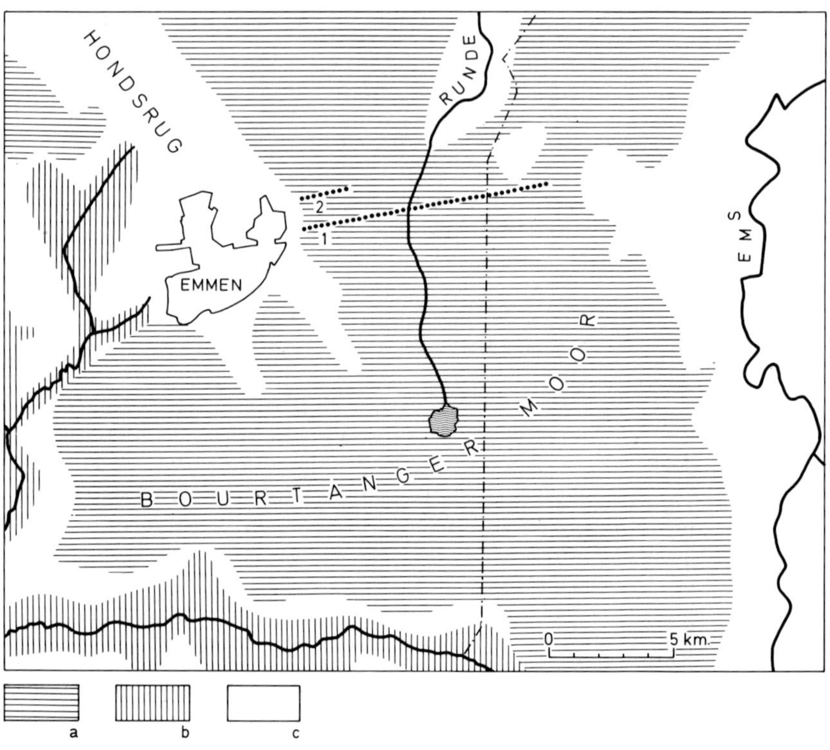

Fig. 1. The original peat-bog situation in Southeast Drenthe, indicating the Iron Age trackways. a. raised bog; b. fen peat; c. higher sandy soil (sand and boulder-clay). 1. The northern plank footpath XV(Bou). 2. The hurdle trackway XIV(Bou).

Fig. 2. Northern plank footpath XV(Bou), trench 3, from 96 m towards the west, June 1963. Photo C.F.D. (= Central Photo Service, State University Groningen).

been occupied by agrarian cultures since c. 3000 BC (Casparie, 1982; 1984).

Groenman-van Waateringe & Smith (1984) describe a leather shoe found in this bog region, which they date to the LBA. The findspot can be indicated rather precisely (fig. 5); this does not correspond exactly to the spot pin-pointed by the authors. Their publication includes a photograph that shows that right next to the spot where the shoe was found part of a bog trackway or footpath is present, although the shoe was found 60 cm deeper in the peat. In spite of the extensive enquiries in this region it was not possible to obtain any definite information about this trackway (pers. comm. Mrs. Smith). In any case the pathway is not part of the northern plank footpath XV(Bou). Nor can it possibly have been a trackway built from the Hondsrug; if that had been the case then we would certainly have found the trackway in the course of the fieldwork of our peat-bog investigations (Casparie, 1972). In view of the height of the pathway in the peat profile it is very well possible that it dates from between 500 BC and the beginning of the Christian Era. It is not inconceivable that it was a particular construction that belonged to the hurdle trackway XIV(Bou); this will be discussed in more detail in 3.2.9.

An extensive investigation based on interviews with the local population, carried out by Mrs. A.F. Smith and Mr. E. Heuving, has shown that another bog trackway is possibly present in this region; this would have run in a direction completely different from that of XIV(Bou) and XV(Bou). On the basis of our knowledge of this bog region, we consider

Fig. 3. Hurdle trackway XIV(Bou), trench 1, towards the east, October 1962. Photo C.F.D.

the existence of such a bog trackway as highly improbable. Consequently this hypothetical trackway is disregarded here.

The ^{14}C-datings used here are non-calibrated values, given in years BP (before present = AD 1950). The datings used regularly in this article are derived from these ^{14}C-datings, and are expressed in years BC, as has been usual in archaeological publications. Now it is preferable to express the non-calibrated datings only in years BP, as was agreed in the last radiocarbon congresses, and to use for calibrated datings cal BC and cal AD. As a consequence of the irregularities in the Radiocarbon calibration curve between 2900 and 1900 BP (radiocarbon age), and using the 1 sigma deviation, the ^{14}C-datings in this paper can be corrected to:

XV(Bou): 800-550 cal BC or 430-400 cal BC
XIV(Bou): 200-50 cal BC
plank (3.2.9): 800-760 cal BC or 680-550 cal BC

The bog burst of c. 500 BC (2.3) is not a radiocarbon date, but the interpretation of 'shortly after 540±60 BC' (Casparie, 1972); this means shortly after 810-560 cal BC or 440-410 cal BC (pers. comm. J.N. Lanting). For the comparison with the radiocarbon dates in the preceding papers on the wooden trackways of Southeast Drenthe, we use uncalibrated ^{14}C-dates in this paper, expressed as BC.

1.4. Acknowledgements

Also this article is mainly the result of the peat-bog research and the excavations carried out by the author together with G. Delger and K. Klaassens and Dr. W.H. Zimmermann (Wilhelmshaven). In addition the following staff members and former staff members of the Biologisch-Archaeologisch Instituut (B.A.I.) cooperated: Mrs. L. van Duinen, Drs. O.H. Harsema, Y. Jongema, Dr. P.B. Kooi, A. Meijer, Drs. J.N. Lanting, Mrs. I. van Otterloo, H. Praamstra ; H.R. Roelink, Prof. J.D. van der Waals, Prof. H.T. Waterbolk, Prof. W. van Zeist, H. Zwier. Assistance was also provided by Mrs. A.F. Smith of Exloo, and E. Heuving of Klazienaveen (in tracing reports of finds made) and Prof. J.C. Vogel (^{14}C-datings). Ir. W.D.J. Tuinzing (Bennekom/Wageningen) provided extensive information about willow-bed cultivation and hurdle-making.

The excavations of the footpath XV(Bou) and much of the peat-bog research took place on terrain belonging to the 'Klazienaveen' firm in Klazienaveen, managing director Ir. K. Meinders and supervisor of the firm G. Hekman. The Koninklijke Nederlandse Heidemaatschappij and Mr. L. van Dijk, State Forestry Service (S.B.B.) in Emmen, provided technical assistance. The Districts Arbeidsbureau (i.e. the local Employment Exchange) at Emmen was responsible for supplying a few labourers on several occasions to help with the fieldwork. With Dr. H. Hayen, of the Staatliches Museum für Naturkunde und Vorgeschichte in Oldenburg, many aspects of the bog trackway research were discussed.

The large amount of drawing was undertaken by G. Delger; H.R. Roelink also did some of the drawing work. Mrs. G. Entjes-Nieborg, B.A.I., typed out the manuscript. Mrs. S.M. van Gelder-Ottway translated the text into English. The author is very grateful to all of these persons for their cooperation.

1.5. Abbreviations

EBA	Early Bronze Age
MBA	Middle Bronze Age
LBA	Late Bronze Age
IA	Iron Age
(Bou)	Bourtanger Moor (= bog); used in the classification of prehistoric (wooden) trackways, according to Hayen (1958).

2. VEGETATION DEVELOPMENT AND PEAT STRATIGRAPHY

2.1. Pollen analysis

2.1.1. *Introduction*

This section gives an outline description of the development of forest between 800 BC and AD 500, as a sequel to the corresponding discussion for the period 3000-1700 BC (Casparie, 1982) and the period 1700-800 BC (Casparie, 1984). Up until 1300 BC a number of *landnams* can be distinguished, that we regard above all as instances of extra emphasis on the reclamation activities and not so much as isolated instances of interference with the vegetation. As a result of the continuing reclamation that went on from c. 3000 BC, deforestation was so advanced by c. 1300 BC that *landnams* can no longer be clearly distinguished in the pollen diagrams. Evidently from this time on a large part of the higher sandy soils in Southeast Drenthe was permanently in use.

The points of departure for the discussion below have been stated in the previous studies on the prehistoric bog trackways (Casparie, 1982; 1984); besides the pollen diagrams mentioned in those studies also the pollen diagram for Bargeroosterveld III is made use of (fig. 6), that was prepared by Y. Jongema. The pollen samples on which this diagram is based were taken in 1967 during the excavation of the MBA footpath XVII(Bou); for the situation see figure 5. This figure also indicates the locations of sampling sites used for the pollen diagrams Emmen I and Bargeroosterveld II, which are referred to here. These pollen diagrams appear in the article on the MBA footpaths (Casparie, 1984: figs 3 and 4); they were prepared by van Zeist (van Zeist, 1959; Waterbolk & van Zeist, 1961).

In the pollen diagrams the *Quercetum mixtum sensu lato* (van Zeist, 1959) is consistently taken as the basis for estimations; broadly speaking this consists of the components of the natural forest on the higher ground. In the Bargeroosterveld III diagram (fig. 6), besides these trees only a few herbs are illustrated, as well as *Alnus*, *Betula* and *Corylus*. The discussion of the development of forest is mainly restricted to the trees. For the situation of the types of forest we have distinguished, see Casparie (1982: fig. 7; 1984: fig. 5); these were severely degraded already in Iron Age times.

On the left side of the pollen diagram (fig. 6) the time scale is divided into zones designated by the capital letters E-J, that indicate the phases of forest reclamation or changes in the pollen picture that have been discerned. E-G have been discussed previously (Casparie, 1984); these concern the Bronze Age. The time spans indicated are approximations of recorded reclamation activities rather than exact datings. By using *Quercetum mixtum s.l.* (van Zeist, 1959) as the basis for estimations the pollen diagram does not give a clear picture of the degree of deforestation, all the more so since only the trees and just a few herbs are illustrated. It is certain, in any case, that the deforestation was considerable already in the MBA.

2.1.2. *The forest history of Southeast Drenthe, c. 800 BC-AD 500* (fig. 6)

H (c. 800-500 BC)

The considerable increase in values for *Fagus*, present in all pollen diagrams for this region, indicates a marked expansion of beech. This almost certainly took place in various places where previously forest had been cut down, but possibly also in abandoned fields. The lower values for *Quercus* are probably partly the result of the way of estimating the pollen percentages, but are also partly due to the fact that oak too was felled

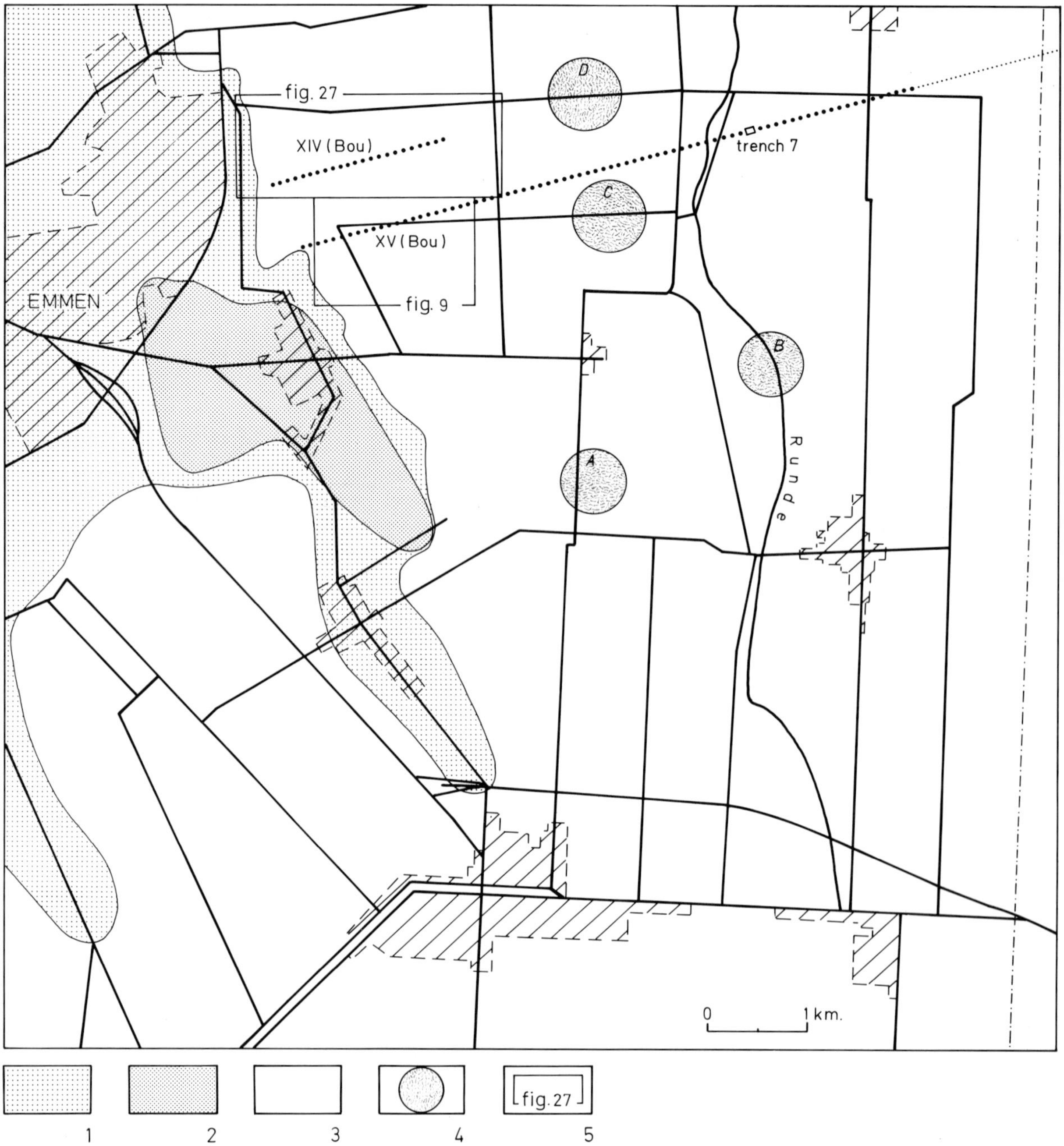

Fig. 4. The peat-bog region of Southeast Drenthe, showing the course of the trackways. 1. sand; 2. boulder-clay (almost) up to the surface; 3. raised bog; 4. bog iron-ore areas (seepage centres A, B, C, D); 5. rectangles of figures 9 and 27. The excavation trench 7 of XV(Bou) is also indicated.

regularly. Besides *Fagus* also *Carpinus* will have been able to become established in suitable places.

Although the pollen diagrams Emmen I and Bargeroosterveld II (Casparie, 1984: figs 3 and 4) indicate that the intensity of occupation may have been less than in the preceding period G (lower values for e.g. *Plantago lanceolata*), this population decline would only have been slight. The presence of urnfields in this region (fig. 5) is indicative of continuing occupation. Also there is no suggestion of any such population decline in the pollen diagram Bargeroosterveld III (fig. 6). In this diagram the *Fagus* maximum is followed by a considerable lower value, that can be dated to 450-500 BC. This general picture is evident in other diagrams for this region (e.g. Casparie, 1972: Emmen 17). This may

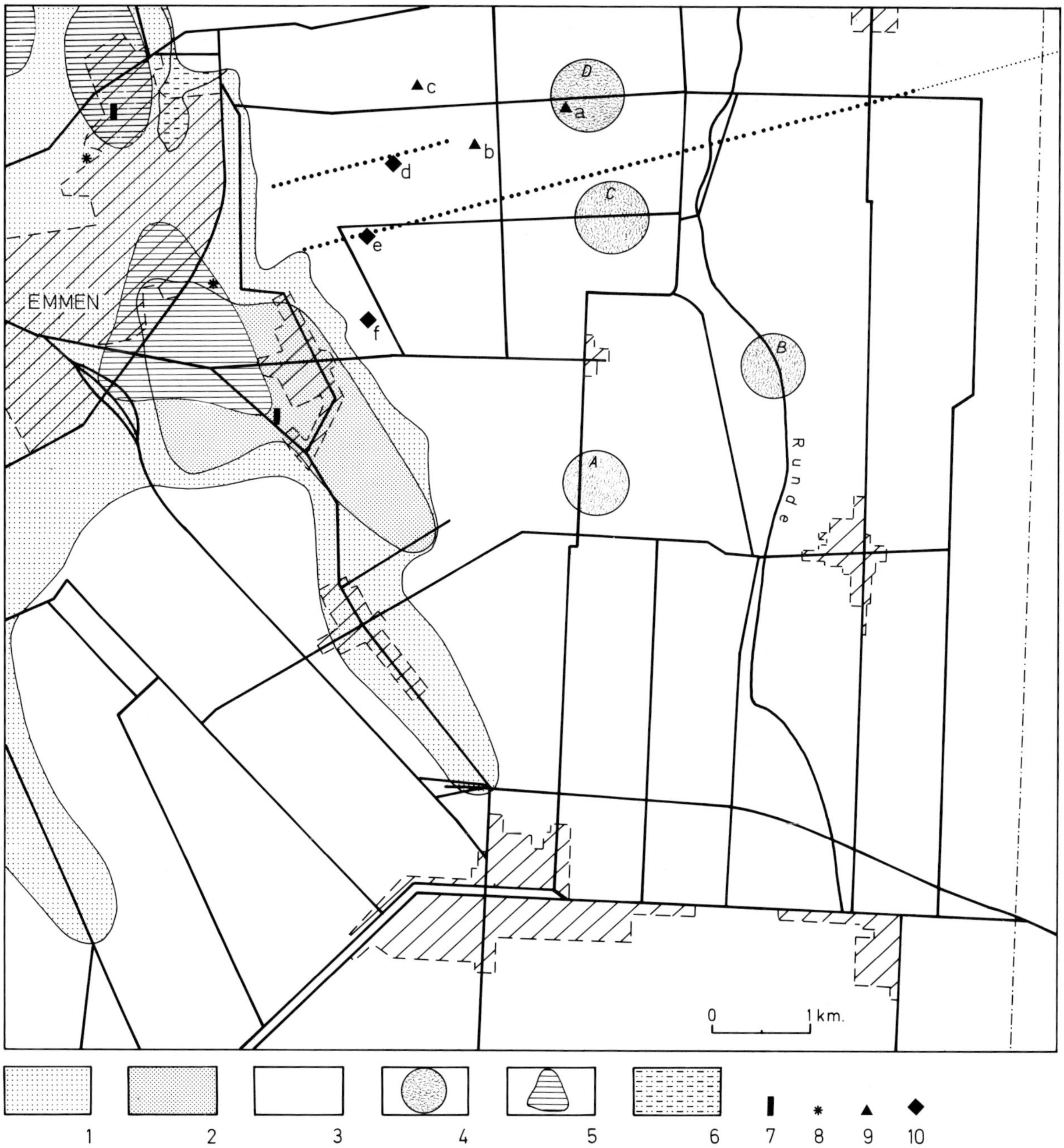

Fig. 5. Some archaeological information relating to Southeast Drenthe and the sampling spots for the pollen diagrams. 1. sand; 2. boulder-clay; 3. raised bog; 4. bog iron-ore areas (seepage centres A, B, C, D); 5. Celtic Fields (after Brongers, 1976); 6. the wet side-valley, the (supposed) area of the willow bed culture; 7. groups of cremation mounds (pers. comm. P.B. Kooi); 8. lumps of iron slag (Modderkolk, 1970); 9. bog finds: a. oaken plank at the bottom of iron-rich seepage peat (Casparie & Smith, 1978), b. plank in poorly humified *Sphagnum* peat (Groenman-van Waateringe & Smith, 1984), c. remains of a peat body (not discussed in this paper); 10. sampling spots for pollen diagrams: d. Emmen I (Casparie, 1984: fig. 3); e. Bargeroosterveld III (fig. 6); f. Bargeroosterveld II (Casparie, 1984: fig. 4). In this study only the diagram Bargeroosterveld III is illustrated (fig. 6).

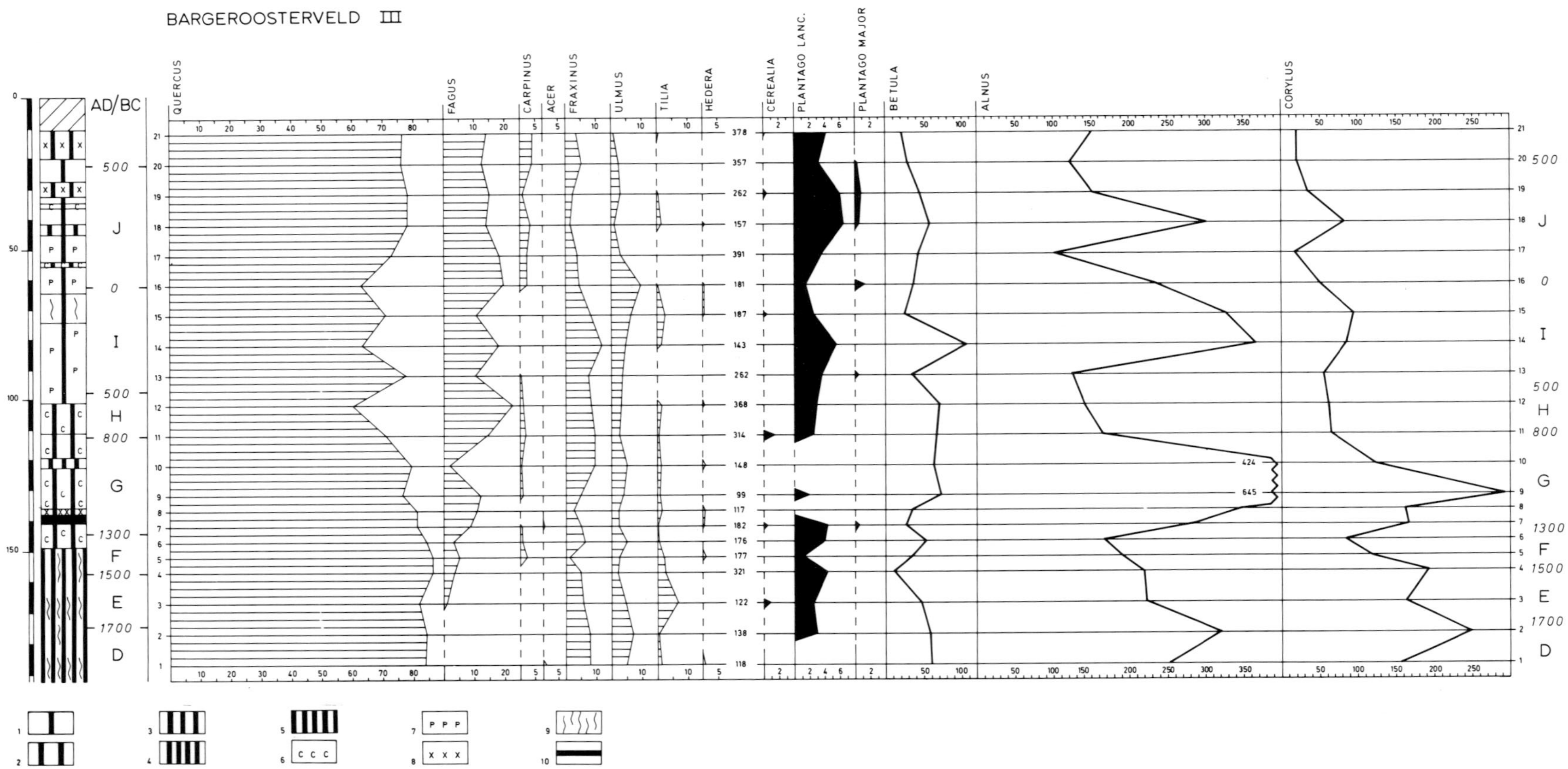

Fig. 6. Pollen diagram Bargeroosterveld III. Situation, see fig. 5. Key (see also Casparie, 1972): 1-5. ombrogenous *Sphagnum* peat; 1. fresh; 2. poorly humified; 3. moderately humified; 4. highly humified; 5. very highly humified; 6. *Sphagnum cuspidatum*; 7. *S. papillosum*; 8. *Calluna*, 9. *Eriophorum vaginatum*; 10. level of the MBA 'southern plank footpath' XVII(Bou).

indicate the selective felling of beech, that would have favoured the expansion of e.g. oak. The proportion of *Tilia* in the forests has become very small in the meantime. The values for *Fraxinus* and *Ulmus* have not changed to any great extent; there is no evidence to suggest that there was any decrease in extent of particularly those types of forest in which these trees dominate.

I (500 BC-0)
The beech remains an important tree, the hornbeam has still not yet expanded to any great extent, and the elm and the ash appear to maintain themselves. The expansion of *Betula* and *Corylus*, together with the increase in *Plantago lanceolata* in Bargeroosterveld III (fig. 6), in 300-200 BC, we associate with the laying out of Celtic Fields in this region (fig. 5). This involved not only the re-use of (old) arable land, but almost certainly also more felling of the remnants of forest still in existence.

The high values for *Alnus* are indicative of a rather extensive bog-fringing forest. It is remarkable that the intensive use of *Salix* for trackway XIV(Bou) (3.2.5, 3.2.7) is not demonstrable in the many pollen diagrams that we have consulted.

J (0-A.D. 500)
In this period *Carpinus* expands, but the hornbeam does not become as common as the beech. *Fraxinus* and *Ulmus* have lower values, which is almost certainly indicative of continuing forest clearance. In the 2nd and 3rd century AD *Plantago lanceolata* has relatively high values. In this period *Betula* and *Corylus* expand once again. The land use in this region is intensive. The occurrence of *Plantago major* in a continuous curve (fig. 6, spectra 18-20) may be indicative of some change in agricultural practice.

2.2. The peat-bog: stratigraphy

The bog trackways XIV(Bou) and XV(Bou) were found in the same ombrogenous peat-bog region as the MBA footpaths XVII(Bou) and XVIII(Bou). The peat stratigraphy of this region has been discussed by us previously in detail (Casparie, 1984: 2.2); for this reason a brief, supplementary discussion will suffice here. The peat stratigraphy near the Iron Age bog trackways is illustrated in figure 7. For the peat profile of XIV(Bou), part of the vertical peat-face C was taken (Casparie, 1972), where the bog trackway intersects with it (3.2.2, trench 3); for XV(Bou) part of the vertical peat-face E was taken, where the pathway intersects with it (3.1.2, trench 4). The peat concerned has shrunk considerably as a result if the artificial drainage since the 16th century AD. The original thickness of the peat deposits could be reconstructed approximately. Generally speaking, the peat was 5-8 m thick below these Iron Age trackways. The thickness of the ombrogenous peat, on top of which the trackways were laid out, varied from 3-5 m. Below this fen peat is present or ferruginous seepage peat (not in the peat sections illustrated, fig. 7), overlying Late Glacial *Braunmoostorf* (Hypnaceae peat) or fluviatile sand.

Both trackways lay in systems of hummocks and hollows, although this cannot be deduced from the peat sections (fig. 7). This is the result of the method used for measuring and drawing the vertical peat-faces, which involved distinguishing three categories of ombrogenous peat: highly, moderately and poorly humified *Sphagnum* peat; due to the exaggeration of the vertical scale in the illustrations of the vertical peat-faces it was not possible to indicate hummocks and hollows or hummock-hollow systems. XIV(Bou) is here situated right next to a hummock (Casparie, 1972: Emmen 12, fig. 44); XV(Bou) lay here in poorly humified hollow-peat.

From the description given by Wesselink (1924) it can be deduced that at the easternmost spot where XV(Bou) was found, on top of the ferruginous seepage peat a layer of fen peat was present, that could correspond to the *Menyanthes-Betula* peat deposit that we have distinguished in this locality (Casparie, 1972); however, the information available is too scanty to permit any meaningful conclusions to be drawn on this matter.

2.3. The peat-bog surface 600-100 BC

The surface of the peat-bog in Southeast Drenthe in c. 1200 BC has been described in detail in our study on the MBA footpaths (Casparie, 1984: 2.2.3). For later periods too the spatial distribution of the types of peat-bog distinguished – especially hummock-hollow systems – can be clearly outlined: in this case for the period 600-100 BC. Here we have chosen the time of c. 500 BC, because then an extensive bog burst occurred in this region; this hardly changed the spatial pattern of bog types, but it nevertheless had a great influence on the hydrology of the bog and consequently also the possibilities of access for people on foot. Figure 8 shows the spatial pattern of bog types; it is almost identical to that of c. 1200 BC. For this pattern and for the description of bog types see Casparie, 1984: 2.2.3 and figure 7. The discussion below is intended as a sequel to the information given in that article. The hummock-hollow systems have already been discussed previously (Casparie, 1972).

Domed complexes. Between 1200 and 500 BC the peat-bog surface grew ±60 cm upwards. The formation of highly humified peat in hummocks was largely superseded by moderately humified *Sphagnum imbricatum* peat, which involved a slight decrease in the proportion of *Calluna* and *Eriopho-*

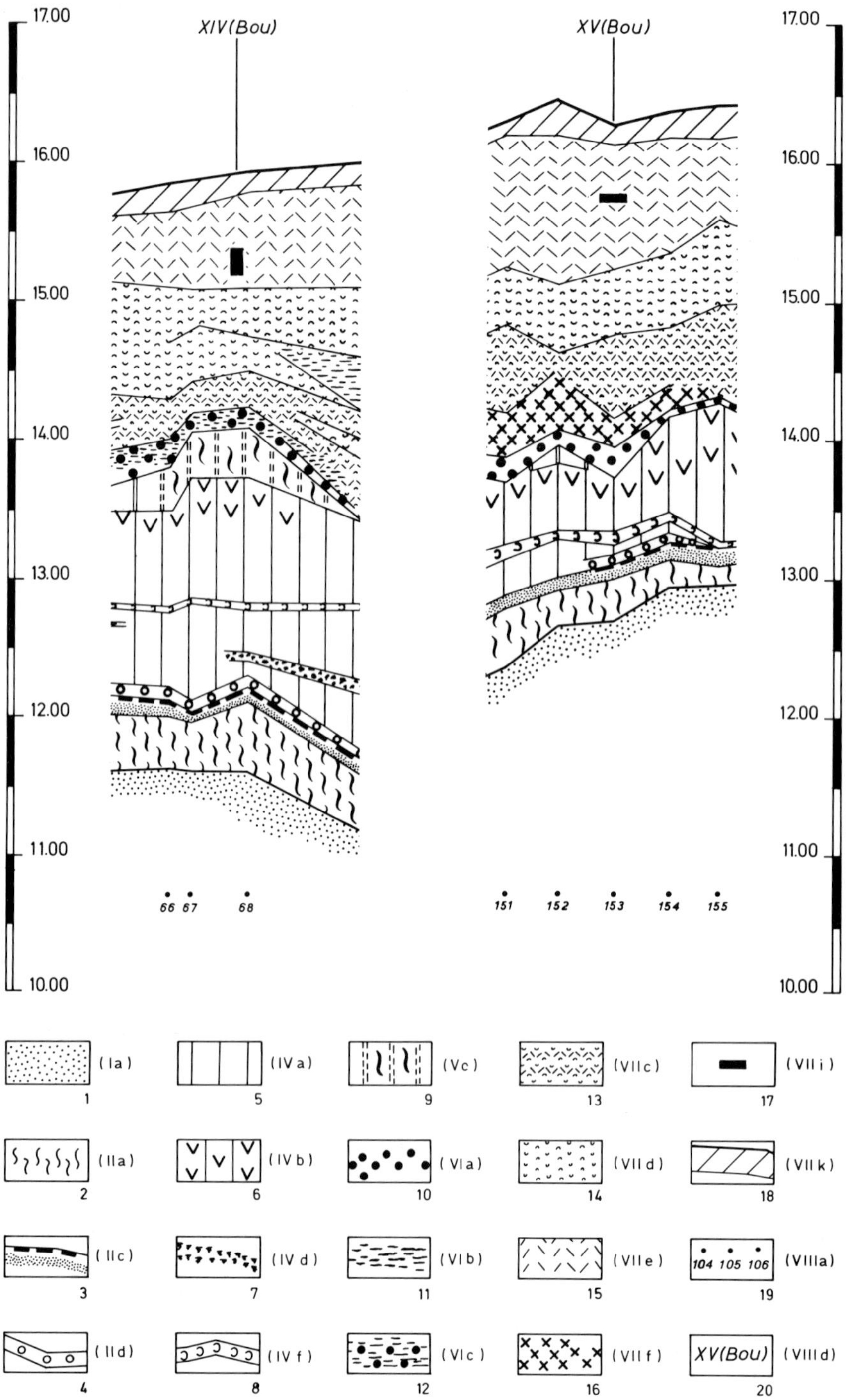

Fig. 7. Peat sections close to the hurdle trackway XIV(Bou) in trench 3, at 3 m (left) and footpath XV(Bou) in trench 4, at 10 m (right), derived from the vertical peat-faces C and E, respectively (Casparie, 1972). Key (the numbers given between brackets refer to Casparie, 1972): 1. sand; 2. *Braunmoostorf*; 3. loess layer; 4. layer of *Betula* twigs; 5. fen peat; 6. fen-wood (carr) peat; 7. charcoal layer; 8. early Atlantic charcoal layer; 9. ferruginous fen peat with siderite; 10. pine stump layer; 11. *Scheuchzeria* peat; 12. *Pinus-Scheuchzeria* peat; 13. highly humified *Sphagnum* peat; 14. intermediate deposits; 15. fresh *Sphagnum* peat; 16. *Calluna-Sphagnum* peat; 17. level of the trackway; 18. secondarily weathered layer, disturbed peat; 19. points examined ('pits'); 20. code for wooden trackway numbering.

rum vaginatum. *Sphagnum rubellum* decreased significantly. In the hollows there grew especially poorly humified *Sphagnum papillosum* peat and to a much lesser extent *S. cuspidatum*. About 100 BC the peat-forming level was ±1 m above that of 1200 BC.

Contact zones. Between 1200 and 500 BC the peat-bog surface grew 40-50 cm upwards; slightly less than in the domed complexes. This applies notably to the hummocks, where the formation of highly humified *rubellum* peat continued to a considerable extent. In the hollows *cuspidatum* peat formed, predominantly in open water. Hummock-hollow systems in contact zones include Emmen 17, Emmen 22+23 and Emmen 32.

The Bargeroosterveld III profile (fig. 6) was

Fig. 8. Outline sketch of the peat-bog surface c. 500 BC, indicating the hummock-hollow complexes distinguished, as described in Casparie (1972). 1. type Emmen 9; 2. type Emmen 17; 3. type Emmen 22+23; 4. type Emmen 32; 5. raised bog lake Emmen 19; the arrow indicates the direction of the erosion of 500 BC; 6. bog-marginal forest. For the meaning of the symbols on the Hondsrug (sand, boulder-clay), see fig. 4: nos. 1 and 2. A, B, C and D: bog iron-ore areas.

sampled through a hollow of the type Emmen 17; also the hollows of type Emmen 32 are characterized by fairly thick layers of *cuspidatum* peat, in which the effects of slight erosion are detectable.

The water content of the contact zones steadily increased, mainly because of increasing precipitation (Dupont, 1986). The water level would have been hardly lower than the hummocks. The hollows were virtually impassable. As a result of the bog burst this situation changed significantly around 500 BC (see below).

The raised-bog lake and the shore zones. This region had become, broadly speaking, one vast body of water, in which extensive patches of floating *Sphagnum cuspidatum* and *Scheuchzeria* vegetations were present. The depth of water in the raised-bog lake Emmen 19 was possibly almost 1 m. In the hollows of type Emmen 22, that had a slightly higher bottom, the depth of water was probably about 60 cm. The peat that had formed was mostly poorly humified. The moderately to highly humified hummocks, type Emmen 23, hardly projected above the water.

The siderite areas. In c. 1200 BC these still projected slightly above the ombrogenous bog surface; around 500 BC this was probably no longer the case. Concerning the siderite area D (fig. 8), it is certain that bog iron-ore was exploited here already before 500 BC (Casparie & Smith, 1978). The siderite area C seals off, as it were, the west-east contact zone to the west of this area; it may have impeded the discharge of water from this contact zone. In any case the raised-bog lake Emmen 19 had no run-off; the water level rose continually until 500 BC.

The bog burst of 500 BC. The sudden drainage of the raised-bog lake Emmen 19 and its shore area Emmen 22+23 has been described by me in detail (Casparie, 1972). The erosion gulleys that formed as a result were more than 1 m deep, possibly up to almost 2 m. We assume that as a result of the increasing water pressure of the raised-bog lake, with an increasing water surplus on the bog surface, the weakest bank of the lake collapsed. The winning of bog iron-ore in the siderite area C may have been partly responsible for this catastrophic discharge. There are, however, no concrete indications of this.

The raised-bog river, the Runde. This raised-bog rivulet developed in the south-north situated contact zones (Casparie, 1972). In view of the large size of the area with erosion phenomena (several square kilometres), the water discharge in 500 BC must have been enormous. We consider it very likely that it was precisely because of this bog burst that the actual valley of the Runde became cut out in the peat; if this happened then any bog trackways that may have been present here would have been washed away.

Bog marginal area. In this belt of forest around the treeless raised bog *Alnus* and *Pinus* were the most important trees, while *Betula* and (in the drier parts) *Quercus* also occurred. The choice of wood for XIV(Bou) and for XV(Bou) indicates that this bog-fringing forest was exploited hardly or not all for the building of the trackways.

The development of the abundance of water in the bog. If we compare the method of construction of the MBA plank footpath XVII(Bou) with that of the IA plank footpath XV(Bou), it is evident that the bog region became considerably more difficult to get across between 1200 and 500 BC. This was undoubtedly the result of the increasing abundance of water that caused a decrease in the carrying capacity of the bog surface.

The footpath XVII(Bou) has for each longitudinal plank (±2.75 m long) one transverse timber, that gives the path sufficient stability. Footpath XV(Bou) has for each longitudinal plank two (fig. 10) or even three transversals (fig. 14): one with holes for the pegs and either one or two intermediate transversals, see also 3.1.4. This last-mentioned situation, that has been found only by Wesselink (1924), was present very close to the Runde (fig. 5), that at that time almost certainly functioned as a contact zone. Possibly the bog surface was so weak here that the longitudinals needed the support of transversals every 90-95 cm. The method of construction of the hurdle trackway XIV(Bou) could indicate that the bog became more consolidated after 500 BC (fig. 28).

After the bog burst of 500 BC. The sudden discharge of water from the raised-bog lake Emmen 19 and other raised-bog lakes was accompanied by the loss of many minerals from the bog. The bog in its entirety became considerably more oligotrophic. The desiccation patches were only of limited extent and were certainly not very intensive. The result of these changes in the bog surface was a marked increase in the growth of fresh and poorly humified *Sphagnum* peat: *S. papillosum* in the erosion gulleys, in which water stagnated again and especially *S. imbricatum* grew on the uneroded bog surface. It is not fully clear how the contact-zone areas now became passable; it is certain, however, that the domed complexes in their entirety contained less water as a result of the drainage by the bog burst. Consequently the somewhat more consolidated peat-bog surface have become more easily passable. This lasted in any case until a few centuries after the beginning of the Christian Era.

3. THE BOG TRACKWAYS - METHOD OF CONSTRUCTION AND THE CONSTRUCTION ELEMENTS

3.1. The northern plank footpath XV(Bou)

3.1.1. *Overview of the field research*

The footpath was possibly known to exist already at the beginning of the 19th century (Westendorp, 1819). The reference to 'a wooden path in the Emmererfscheidenveen' is too vague to permit us to identify this as the footpath XV(Bou). In the 20th

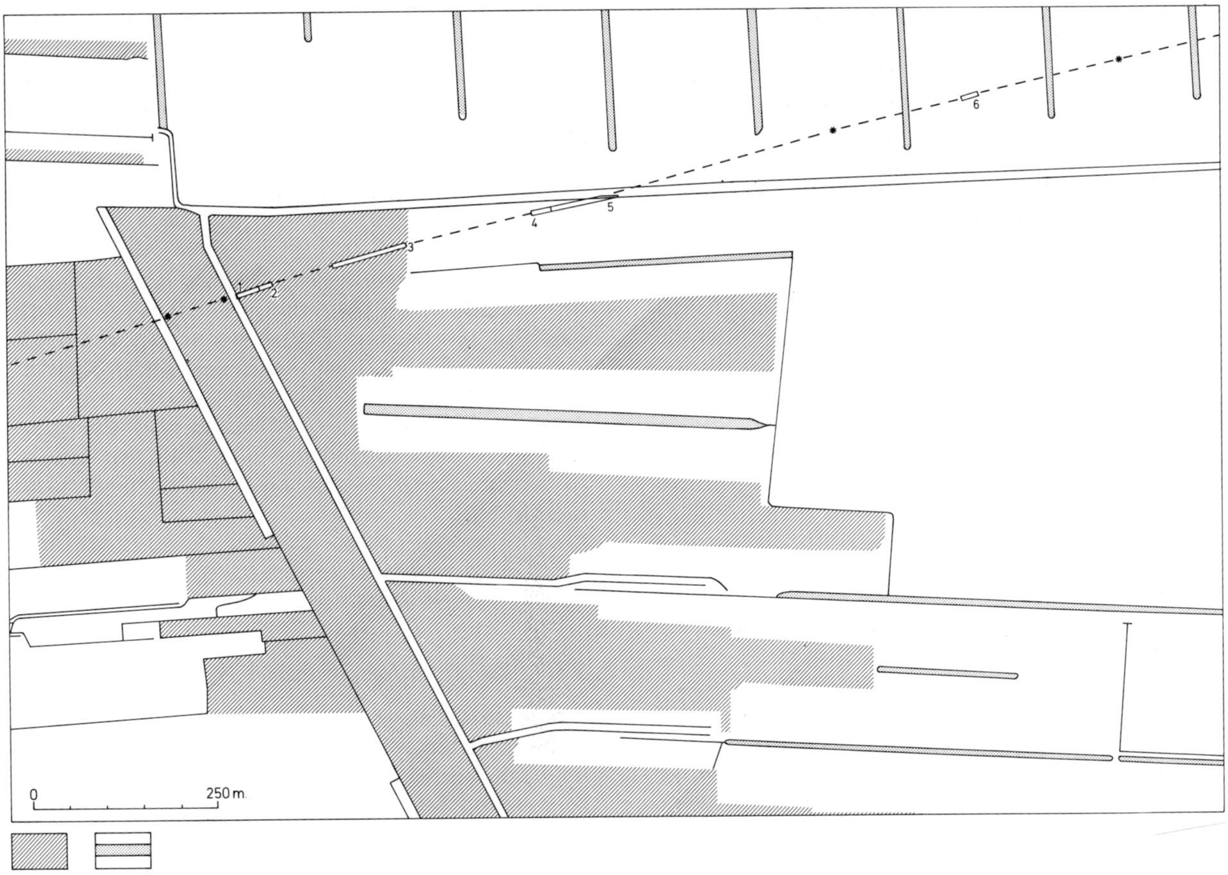

Fig. 9. Northern plank footpath XV(Bou). Location of the excavation trenches 1-6 and the observations points, indicated by an asterisk. The definite, unexcavated part of the course is indicated by a broken line; the path continued further to the west and to the east. For the precise location of this map and of trench 7, see fig. 4. Key: 1. peat remains; 2. water.

century there have been many reports of this pathway; only those that have been followed up by closer investigation (excavation) are discussed here.

The first excavation took place in 1910 (Wesselink, 1924); in the course of this excavation the remarkable method of construction could be ascertained (fig. 14). The situation of the excavation trenches and the course of the path are shown in figure 9. The other parts of the pathway that have been excavated are shown in figures 10-13. The construction had been disturbed in large stretches of the pathway.

The exposure of those parts of the footpath that we excavated took place as follows. The topmost peat was dug away using a shovel or mechanical digger, until the path had almost been reached. Then the remaining peat was removed with a trowel or by hand, and the elements of the path were cleaned up with a paint scraper and brush. After the elements had been measured and drawn, samples were taken, where necessary; the wood was then removed and the underlying peat dug away down to a depth of c. 30 cm. Throughout the entire procedure photographs were taken of parts of the path.

In trenches 2, 4 and 5 the topmost peat layer had already been removed by machine, by the Maatschappij Klazienaveen, owner of this peat area concerned. As a result of this mechanical digging the wood of the pathway was damaged in many places.

During the previous excavations (1910, 1952) a working method was used that did not differ in principle from the method described above.

1910, trench 7. Wesselink exposed the path in two places, situated ±65 m apart. On the basis of Wesselink's data it is possible to establish the location of these two trenches quite well (fig. 4). The eastern trench was photographed and the part of the trackway exposed in it was drawn schematically (fig. 14). Each trench was about 10 m long.
Aim: to establish the construction method, documentation, information about the peat stratigraphy. Wesselink assembled a lot of information, that

will be made use of in the discussion below.
Wesselink, 1924.

1952, trench 6. Excavation carried out May 7th-12th 1952 by H. Praamstra and W. van Zeist, following reports that the path had been found in the course of peat-digging activities. At that time it was not yet known that the MBA footpath XVII-(Bou) was present in the vicinity. Length of trench 28.00 m (fig. 13). While the excavation was in progress the course of the trackway was established, by means of reconnaissance, over a considerable distance (3.1.5).
Aim: to establish the construction method and the course of the trackway; sampling for palynological research. The construction method could not be established with certainty.
Dating: see 3.1.3.
No finds were made.
Van Zeist, 1955a; 1955b.

1961, trench 4. Excavation carried out September 24th-28th 1961 by H.R. Roelink and W.A. Casparie. In the course of peat research in this region (Casparie, 1972) it became evident that the path had already been destroyed over a considerable length as a result of peat-digging. The part of the path that had not disappeared totally was subsequently excavated. Many longitudinals had already been severely damaged. Length of trench 24.30 m (fig. 12).
Aim: documentation; the construction method could not be established.
No finds were made.
Casparie, 1972.

1963, trenches 2, 3 and 5. Excavations in the months May-July by G. Delger, K. Klaassens, W.H. Zimmermann and W.A. Casparie, during the campaign of the peat research from May-October 1963. The area of peat-land which contained these parts of the footpath as well as portions of the footpaths XIV(Bou) and XVII(Bou) was dug away in the course of 1963. Lengths of the trenches: trench 2: 18.50 m, trench 3: 102.70 m, trench 5: 80.60 m (figs 10-12). Between the trenches 2 and 3 (87 m) the footpath had been severely disturbed.
Aim: to gain insight into hummock-hollow systems of the fossil bog surface, the construction method, traces of use, dating.
Finds: in trench 3 at 83 m a wooden mallet was found, which had undoubtedly been used for driving the pegs into the bog surface (fig. 26).
Dating: a ^{14}C-dating was carried out for a plank of trench 3 at 93-94 m, see 3.1.3.
Casparie, 1972.

1967, trench 1. Excavation carried out June 5th-9th 1967 by K. Klaassens, H.R. Roelink and W.A. Casparie. Length of trench 33.10 m (fig. 10). This trench is adjacent to trench 2.
Aim: to gain insight into hummock-hollow systems of the fossil bog surface and to collect more detailed information on the construction method and the use of the path.
Dating: see 3.1.3.
Casparie, 1972.

1978, a few reconnaissances. Reconnaissances made on November 6th and 7th and on December 20th 1984 by G. Delger, K. Klaassens, L. van Dijk (S.B.B.) and W.A. Casparie in the course of field-work of S.B.B.
Aim: to assess the state of preservation of the wood and the peat for the purpose of enabling the S.B.B. to take measures to look after this archaeological monument.
Dating: samples of the wood were collected for a dendrochronological dating, see 3.1.3.
The reconnaissances are not discussed separately.

3.1.2. *The situation per trench*

The method of construction of the path was found to be consistently unclear throughout all of the field research. Some of the disturbances that were come across had definitely been caused intentionally. These disturbances are of prehistoric age (3.1.6). Without the investigation by Wesselink (1924), it would probably have been impossible for us to fully understand the method of construction.

The path showed a rather pronounced relief, as is usual with prehistoric bog trackways. The variation in height, measured at the transverse timbers, amounted to at most 80 cm per trench. We may attribute about 30 cm of this to the relief of the hummock-hollow surface of the bog. This relief was originally more moderate; it is pronounced due to the shrinkage of the peat as a result of the systematic drainage of the bog since the 16th century AD. Owing to this shrinkage, in the middle of the peat-bog region particularly, approximately from trench 3 on, the level of the footpath has come to lie a few metres lower.

Local subsidence, for example of extremely desiccated old vertical peat-faces, resulted in a yet more pronounced relief, e.g. trench 1, the western 3 m. The heights that were measured do not add anything to our understanding of the footpath; consequently they are not mentioned on the excavation drawings.

In the following descriptions of the trenches the mean thickness of the peat above the footpath is mentioned per trench; variations in thickness per trench are not given.

Trench 1, fig. 10. One of the major objectives with this excavation was to acquire new data on the

Fig. 15. Northern plank footpath XV(Bou), trench 1, towards the west. Most of the longitudinal planks are missing. In the foreground a repaired transverse timber; to the right of it a plank driven into the peat. In the background, close to the farmhouse, is situated the starting-point of the path. June 1963. Photo C.F.D.

method of construction. The path lay ±40 cm below the peat surface. The western end of the trench had subsided over a length of c. 2 m.

It is remarkable that longitudinal planks are almost absent from this trench (fig. 15); over a length of 33 m in this trench only 2 planks were found, both of which had been driven vertically into the peat. The state of preservation of the wood that was found was good. The hummock-hollow system, on which the path was laid out, could be clearly discerned.

Trench 2, fig. 10. The path lay about 40 cm below the peat surface, that showed traces of digging activities in many places. As a result the wood of the path had been damaged to such an extent that accurate observations could no longer be made. Here, too, most of the longitudinal planks were missing; no vertical planks (i.e. planks driven vertically into the peat) were found, either. The hummock-hollow pattern was difficult to see.

Trench 3, fig. 11,. Here the path lay 80-100 cm below the peat surface. In the excavation trench a number of modern drainage ditches are visible. The damage caused by these to the path was relatively slight. No additional (recent) disturbances were present.

In the western half of the trench most of the longitudinals were missing, nor had they been driven into the peat. The eastern part of the trench

Fig. 16. Northern plank footpath XV(Bou), trench 5, from 4.80 m towards the east. Most of the longitudinal planks are missing. July 1963. Photo C.F.D.

contained relatively many longitudinals; most of them were still present in situ (fig. 2). The state of preservation of the wood was fairly good. The hummock-hollow pattern could be clearly discerned.

Trench 4, fig. 12. The presence of the footpath was evident from wood remains found on the vertical peat-face (Casparie, 1972: fig. 24, peat-face E, pit 153). The exact situation and the peat stratigraphy at this spot are given in this way.

When the excavation was carried out the overlying peat had already disappeared; the serious damage was caused by the recent peat-digging activities. Therefore it can only be established that fewer longitudinals were present that one would expect to find on the basis of the number of transversals. No vertical planks were found. The hummock-hollow pattern was not investigated.

Trench 5, fig. 12. The path lay c. 60 cm below the peat surface. The excavation took place during the peat-digging operations; this resulted in some extra damage to the path.

Here too the most of the longitudinals are missing (fig. 16); in fact relatively many vertical longitudinal planks (fig. 17) or transversals were found. Between 64 and 67 m bits of wood refuse were found (figs 12, 18). The method of construction was impossible to determine. The hummock-hollow pattern was clearly visible. Here the path lay right next to a contact zone (2.3).

Trench 6, fig. 13. The path lay 60 cm below the peat surface. On the field drawing the possible presence of a disturbance is indicated in two places. Just what kind of disturbances these are is not known, since no further details are available. Predominantly vertical planks were found. Therefore the method

Fig. 17. Northern plank footpath XV(Bou), trench 5, from 38-53 m towards the west, with a few longitudinal planks driven into the peat. July 1963. Photo C.F.D.

Fig. 18. Northern plank footpath XV(Bou), trench 5, secondarily used wood, found between 64 and 67 m, partly charred. The piece on the right shows numerous traces of cutting. Photo C.F.D.

of construction could not be established.

Having re-examined the field drawing, we think it possible that this excavation revealed also part of the MBA footpath XVII(Bou), about 7 m south of the course of the path excavated here, XV(Bou). This would then be the easternmost point at which the MBA footpath has been encountered (Casparie, 1984).

Trench 7, fig. 14. According to Wesselink (1924), here the path lay about 1 m below the peat surface. The method of construction is clearly discernible here; the stretch of path found here clearly shows the construction method (fig. 19). The state of preservation of the wood is poor according to Wesselink; a photograph published by him is clearly indicative of this.

3.1.3. *Dating*

The path, as it was found in 1952 (fig. 13), was dated by means of pollen analysis (van Zeist, 1955a: pp.

48-49, fig. 16). The age that was then established, 3rd-1st century BC, has turned out to be too young. From trench 3 of the 1963 excavation a longitudinal plank between 93 and 94 m (fig. 11) was selected for a ^{14}C-dating of the youngest annual rings; the dating obtained was GrN-4622: 530±40 BC (not calibrated).

This Iron Age dating is supported by the research on the peat stratigraphy. In many places the path is situated fairly high in a thick *Sphagnum cuspidatum* deposit, the end of which can be associated with the bog burst that can be dated to c. 500 BC (Casparie, 1972). Dupont (1986) has dated this bog burst more precisely to 500 BC (calibrated). The footpath XV(Bou) is therefore somewhat older than this date.

A few transverse timbers, found during the preliminary surveys of 1984 (3.1.1) have been investigated dendrochronologically by Dr. B. Schmidt, Labor für Dendrochronologie, University of Köln. The wood concerned is sapwood, that cannot be correlated with any one of the standard chronologies (pers. comm. B. Schmidt).

3.1.4. *The method of construction*

Here we are concerned with a footpath two planks wide, with transverse timbers used for extra stability. The reconstruction drawing (fig. 19) is based essentially on the data of Wesselink (1924); during the later excavations observations were made repeatedly to see whether the wood remains found fitted in with this construction method. This actually turned out to be the case; only minor deviations from this construction method have been detected.

From information collected by Mrs. A.F. Smith (Exloo) in the region concerned it can be deduced that east of trench 6 (fig. 9) the path probably had a different method of construction, which has not been established in any excavation and that in the immediate vicinity of the rivulet the Runde (fig. 5) a special substructure was present. This has not been established by further research, either. These two points of information will be considered in more detail at the end of this section.

On the basis of Wesselink's data and our own excavations we arrive at the following picture (fig. 19):

The longitudinal planks, that were laid out side by side, are supported at their overlapping ends by a transverse timber with a square hole, present on either side of the overlying planks. Over this a corresponding transverse timber also with two holes was laid out. The plank ends and these 'double transversals' were consistently fixed in the bog surface by means of two pegs. These pegs, that were driven through the square holes, either had a thickened head or were driven somewhat obliquely through the holes into the peat. In this way the ends of the longitudinals were firmly fixed. In addition transverse timbers without holes were used. These served to support the longitudinals between the 'double transversals'. In trenches 1-5, between these double transversals there was always one transverse timber present without holes (figs 10, 11, 12). In trench 6 the situation was not clear; for trench 7 Wesselink mentions that two such transversals are always present between two double transversals (fig. 14). Also small pieces of wood have been used to give extra support here and there. The usable width of the path was 44-50 cm. The plank width measurements correspond to this (3.1.5). The distance between the double transversals varied, with a few exceptions, between 2.67 and 3.36 m, though particularly between 2.85 and 3.20 m. The mean distance was 2.95 m. In prehistoric times the method of construction was probably adapted slightly here and there; the path was cleared away for the most part, not very long after it had been built, see 3.1.7.

Messrs. L. Kuilder and H. Lasker, both now of Emmen, were familiar with the path in the area 200-700 m east of trench 6, in the 1930's, when they lived there. This area was then in the process of being exploited by means of peat digging; they saw the wood remains on a number of occasions. The

XV (Bou)

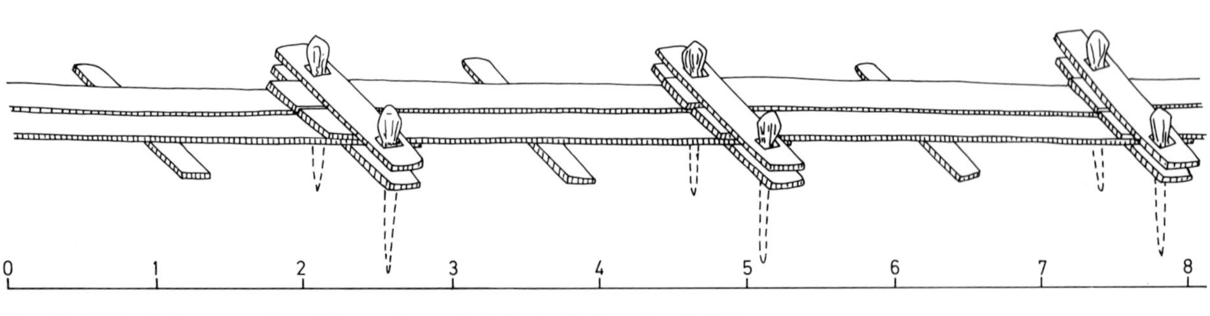

Fig. 19. Method of construction of the northern plank footpath. Drawing G. Delger.

path concerned consisted of planks laid out transversely, each measuring about 1.25 m. At both ends holes had been cut out, through which pegs were fitted, that were driven into the bog. The planks were of oak wood, measuring 25-30 cm in width and 3-5 cm in thickness. Possibly a substructure was present consisting of round wood laid out longitudinally; only one of the eye-witnesses was able to remember this.

From the information they provided it is evident that the bog here was exceptionally wet; this may have been the shore zone of the raised bog lake Emmen 19 (2.3). Mrs. Ottens-Ottens of Emmer-compascuum informed Mrs. Smith that in the 1930's she had seen vertically fixed posts in the bog near the Runde, the pointed ends of which were sticking into the mineral subsoil. These posts stood about 1.5 m apart, and their top ends were connected by a horizontal log. The level of this horizontal part of the construction closely corresponds to the level at which the path XV(Bou) was found in the peat. The direction of the repeated system is thought to correspond to that of footpath XV(Bou). Its precise location could not be ascertained.

We must certainly assume (3.1.8) that near the Runde the path must have had a substantial foundation, e.g. on the basis of the posts driven into the sandy subsoil. For this reason it is not impossible that the information given here above concerns the footpath XV(Bou).

3.1.5. *The construction elements*

The longitudinals and the transversals are all made of cleft wood, namely *Quercus*. After the wood had been cleft the planks and the transverse timbers were cut into shape, where necessary, with the aid of an axe (fig. 20).

The planks were mainly made out of heartwood; some of them had been cleft from the very middle of the tree, as a result of which a longitudinal split had arisen (fig. 21). Only a few planks have remained preserved in their entirety. They varied in length from 3.13-3.82 m. On average they would have measured 3.25-3.50 m. This can be deduced from the distances between the 'double transversals'. The width of the planks varied mostly from 18 to 26 cm, occasionally from 28-32 cm. The thickness of the planks showed considerable variation. Some planks were 1-3 cm thick, some 2-5 cm thick, while others measured 7-9 cm in thickness.

The planks had not been neatly finished off. It is true that irregularities, produced when the wood was cleft, had been cut away with the aid of an axe, but the ends of the planks, for instance, showed little uniformity. In a number of cases notches had been cut in the planks to make them fit between the pegs (fig. 22).

The transverse timbers were also mostly of *Quercus*, as a general rule sapwood, sometimes semicircular cleft tree-trunks of small diameter. In a few cases *Alnus* was used. The transverse timbers with squared holes (fig. 22) were mostly between 100 and 150 cm in length; sometimes they were considerably longer, up to about 2.50 m. The width was generally between 14 and 22 cm, in many cases 15-16 cm. The thickness varied from 1-8 cm. Thicknesses of 1-4 and 5-8 cm occurred regularly. In many cases these transverse timbers consisted of slabwood, the outermost planks of the tree trunk, often with the bark still present. With these slabs

Fig. 20. Northern plank footpath XV(Bou). A few planks driven vertically into the peat, found in trench 5 between 77 and 80 m. Length and width: foremost plank 3.05 x 0.29 m; middle plank 3.45 x 0.24 m; rearmost plank: 3.15 x 0.21 m. The planks showed no traces of use; the cleft planks had been trimmed into shape with the aid of an axe or adze. Width of cutting edge of the axe: at least 7 cm. Photo C.F.D.

Fig. 21. Northern plank footpath XV(Bou). Cutting traces on the transverse timbers found in trench 5. Above: an unfinished hole; below: an incompletely shortened plank.

Fig. 22. Northern plank footpath XV(Bou), trench 3, between 39 and 43 m. Notch in a longitudinal plank to facilitate the fixture of the peg through the underlaying transverse timber. Photo C.F.D.

the flat side of the lower transverse timber usually rested on the peat; the round side between the cut-out holes was then sometimes trimmed flat to make these transversals support the planks more effectively. No essential differences in size, finishing off, etc. could be discerned between the lower and upper transverse timbers. The holes that had been cut out mostly measured 7 x 7 cm (fig. 21). Those transverse timbers that had not been provided with holes had roughly the same characteristics as the perforated transverse timbers; they were, however, more diverse in form. The length varied mainly from 1.07 to 1.80 m; a transverse timber of 3.96 m (fig. 12, trench 5) undoubtedly originates from a longitudinal plank that must have split accidentally. The width was usually 14-20 cm; the thickness varied from 1-8 cm. Some were triangular or semicircular in cross-section.

The smaller pieces of wood, that were used incidentally for providing extra support, could only be recognized as such when they were still in their original position, as in trench 3, at 95 and 96.50 m. These pieces of wood were usually only 50 cm long, 5-10 cm wide and 1-5 cm thick.

Fig. 23. Northern plank footpath XV(Bou). A few transverse timbers from trench 5. Photo C.F.D.

Fig. 24. Northern plank footpath XV(Bou). Pegs from trench 5. Their length varies from 1.10 to 1.70 m. With many of the pegs the thickened head - on the right - is still visible.

Some of the pegs were made of roundwood, especially *Alnus*, with diameters of 5-7 cm. These were found in trench 1 in particular. Most of the pegs were made of *Quercus*, however, especially sapwood. They were cut into their four-sided to seven-sided shape by means of striking blows longitudinally with an axe (fig. 24). The length of the pegs was mainly 85-115 cm; the average length was just over 100 cm. In trench 1 the average length was ±75 cm. The pegs that had been cut into shape with an axe either had a thickened head (±9 cm in diameter) or they gradually tapered towards the point. Thus the pegs could be used effectively for fixing the longitudinals between the transversals. The heads of the pegs probably stuck out 15-30 cm above the footpath.

Freshly cut wood was always used for these construction elements; any secondary wood that may be present does not include construction elements.

3.1.6. *The rest of the wood*

A fairly large number of undefinable pieces of wood were found, that undoubtedly originate for the most part from the construction elements discussed above (3.1.5). These pieces broke off the construction elements at the time when the pathway was purposely destroyed (3.1.7). A few pieces may have broken off as a result of use of the pathway, before it was destroyed. There is little sense in devoting any special attention to these fragments. In addition, as mentioned already above (3.1.5), a few pieces of wood were found here and there that were used to

Fig. 25. Northern plank footpath XV(Bou), trench 3, from 72 m towards the east. In the foreground a slightly damaged double transversal, still holding in place one longitudinal plank. Two *Alnus* branches, one on each side of the path, possibly deposited after the destruction of the path. July 1963.

Fig. 26. Northern plank footpath XV(Bou). Wooden mallet, *Quercus*, found in trench 3 at 83 m. Length of the mallet: 103 cm. Photo C.F.D.

provide extra support (e.g. trench 3, fig. 11, between 62 and 64 m), or that served to fix a peg in position. These pieces of wood can hardly be distinguished from the above-mentioned fragments of construction elements; in view of their irregular occurrence it is clear that they did not consistently form part of the basic structure. They are mainly of *Quercus*; also *Alnus* was found in a few cases. The builders of the pathway would have taken these pieces of wood from the higher sandy soils. These pieces of wood were probably refuse left over after trees had been felled and cut up (3.1.9).

In trench 3 (fig. 11), at 36.40 m, three *Alnus* sticks were found that had been driven vertically into the peat, about 80 cm away from the path. It is not known how far they stuck out above the peat-bog surface, but in view of the limited depth of the stick remains, relative to the level of the path, and in view of the length of the other *Alnus* branches lying in this trench, this was probably 60-90 cm. In trench 3 between 69.0 and 70.10 m an *Alnus* branch lies close to the path. Between 74.0 and 76.10 m, in trench 3, two *Alnus* branches are present, one on each side of the path (fig. 25). The one on the south side lies on top of the broken transverse timber. This can be seen as a weak indication that the branches were deposited here after the destruction of the path. One gets the impression that these branches were used to mark out a route. Small pieces of *Alnus* branches were found in different places in this trench. The wooden mallet of *Quercus* that was found in trench 3 between 82.80 and 83 m (fig. 26) has a more or less round handle and a striking part that is triangular in cross section. Length of the striking part: 50.5 cm; the three sides are 13.0, 13.0 and 11.5 cm wide, respectively. Only on one striking surface the wood has mouldered somewhat as a result of use. The handle is 52.5 cm long; the diameter varies from between 7.0 and 8.5 cm in the thick part, to 5.0-7.0 cm at the thin end.

Also in trench 5 (fig. 12) branches have been found in a few places, for example between 44 and 46 m. Between 64 and 67 m a few pieces of wood were found (fig. 18) that cannot be regarded as construction elements of the pathway, but that nevertheless belong to the path. The three plank fragments of *Quercus* (64.20-64.60 m, 65.80-66.10 m and 66.10-66.40 m) are partly charred. The process of carbonization did not occur on the spot. It is almost certain that this was secondarily used wood; the function of this wood is not at all clear. The same applies to the pieces of a beam with notches (fig. 18), that measures 77.0 cm in length, and in cross-section 9.5 and 5.5 cm. The notches were made using an axe. These pieces of wood were all found lying horizontally.

3.1.7. *The destruction of the path*

Over large stretches of the pathway the longitudinals and the upper transverse timbers are missing. Many pegs have been pulled out. We interpret this as evidence of the intentional destruction of the path by means of removing the longitudinal planks, as is clearly evident in trench 3, or driving the planks vertically into the bog, as was evident in trenches 5 and 6 in particular. Only the eastern part of trench 3, between 92 and 102 m, appears not to have been destroyed in this way.

The planks that have disappeared completely were probably taken away to the settlement on the Hondsrug. Unusable planks and the wood from more distant parts of the path were simply driven into the bog. As a result it became impossible to use this wood again.

In view of the good quality of the wood that was driven into the bog (fig. 17), we may assume that the intentional destruction took place relatively soon after the path had been laid out. These vertical planks showed no signs of weathering on the surface.

We see two possible reasons for this destruction. First of all, it may have happened that the need was no longer felt to communicate with inhabitants on the other side of the bog (3.1.8), or that the destination in the bog no longer required a means of access, so the path could be dispensed with. The other possible reason is that the bog became too dangerous to enter, so that destruction of the pathway was necessary to prevent access to the dangerous spots. The latter we consider the more likely and more appealing explanation. The path lay alongside a very wet contact zone; after c. 1200 BC, when a large lake formed here, the bog region concerned became steadily wetter (Casparie, 1972) and thus more difficult to pass across (2.3). The bog burst of 500 BC, that put an end to this lake, caused so much erosion precisely in this stretch of the route of the path that in the bed of the Runde (fig. 5, 3.1.8), for example, it must have been extremely difficult to pass. The erosion gulleys that formed in the bog were probably almost two metres deep; the width of the eroded area certainly measured a few hundreds of metres. Whether the destruction of the path can be dated to before or precisely after the bog burst remains unknown.

3.1.8. *Direction, starting point, destination, length and function*

From the situation of the excavation trenches and observation points (fig. 9) the direction of the pathway is clearly evident. This is approximately WSW-ENE, like that of the other prehistoric wooden trackways in this region. In the stretch of pathway that is definitely known to have existed, measuring c. 4.5 km in length, there are no appreciable bends. We may therefore assume that the path was laid out as a direct connection, undoubtedly from the Hondsrug (3.1.9, fig. 4). The starting-point must therefore lie ±150 m west of the westernmost spot where the path has been observed, on the east side of the side-valley of the Hunzelaagte (a low-lying area on the Hondsrug), notably at the northern end of this sandhill. This starting-point is about 90 m north of the starting-point of the MBA footpath XVI(Bou) (Casparie, 1984).

On the above-mentioned sandhill no Iron Age occupation has been discerned. To the west of the side-valley the Hondsrug was intensively occupied in the Iron Age, as is evident from the presence of Celtic Fields (Brongers, 1976), from abundant traces of iron-smelting activities (Modderkolk, 1970), from urnfields and groups of cremation mounds (fig. 5; Casparie & Smith, 1978; pers. comm. P.B. Kooi). It is fairly certain that the side-valley, that was a wet, marshy lowland area in Iron Age times, also had some kind of man-made bridging connection. This area was reclaimed so long ago that nothing is known of any such construction (see also 3.2.7).

The extensive interviews made in this region by Mrs. A.F. Smith (Exloo) since 1977, later in cooperation with E. Heuving (Klazienaveen), have made it clear that the path was almost certainly encountered also a few kilometres east of Wesselink's trench No. 7 (fig. 4). Its course could even still be indicated in various places (personal communication Smith and Heuving). The path probably extended as far as the eastern margin of the Bourtanger Moor (pers. comm. H. Hayen). This probable course of the path is shown in fig. 1.

Wesselink's investigation (1924) indicates that the path was certainly intended to provide access to something east of the bog rivulet the Runde. This is also suggested by the information, that is otherwise somewhat vague, about a different, notably more robust building method used in the very wet bog-zone close to the Runde (3.1.4). We think it very well possible that this path was intended as a means of communication with the inhabitants on the east side of the Bourtanger Moor. If this was so, then the path would have measured almost 12 km in length, this being the width of the raised bog area between the Hondsrug and the dune area next to the River Ems (fig. 1).

As this pathway had a walking surface 45-50 cm wide it would have provided a comfortable passage over the domed complexes (2.3); the contact zones and the bog surface near the Runde would have been too wet to allow this construction method as a suitable bridging connection (fig. 18), as is also suggested by the scarce information about the construction method used there (3.1.4).

3.1.9. *The use made of the wood*

From the wood found (planks, transverse timbers, pegs) it can be deduced that oak trees were used measuring ±40 cm, that were 80-140 years old. The sapwood was used for making transverse timbers, as is evident from the many pieces of slabwood, and for making pegs. The heartwood was used for making the planks, which measured 22-30 cm in width.

The oaks undoubtedly came from the higher sandy areas, notably from the Hondsrug. We may assume that the trees used for this pathway were cut down in forests situated fairly close to the settlement. On the Hondsrug the felled trees were cut up into the construction elements: planks, transverse timbers, pegs. Seeing that here and there wood chips were found, it is evident that the construction elements were trimmed to the required size on the actual building site of the path, in the bog. In any case, excessively wide planks were made narrower so as to facilitate placing them between the pegs of the transversals (fig. 22). We have not been able to show with certainty that the square holes in the transversals were cut out during the process of building the path in the bog, but the presence of a few incomplete holes in the transverse timbers are indicative of this.

It has been estimated by Hayen (1979) that for the 3 m wide Iron Age trackway VI(Pr) in the 'Grosse Moor' (Great Bog) near Diepholz (F.R.G.), which was made of transversely laid planks, an area of 42 ha of forest must have been felled in order to build 1000 m of trackway. If we apply his estimation method to the use of wood for our footpath XV(Bou), that required about a quarter as much as wood per unit length, then we find that a footpath 12 km long requires about 120 ha of forest. This is a considerable large area. The construction of this path must have made extremely great demands on the wood supply of this region, that had certainly been deforested to a large extent as a result of the intensive occupation present here for c. 2500 years already. This is an additional indication that the path was built to serve an important purpose, and that its construction could have taken place after very careful consideration and planning.

Particularly in trench 1 pegs of *Alnus* were found (3.1.5). Here and there branches of *Alnus* and *Betula* were come across. It is not certain whether the bog-fringing forest, where both of these trees occur, was used to provide this wood.

3.2. The hurdle trackway XIV(Bou)

3.2.1. *Overview of the field research*

Wesselink (1924) observed this trackway in 1910; subsequently he was able to establish its course. He describes the trackway as consisting of a surface-layer of logs and branches (bundles of rods), supported by a substructure of roundwood (3.2.4). Since the beginning of the 20th century the trackway has been regularly cut through in the course of ditch-digging activities, carried out in connection with the cultivation of buckwheat in this peat area.

During our field research, notably when the vertical peat-face C was being investigated (Casparie, 1972), we came across the trackway. The first excavation took place in September 1962; the trackway surface-layer turned out to consist of hurdles. In October 1962 a second excavation followed (fig. 3). In the summer of 1963 the last remaining part of the road was excavated. The situation of the excavation trenches and the observation points, and the course of the trackway are shown in figure 27. The excavation drawings are given in fig. 28. The method of excavation is the same as that used for XV(Bou) (3.1.1).

1962, trench 3. Excavation of September 7th-12th 1962 carried out by G. Delger and W.A. Casparie. It became evident during the peat research that the trackway in the vertical peat-face C was destined to be cleared away in that month definitely as a result of peat-digging. For this reason this stretch of the trackway was exposed. Length of trench: 3.60 m (fig. 29).
Aim: to establish the method of construction, to take samples for dating purposes, documentation. No finds were made.
Casparie, 1972: figs 22 and 44.

1962, trench 1. Excavation of October 3rd-5th 1962 carried out by A. Meijer, G. Delger and W.A. Casparie. The part of this trackway that was found in vertical peat-face B was destined to be cleared away in the years 1962 and 1963. Here a longer stretch of the trackway was exposed, partly because the first excavation did not permit us to establish the construction method with certainty. Length of trench: 10.70 m.
Aim: documentation, to determine the method of construction, to gain insight into the hummock-

Fig. 29. Hurdle trackway XIV(Bou), trench 3, towards the west. September 1962. Photo H. Kroon.

hollow system of the fossil bog surface.
No finds were made.
Casparie, 1972: fig. 15.

1963, trench 2. Excavation in the months May-July 1963 carried out by G. Delger, K. Klaassens, W.H. Zimmermann and W.A. Casparie, during the campaign of the peat-bog investigation of May-October 1963, shortly before peat-digging at the spot concerned. Length of trench: 38 m.
Aim: documentation, to gain insight into the hummock-hollow systems of the fossil bog surface.

On and next to the trackway-surface various pieces of wood were found, that presumably have something to do with the trackway, either at the time when it was being built or perhaps later, when it was in use. On these pieces of wood no traces of working or other special features are visible. For this reason these bits of wood are not discussed any further.
Casparie, 1972: fig. 40.

3.2.2. *The situation per trench*

Parts of the trackway lay so close to the peat surface that in many places roots had penetrated the trackway structure, notably those of *Molinia caerulea*. Naturally, the presence of roots in the remains of the wooden trackway hampered the excavation, in particular the cleaning of the trackway-surface. This was the case in trench 1. In trench 2 the peat surface had been affected by the recent vegetation to such an extent that the trackway surface had largely decayed into a reddish-brown granular mass, in which many details of the trackway construction could no longer be discerned. This situation was also encountered in two patches of peat, present at c. 180 m and 380 m east of trench 3, respectively. At these places excavation was pointless; they are indicated in figure 27 as observation points.

In trench 3 the state of preservation of the wood of the trackway was excellent; here no serious penetration of roots had occurred into the trackway level.

The road showed a distinct relief, that can be attributed entirely to the hummock-hollow system (Casparie, 1972: figs 40 and 44), although the shrinkage of the peat has certainly made the relief more extreme. The original difference in height between hummocks and hollows would certainly have been some 10-20 cm; locally this difference would have been greater on account of the very weak peat in the hollows (3.2.4). The differences in thickness of the peat above the trackway are indicated only roughly in the descriptions given below. The recent peat-bog surface has originated as a result of its being burnt prior to buckwheat cultivation, which has been practised notably in the vicinity of trenches 1 and 2.

Trench 1, fig. 28. Here the trackway lay 30-80 cm below the peat surface. The wood of the trackway-surface had not been affected by recent penetration of roots to such an extent that observation was difficult. The trackway here was largely undisturbed; the missing hurdle (fig. 3) must have disappeared already before the trackway became overgrown by peat. A distinct pattern of hummocks and hollows was present, that is not shown in the excavation drawing.

Fig. 30. Hurdle trackway XIV(Bou), trench 2, between 21 and 24 m, towards the north. On the northern side of the trackway surface the (broken) *Alnus* plank. The trackway-surface of *Salix* wattlework, situated here on a hummock, has almost completely decayed due to the destructive action of *Molinia caerulea* roots of the recent vegetation. July 1963.

Fig. 31. Hurdle trackway XIV(Bou), trench 2, between 21 and 22 m. Part of the plank shown in fig. 30 with short pegs of *Fraxinus* wood. July 1963.

Trench 2, fig. 28. Here the trackway lay 10-80 cm below the peat surface. In the western part of the trench (0-21 m) the trackway could be observed only incompletely. Nevertheless, it could be established that the trackway had certainly been present here. In the other part of the trench the shape of the construction elements could still be determined satisfactorily. This trench in its totality certainly does not give a complete picture of the trackway. The trackway lay on a distinct hummock-hollow surface, that has been illustrated by us previously (Casparie, 1972: figs 39 and 40).

On the former bog surface, between 25 and 30 m, a peat-bog stream had been present, in which extensive *Sphagnum cuspidatum* vegetations were present. Here the trackway construction had been adapted.

Here too part of the trackway probably disappeared already very long ago; between 25 and 31 m the trackway surface has a width of only one hurdle, while the substructure here is wider.

Trench 3, fig. 28. Here the trackway lay 60-70 cm below the peat surface; it had already been found and drawn during our field investigation (Casparie, 1972: fig. 22, peat-face C, pit 68, and fig. 44, section Emmen 12). The precise location and the peat stratigraphy at this spot are also given in the field

Fig. 32. Hurdle trackway XIV(Bou), trench 2, between 26 and 31 m, towards the north. Trackway-surface hurdle with double stakes, that has subsided between the timbers of the substructure. Below this hurdle lies the foundation of *Salix* bunches. July 1963.

investigation report. The trackway here lay in a hummock-hollow system.

3.2.3. *Dating*

The trackway has been dated by means of pollen analysis (van Zeist, 1955a: p. 48, fig. 15). The age that was thus established, after AD 200, has turned out to be too young. From trench 3, that was excavated in September 1962, a few rods (*Salix*) of a hurdle have provided a ^{14}C-dating: GrN-4147: 170±50 BC (uncalibrated). The rods were 2-4 years old. This Iron Age dating is supported by the stratigraphical peat research (Casparie, 1972). In many places the trackway lies in the poorly humified *Sphagnum papillosum* hollow-peat and over moderately humified hummocks, where *S. imbricatum* is an important peat-forming species.

This dating indicates that this is the latest prehistoric peat-bog trackway in Southeast Drenthe about which information is available. Any peat-bog trackways that were built at an even later date would have disappeared as a result of the burning of the peat surface prior to the cultivation of buckwheat.

3.2.4. *The method of construction*

On a rather irregular horizontal framework of roundwood, hurdles, i.e. mats of wattlework, were laid out (fig. 3). Here and there this method of construction was supplemented in some way, undoubtedly in connection with particular conditions of the bog surface.

The substructure consists of logs, i.e. young tree-trunks cut to the required length, laid out longitudinally and logs laid out transversely; these two sets of logs are not joined together in any way. There is little or no regularity in the distance between these elements of the wooden substructure. In the longitudinal direction 4-8 logs lie alongside one another. This distance between the transversely placed logs varies from ±1.20 m to a few metres. It may be assumed that the logs forming the substructure shifted somewhat with respect to one another during the time when the trackway was in use, though possibly also afterwards. This was particularly clearly evident with a few very weak parts of the peat-bog surface (fig. 28, trench 2, 15-21 m).

The hurdles had been laid out either in a longitudinal direction (in which case two hurdles lay alongside each other) or transversely. It was not possible to discern any regularity in this alternating pattern. It is possible that local circumstances (e.g. hummock with pronounced relief) determined the direction in which the hurdles were laid out. The way in which hurdles are made is discussed in 3.2.6.

The hurdles, which form the trackway surface, had been laid down neatly next to one another. They are not connected to one another, nor are they fixed with pegs to one another, to the substructure or to the peat.

On the trackway-surface there is no covering layer present, e.g. in the form of heather sods.

The method of construction evident in trench 2 has been adapted in two places (fig. 28). Between 21 and 24 m, a plank was found on one side of the

Fig. 33. Hurdle trackway XIV(Bou), trench 2, from 27 m towards the east. The foundation of *Salix* bunches, cut through here in two places, below the hurdles with double stakes, after the removal of the trackway-surface hurdles. July 1963.

trackway, on the short side of two transversely laid out hurdles. Near the ends of this plank holes had been cut out in a simple way (fig. 30), through which short pegs were fixed (fig. 31). This plank probably served to prevent the trackway surface from wobbling on the very pronounced hummock that is present here. In trench 2, between 25 and 31 m, below the hurdles with double stakes present here (3.2.6) and the substructure, filling material consisting of bunches of rods was present, in the bed of the previously mentioned bog stream (3.2.2, subheading trench 2) (figs 32, 33). Evidently the carrying capacity of the peat here was insufficient to allow the hurdles to be supported only by the substructure of logs. These willow rods will be discussed in more detail in 3.2.6.

The hurdles varied in length from 2.60 to 2.80 m; their width varied from 1.10 to 1.40 m. The length:width ratio is roughly 2:1. On either side the trackway-surface projected 20-40 cm beyond the substructure. The trackway-surface probably lay 15-25 cm above the peat surface; the width of the trackway was ±2.70 m. This construction would permit a comfortable passage over the bog, though the trackway surface would be rather springy, as it were.

The previously mentioned (3.2.2) absence of parts of the trackway surface in trench 1 (between 5 and 8.50 m) and trench 2 (between 25 and 31 m) we do not regard as an adaptation of the construction method. In the first place we assume that a hurdle (for some unknown reason) was removed; in the second place it is possible that two hurdles became washed away, after they had subsided into the bog stream.

It is not certain whether the construction was strong enough to take wheeled traffic. No vehicle tracks have been found. Two damaged patches in the trackway surface, in the form of oval holes (trench 2, 2.50 m and 3.70 m, respectively), indicate that it was trodden by animals.

3.2.5. *The construction elements*

For the longitudinally placed substructure elements logs were used consisting of 5-m lengths of tree-trunks of different species of wood. *Alnus* predominates; in addition *Betula*, *Fraxinus* and *Pinus* occur. The logs used were 8-18 cm thick at the base end. The trees were 10-20 years old; as a general rule straight tree-trunks were used. The transversely placed logs of the substructure are more irregular in shape. The length varies from 1 to 3 m; the diameter of the base end is 8-15 cm. The species of wood that have been found are *Alnus* and *Betula*. Also these trees were 10-20 years old. The trees that were used for the substructure would undoubtedly have been present in and near the settlement and along the bog margin.

In trench 2, between 25 and 30 m (fig. 33), an extra foundation of bunches of rods has been found, as mentioned above (3.2.2). The bog stream present here is filled up with *Salix* rods, of the same type as those used for making the hurdles; this is described in more detail in 3.2.6. The branches used were 3-4 cm thick, and mostly 3-4 years old, with lengths varying from 2 m to almost 5 m.

The hurdles were made out of rods of *Salix* measuring 2.5-4 cm in diameter; these rods were 3-4 years old, and measured up to almost 5 m in length.

Fig. 34. Hurdle trackway XIV(Bou), trench 1, general view towards the west. October 1962. Photo C.F.D.

Fig. 35. Hurdle trackway XIV(Bou). The ends of the *Alnus* plank from trench 2, between 21 and 24 m, with the holes cut through them.

The rods had not been peeled; for the making of the hurdles see 3.2.6.

The length of the hurdles varied from 2.60 to 2.80 m, their width from 1.10 to 1.40 m, as is already mentioned (3.2.4). The thickness of these hurdles varied from 5 to 8 cm. When the excavations were in progress it turned out that it was not possible to draw the hurdles in detail. We have had to restrict our attention to the stakes, that form the warp of the wattlework, and the surrounding of the hurdles. The way in which the woven network holds together is clearly visible in figures 29 and 34. A few

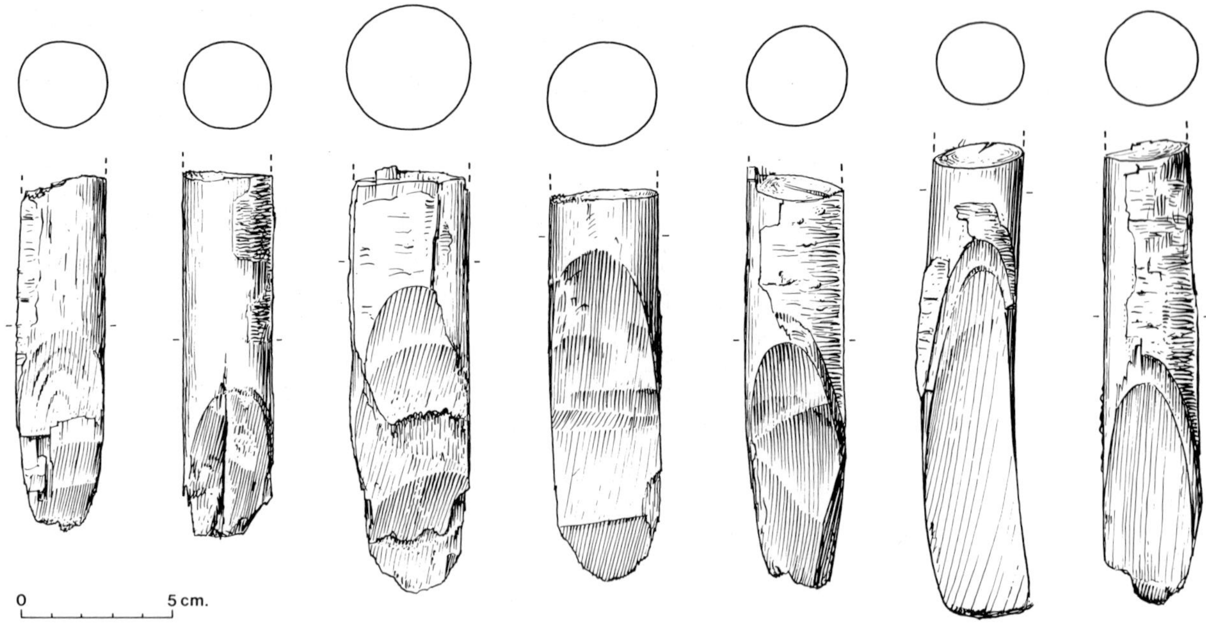

Fig. 36. Hurdle trackway XIV(Bou). Basal ends of the *Salix* rods from the foundation of trench 2, between 26 and 31 m. It can clearly be seen that the branches were cut downwards from the willow stools. Drawing H.R. Roelink.

hurdles had double stakes serving as the warp; these more substantial hurdles were used as a trackway surface-layer over the bog stream (trench 2, 25-30 m), among other things.

The (broken) plank in trench 2, between 21 and 24 m (fig. 30) on the northern side of the trackway had been cut tangentially out of an *Alnus* trunk. It is 2.80 m long, 22 cm wide, and 3-7 cm thick. Close to each end a hole has been made (fig. 35) by means of simply striking a few cutting blows from each side. Sticking through these holes were 4 or 5 short pegs of *Fraxinus*; these pegs were spindle-shaped, 12-15 cm long and 2-3 cm in diameter. They were in an advanced state of decay. Although during the excavation it could not be demonstrated that the plank had been fixed to the adjacent hurdles by means of these pegs, it seems likely, nevertheless, that the plank served to provide support to keep the two hurdle ends effectively in place on the very uneven hummock present here. The plank itself had not been fixed in the peat.

3.2.6. *Hurdle-making*

The hurdles that were used for this trackway are identical to present-day hurdles of *Salix* wood as regards the kind of wood used, the weaving technique and overall dimensions. Therefore they must have been made in the same way. Ir. W.D.J. Tuinzing of Bennekom, an expert on the cultivation of willow beds and the processing of willow rods, provided us with extensive information on hurdle-making. Although we shall refrain from giving a full account of the many aspects of this technique, a detailed description of the way in which the Iron Age hurdles were made is appropriate here. For this purpose we have gratefully made use of Tuinzing's information (in: Westendorp et al., 1985). By studying the rods that were used for the foundation in trench 2, 30-35 m (fig. 33), we are able to obtain a clearer picture of the hurdle-makers of 170 BC, since the same starting material was used for this foundation and for the hurdles.

For making the hurdles four-year-old willow rods (i.e. branches of *Salix*) were mainly used. The diameter at the bottom end varied from 2.5 cm to almost 4 cm. The hurdle consists of an interwoven network with a warp and a weft. The average thickness of the weft elements was almost 3 cm, that of the warp elements about 3.5 cm; the rods used for the foundation measured over 3.5 cm in thickness. From the cut surfaces of the branches it can be seen that these were cut off from above (fig. 36); the branches were held under tension, so that they curved slightly to one side. Then, with one to three blows of an axe or billhook, applied at an angle of 15-25°, each branch was cut through to such an extent that the last part of the cut surface broke off due to the tension. This technique is indicative of the cultivation of willow coppice with low willow stools, that were cleared of their shoots in a four-years' rotation (3.2.7). The wood was not stripped.

Fig. 37. Hurdle trackway XIV(Bou), trench 1, between 4.50 and 8.0 m, towards the south. In the foreground the pointed basal ends of the stakes, that were stuck into the ground when the hurdles were being made. October 1962. Photo C.F.D.

Fig. 38. A modern hurdle of *Salix* rods being made by A. van Weverwijk of Culemborg, near the River Lek in the riverine region of the Central Netherlands. Photo taken on April 14th, 1949 by Ir. W.D.J. Tuinzing. In the foreground the tool kit of the hurdle-maker: wood-cutting knife, a kind of wooden hammer and billhook.

Fig. 39. Modern hurdle as a windscreen for livestock, near Culemborg. Photo taken on July 20th, 1962 by Ir. W.D.J. Tuinzing. Length of the hurdle: 3.55 m; height from the ground to the top of the stakes: c. 1.60 m.

Fig. 40. Hurdle trackway XIV(Bou), trench 1, 8-9 m, towards the east. End of a hurdle with weft rods, bent back 180°, visible (arrows). By twisting these rods around their longitudinal axes the wood fibres can be separated from one another, so that the rods can be turned 180° without breaking. As a result of this the wood decays more rapidly. October 1962. Photo C.F.D.

The rods were cut off in the autumn or winter, as is clear from the consistently well developed last annual ring of the wood. The side branches had been cut away; it could not be ascertained how this had been done.

For the extra foundation of trench 2, 25-30 m, bundles of 30-40 such branches were used. No traces have been found of any binding material that may have held the bundles together.

For each hurdle first of all 12 branches measuring 3-4 cm were selected. These were sharpened at the bottom end, while the top end was trimmed so as to produce stakes 1.40-1.60 m long. These were stuck or driven vertically into the ground at intervals of 22-25 cm, to a depth of 10-20 cm. The pointed base ends had usually been sharpened in such a way that they were facetted, usually with 3 or 4 sides (fig. 37).

The actual weaving of the horizontal rods through the vertical stakes is done as shown in fig. 38. To make a flat hurdle (fig. 39) out of springy willow rods requires rather a lot of skill. At the stakes at the end of the hurdle the weft element is twisted around its longitudinal axis, so that the wood fibres become separated, though they can still slip over one another. The weft element can now be bent back 180°, without it breaking. The wood does indeed become damaged as a result of this, and therefore is more susceptible to rotting (fig. 40).

Hurdles that are exposed to some kind of heavy pressure, for example being used as a trackway-surface, are made to be more solid. If thicker rods are selected as stakes then there is a danger of the weft elements breaking; this can be avoided by using warp elements of two stakes together. Another method is to use thicker weft elements, namely with a diameter of 3.5-4 cm at the bottom end. Both these ways of reinforcing were applied to the prehistoric hurdles.

Many hurdles measure c. 2.80 x 1.40 m; this could be called a standard size. When the interweaving of the network has been completed, the projected ends of the weft elements are removed with the aid of a chopping knife or billhook, after which the hurdle can be loosened from the ground. It is then ready for use.

Hurdle-making is an activity that is carried out mainly in the second half of April and the first half of May, because then the wood is more pliable on account of the higher temperature compared with the winter situation.

Hurdles were used in the past for various purposes: not only as surface coverings of trackways and bridges, but also for fences, walls, ceilings, doors or gates and as windscreens. This was probably also the case already in Iron Age times.

3.2.7. *The use of the wood and the location of the willow coppices*

For the trackway, that was over 3000 m long (3.2.8), about 2000 hurdles were used. How much manpower was necessary for making the trackway?

Hurdle-making is rather labour-intensive. A good hurdler can make 4 hurdles a day. The most suitable period for hurdle-making – between mid-April and mid-May – lasts at most 30 days. Assuming that the hurdles were made in one season at the rate of production of 4 per man per day, for 2000 hurdles the number of people needed would be 17.

The cutting and trimming of the rods is not very time-consuming; in one day one person can prepare enough wood for about eight hurdles (55-65 rods per hurdle). The most usual period for doing this (between the beginning of November and the end of February) lasts c. 120 days. Seventeen people would have been able to cut enough wood for the hurdle trackway in about two weeks.

The amount of time invested in transporting the wood to the place where it was made into hurdles (probably the settlement), and in taking the hurdles to where the trackway was being built, is more difficult to estimate. Both bundles of rods and the hurdles were probably transported by wagon. We assume that the willow coppices, the settlement and the starting-point of the trackway were situated relatively close together. For this reason the transport required would not have been the limiting factor; the best time for building the trackway would have been between April and mid-September, a period of about 150 days. A group of 17 people could easily transport 2000 hurdles over a distance of at most 3-3.5 km in 35 days.

In addition to these activities, the work that was put into making the substructure has to be taken into account (see below) to obtain a good idea of the total amount of time that was invested in building this trackway.

The area of willow coppice that was exploited for making the hurdles of this trackway can be roughly estimated, on the basis of a few assumptions. The information is based to a large extent on present-day willow-bed cultivation. At the present time this is practised exclusively on the very heavy though extremely fertile clay soils of the riverine region of the Central Netherlands. The yield of wood obtained there will be higher than what we may assume for the willow coppices of Southeast Drenthe in 170 BC (see below), partly because of the use of specially selected varieties of *Salix* nowadays. The value estimated here for the area of willow coppice therefore has to be regarded as a minimum value.

The wood is cut in a four-years' rotation; we assume that a quarter of the coppice area always had four-year-old wood growing on it. That is to say, we assume that of the total area of coppice every year a quarter was cut. A productive willow stool, that is stripped of all its branches every for years, has – very approximately – 40 shoots, of which 20 will be straight, of the right length and thickness. Thus for each hurdle (55-65 shoots) three willow stools will be needed, on average. Therefore the 2000 hurdles of the trackway required 6000 stools. If we assume that the willow stools stand 3 m apart on average (in practice this distance varies from 1 to 6 m), then each stool occupies 7.5 m^2. Thus for this trackway an area of 4.5 ha of willow coppice would have been used. The total area of willow coppice, under conditions of constant exploitation, would then have been (at least) 18 ha, of which a quarter was harvested each year.

Hurdles are of very limited durability; usually they last for only a few years. It is therefore certain, in view of the great skill of the hurdlers, that willow coppice was harvested every year to obtain sufficient raw material for this and other wattlework. It is conceivable that for the purpose of such a large-scale activity as the building of this trackway willow rods were cut exclusively for these hurdles, while all the other wattleworking activities had to be postponed for a year. In that case the willow coppices of these trackway builders need not have been larger than c. 18 ha.

As we have argued above, we may assume that the willow coppice was present in the neighbourhood of the trackway. The location of this coppice can be pin-pointed: an area less than 30 ha in size, namely the side-valley of the Hunze depression, c. 1 km northwest of the starting-point of the trackway (fig. 5). The bog itself and its marginal area could not have been used as coppice, as they are not suitable for the growth of *Salix*. The extensive field complexes, the Celtic Fields (Brongers, 1976), that take up a considerable part of the higher soils near the trackway (fig. 5), could not have been used for *Salix* cultivation, either. This also applies to the highest parts of this region; the settlement where the trackway builders lived was almost certainly here (4.6).

The side-valley, with its subsoil of partly ice-pushed boulder-clay, became steadily wetter as a result of the upward growing bog surface in the Hunze depression, all the more so because there it has a very limited amount of slope. The bottom of the side-valley projected barely 2 m above the peat-forming level in the Hunze depression, while the length of the valley was 800 m.

The catchment area for the supply of water to this side-valley is small: at most 150 ha. Consequently the side-valley would have been very susceptible to drying out in periods when no rain fell. This drying-out would have been especially noticeable during the summer months on occasions when the bog surface subsided somewhat, with an accompanying increase in the drop in the side-valley. It is possible that this accounts for the remarkably large number of growth interruptions in the *Salix* wood that was used for this trackway. A large proportion of the stakes that were analyzed showed so-called double or false annual rings, indicative of a cessation in growth occurring nearly every year. The obvious conclusion is that this represents an annual drying-out of the side-valley during the summer period.

In and around such willow woods also *Fraxinus* and *Alnus* were undoubtedly present.

The number of logs that were used for the substructure cannot be determined so readily. On the basis of the excavation drawings (fig. 28), it is possible that the substructure required c. 5000 logs measuring about 6 m in length, that were partly used cut up into shorter logs (as transverse timbers of the substructure). Such logs with a diameter at the bottom end of 8-15 cm have a volume of about 40 litres. The total quantity of wood used for the substructure is thus 200 m^3. On the basis of our calculations relating to the timber requirements for the Neolithic trackway of Nieuw-Dordrecht (Casparie, 1982: pp. 155-156), we arrive at a total area of woodland of 3-5 ha. The trees were 10-20 years old when they were felled. If we assume that one man can fell 16 such trees in a day, and that the further processing of the wood (cutting off side branches and transport) costs just as much as time as the felling, then one person should be able to produce 8 such logs per day. The 5000 logs thus represent more than 600 man-days; for a group of 17 people (the same group size as was assumed for the hurdle-making) this work could have been done in 35 days.

From our rough estimations it is evident that the cutting of the willow rods, the making of the hurdles, the felling of the trees for the substructure and the transport of the construction elements, together with the building of the trackway over a total length of c. 3000 m (3.2.8), altogether meant at least four months work for 17 people.

The area of willow coppice required was at least 18 ha; of this the yield of one year – 4.5 ha – was utilized for this trackway. In the following year the greater part of this coppice area would have been able to provide plenty of wood once again for wattlework required for various other purposes. In view of the point of departure for the calculations, namely that these are minimum values of areas, it is likely that the whole side-valley, extending over an area of less than 30 ha, was used for this willow coppice. The forest that provided the logs for the substructure could have been exploited again for its timber after 10-20 years. It is fairly certain that this was not undisturbed, dense forest; thus the areas size of 3-5 ha should be regarded as a minimum value. In view of the extent of the area of willow forest, and of the size of the area of land under cultivation (c. 250 ha of Celtic Fields, see fig. 5), it is clear that we are concerned here with a landscape that was being used intensively already in this period, where forest was only sparsely present. It is possible that the people living here had to cut down all the trees of the right size that were available in order to build the substructure. This would also explain why the substructure contained species of wood that came from different places.

3.2.8. *Direction, starting-point, destination, length and function*

From the location of the excavation trenches and the observation points (fig. 27) it is clear in which direction the trackway ran; this was SWS-ENE, just like the other prehistoric bog trackways in this region. Some of the observation points were given by Wesselink (1924), others we established ourselves. The westernmost observation point is based on information from the owner of this terrain, W. Welling of Emmerschans, that he acquired in 1963. At this spot (fig. 27), a few decades earlier, a lot of wood had been found in the peat, precisely at the transition from sand to peat. We have interpreted this as the starting-point of the hurdle trackway or at least part of the trackway close to where it began.

The trackway was laid out as a direct connection; it was certainly at least 1500 m long (fig. 5, between excavation trench 1 and the easternmost observation point). It definitely could not have been intended to extend across the whole raised bog (a distance of 11 km); for this purpose its method of construction is unsuitable. We are convinced that it served to provide access to the bog iron-ore (siderite) centre D (fig. 5). If this was indeed the case, then we are concerned with a trackway that was 3000 m long or slightly longer.

The presumed function of the trackway, namely to permit the transport of iron-ore, is suggested not only by the presence of lumps of iron slag in the area of the (presumed) settlement (Modderkolk, 1970) and by the direction of the trackway (towards a bog iron-ore area), but also by a few finds from the peat, notably the piece of wood with a rectangular hole (3.2.9), that was found in the bottom layer of this siderite area D (Casparie & Smith, 1978). The iron-ore (van Heuveln, 1958) formed between 5200 and 3100 BC (Casparie, 1972; 1984). It had probably become overgrown by poorly humified *Sphagnum* peat already at the time when this trackway was built. We do not know whether the spot was still visible from a distance, as it had been in the Middle Bronze Age (Casparie, 1984).

3.2.9. *A few wood finds*

During the excavations of this trackway no finds were made that give any indications of its function. With many finds from the peat it is not clear whether they are related to activities in the bog, or whether they are objects that were accidentally lost, for example. Two wood finds in the area of the trackway are indicative of exploitation of the bog close to the trackway XIV(Bou) during the Iron Age.

1. In the siderite area D, at the transition from sand to peat, a plank was found (fig. 5:1), that has been dated to 575±35 BC. This find – precisely in the course of trackway XIV(Bou) – indicates that bog iron-ore was dug here in the Early Iron Age (Casparie & Smith, 1978). Close to the plank, in which a rectangular hole had been cut out, various rows of small poles were found (pers. comm. H. Kuilder, Emmen), also at the base of the peat, that are indicative of relative extensive digging activities in the bog.

The plank, *Quercus*, measures 1.14 m in length, 20-21.5 cm in width, 3.5-5.4 cm in thickness; it was prepared entirely by chopping, i.e. it had not been sawn. The chopping traces indicate that an axe was used with a cutting edge of about 6 cm. The

Fig. 41. Oaken plank, found at the bottom of the peat in seepage centre D (fig. 5.). The plank has been dated to 575±35 BC. Length of the plank: 114 cm. Photo C.F.D.

Fig. 42. The peat-digger J. Velema indicates the plank, in the peat near XIV(Bou) (fig. 5). The plank (not dated) may form part of a construction of some sort close to the hurdle trackway. Photo c. 1933 by J.H. Botterweg.

rectangular hole that had been cut out measures 64.5 cm in length, while its width varies from 6.5 to 8 cm (fig. 41).

The level at which the find was made, 2.75-3 m deep in the peat, can be dated to c. 5000 BC (Casparie, 1972). We are therefore concerned here with an object that had been deeply dug in. Its precise function is not known, but in view of the quality of the peat at this spot - rich in iron-ore - we can be almost certain that it had something to do with exploiting bog iron-ore.

2. Groenman-van Waateringe and Smith (1984) describe a leather shoe found in the peat, that can be dated to the Late Bronze Age. Their illustration Plate 1a (fig. 42) shows the finder of the shoe, at the spot where it was found in the peat. The finder is pointing to a plank of a bog trackway. The photo was taken in about 1933 by J.H. Botterweg (Drachten). In figure 5 this spot is indicated as No. 2. The level of the plank can be roughly indicated: between 500 BC and the beginning of the Christian Era. In view of the situation (fig. 5:2) and the level of this piece of wood (undoubtedly a plank, though no measurements are known), it is possible that this wood can be associated with the trackway XIV-(Bou). In any case we can be certain that it was connected with some kind of activity that can be dated in the Iron Age.

The location given by us (fig. 5) deviates from that mentioned by Groenman-van Waateringe and Smith (1984). The location they mentioned was put on their map in the wrong place, as the result of a misunderstanding.

4. INTEGRATION AND DISCUSSION

4.1. Introduction

In previous studies on the bog trackways of South-east Drenthe (Casparie, 1982; 1984), we focussed our attention on many aspects in order to explain as well as possible the Neolithic bog trackway of Nieuw-Dordrecht XXI(Bou) and the MBA footpaths XVI(Bou), XVII(Bou) and XVIII(Bou). Many of the points of departure discussed in those studies are also applicable to this discussion of the Iron Age bog trackways, that forms a sequel to the previous studies. Those points of departure are not repeated here.

4.2. The accessibility of the peat-bog area

The development of the surface of the raised bog from the time when the MBA footpaths were built (1200-1100 BC) until the bog burst of 500 BC is mainly a matter of increasing abundance of water and consequently greater problems for anyone wanting to pass across the bog. For the domed complexes it is difficult to assess whether the bottoms of the hollows could be safely trodden in the summer period without the aid of planks or hurdles; the contact zones and the area of the raised-bog lake were in any case impassable throughout the whole year (2.3). The raised-bog rivulet the Runde was probably a rather insignificant stream until the bog burst; nevertheless the footpath XV(Bou) must have had a very substantial construction in the contact zone of the Runde. The accessibility of the bog iron-ore areas, notably area D (fig. 8), that was evidently being exploited in the Early Iron Age (3.2.9), is not clear. In about 500 BC the siderite areas had already become overgrown by (poorly humified) *Sphagnum* peat.

After the bog burst the peat surface of the domed complexes possibly became slightly more easily passable. A few centuries later this was no longer the case, as can be deduced from the need for an extra foundation of bunches of rods for the trackway XIV(Bou) (3.2.4). After the bog burst the water supply in the contact zones decreased considerably; nevertheless these zones almost certainly remained hardly accessible at all. A lot of water soon accumulated once again, and these zones were marked by deep erosion gulleys. The absence of proper wooden trackways, i.e. trackways substantial enough to take wheeled traffic, can possibly be interpreted as an indication that this peat-bog region still had a very limited carrying capacity and was thus very difficult to pass across. In view of the scarcity of peat-bog finds dating from the Iron Age in this region (pers. comm. W.H. Zimmermann, Wilhelmshaven), it is evident that the bog surface was trodden less in the Iron Age than in the preceding periods.

4.3. The reason for the existence of the wooden trackways

The most important reason for building the footpath XV(Bou) was probably the need for contact with inhabitants of the east side of the Bourtanger Moor (3.1.8). The way in which this need became expressed raises a number of questions. Why was a footpath built and not a trackway for wheeled traffic, that offers more possibilities for transport? It is very probable that the wood supply in the region where the path-builders lived was too small to permit a trackway to be built; for this purpose about 500 ha of forest would have been required (3.1.9), which was no longer present as a result of the intensive occupation (4.6). Evidently this path was not intended for transporting large quantities of material. Therefore such a path cannot be regarded as a trade route; the practical possibilities were too limited. Why was so much energy put into establishing a connecting route, that could only have served a limited purpose and that was not intended to take intensive traffic? It is fairly certain that the hardly passable state of the bog surface in c.

530 BC (4.2) necessitated the building of this path, but it is not known whether this is the successor of a previous route, without any wooden walking surface, across the bog. We may assume that this peat-bog surface had been very wet for a long time already; with this path it therefore seems that a new need for contact arose rather than than the gradually increasing inaccessibility of the bog necessitated the improvement of an already existing route. It is also certain that at about that time people did indeed make their way across the bog (3.2.9), notably for the purpose of winning iron-ore.

It is not clear whether the walking surface of this path projected high enough above the bog surface to permit people to walk along the path dry-shod. Although the path was well supplied with transverse timbers and had a more substantial construction in the exceptionally wet parts of the bog (3.1.4), it was probably usable as a comfortable passage only in dry, rainless periods (3.1.8).

Between the time of construction of this path and the bog burst (2.3) there was probably hardly any difference in time; the ^{14}C-datings are also indicative of this (2.3 and 3.1.3). In view of the quality of the wood of the planks that were driven into the peat (3.1.7), it is not impossible that the bog burst surprised the path-builders while they were at work. We may assume that they had not foreseen this bog burst; the effect of this catastrophe on the passability of the raised bog was so drastic that it was decided to demolish this path (3.1.7). Partly because no traces of use are evident and no finds relating to the path were made, it cannot be clearly established for what kind of communication this path XV(Bou) was intended.

The hurdle trackway XIV(Bou), that was almost certainly intended to reach the iron-ore centre D (fig. 5; 3.2.8), is remarkably wide, wide enough to take wagons. The MBA wooden trackways in this region, that were also built for the purpose of winning bog iron-ore, are merely footpaths. This can be adequately explained by the fact that in the MBA this material could be processed only in small quantities (Charles, 1984). The technique of iron processing became considerably improved in the Iron Age, as can be deduced from many finds of pieces of iron slag of *Rennfeueröfen* (i.e. furnaces for iron smelting) (see i.a. Hayen, 1968). Also from the inhabited area close to the hurdle trackway sufficient information is available on iron-working in this period (Modderkolk, 1970). Yet here too a few questions arise. Can the remarkable width of this hurdle trackway be attributed to a greatly increased demand for iron-ore, that had to be won in a very short time? In that case the use of wagons or of a large workforce would be plausible. No clear indications have been found that the path was intended to be made use of in either of these ways.

The (possibly) increased demand for iron-ore would probably also have led to an intensive exploitation of the bog iron-ore areas A, B and C (fig. 5). Yet here no wooden trackways were made use of; at least no Iron Age trackways relating to these areas have been found. These areas may have been relatively easily accessible. The question then arises as to why people took the trouble to build a labour-intensive trackway to get to the iron-ore area D, a trackway that could only have been used for a very short time (4.4).

4.4. Building methods, organizational aspects and the duration of the period of use

The northern plank footpath XV(Bou) was clearly planned in advance; in view of the consistent method of construction (fig. 19) over a relatively great length it is fairly certain that the construction was carried out according to specific instructions stringently prescribing the application of a particular building method and choice of material (predominantly oak timber). The construction elements were probably made in the settlement, but they had to be trimmed and cut into shape at the building site in the bog to make them fit together well. In terms of the dimensions of the construction elements this footpath hardly differs from the MBA footpath XVII(Bou) (Casparie, 1984) that is more than 600 years older. With the use of metal axes, evidently enough standard techniques and standard measurements had been developed, that remained in use for a very long time.

The path is made of freshly cut oak timber; only in one place, trench 5, 64-67 m (fig. 18), some secondarily used wood has been found. The significance of this and the previous function of the pieces of wood concerned are unknown.

The wood has been worked more roughly than in the MBA footpath XVII(Bou); this may be attributable to the use of iron tools. A wooden mallet (fig. 26) has been found, that must have been one of the tools used for building the path.

It was the intention of the path-builders to lay out a straight pathway; no bends are present at all. In view of the straight course of the path over a considerable length it is clear that the direction of the path was predetermined in the terrain.

The path had great stability; on account of its surface width of c. 0.5 m it was very comfortable, but it is questionable whether it would have allowed people to walk here dry-shod in the very wet area of bog (immediately before the bog burst). The use of extra transverse timber (3.1.4) compared with the MBA footpath XVII(Bou) would have meant only a slight improvement. The fixing of the transverse timbers by means of pegs into the peat and the way in which the longitudinal planks are firmly held in place between the transversals in footpath XV(Bou) are indicative of a throughout knowledge of the bog and experience with the problems of carrying capacity. Also the more substantial means of construction in the excessively wet parts of the bog (3.1.4) is indicative of this.

The hurdle trackway XIV(Bou) has a very simple means of construction (fig. 3; 3.2.5); the substructure consists of ordinary, general-purpose timbers and the trackway-surface elements are standard items of use (3.2.5, 3.2.6). Therefore it is impossible to say whether this path was also built according to specific instructions prescribing the materials to be used. In view of the straight course of the path it is probable that the route was determined in the terrain beforehand (fig. 5). The making of the willow hurdles (3.2.6, 3.2.7) is a routine task restricted to a particular season; this hurdle-making required great technological skill, and was almost certainly a traditional craft of long standing (4.8).

Like recent willow beds, the willow coppice that was exploited for this trackway (3.2.7) undoubtedly provided willow rods for various other purposes too, such as finer basketry, for making bushels and hampers. The basketry was almost certainly done in the settlement.

The trackway XIV(Bou) had fairly great stability when trodden underfoot, partly on account of the resilient nature of the surface material. The laying out of wattles on a framework of roundwood gives a maximal surface-increasing effect of the hurdles on the bog surface. Such a means of construction (fig. 4), that exerts a minimum pressure on the bog surface, is insufficient, however, in very wet contact zones, as is clear from the (supplementary) use of withy-bed bundles as foundation material in a bog-stream channel (3.2.4). The absence of connections between the trackway-surface and the substructure together with the absence of pegs to fix the trackway to the peat means that the trackway-surface elements can easily slip out of place with respect to the substructure. Evidently the path-builders did not regard this as a serious problem.

The construction of both trackways involved fairly large quantities of wood, that was probably processed within a short space of time. The availability of this wood will be discussed in more detail in the following section (4.5). The wood came mainly from the Hondsrug, the higher sandy areas (3.1.9, 3.2.7). For the transport of the construction elements from the place where they were made to the building site it is presumed that wagons were used, although no direct indications of this have been found. We doubt whether also the hurdles of trackway XIV(Bou) were also used for taking wheeled traffic. In the case of both trackways the transport by wagon is thought to have taken place as far as the bog margin.

The construction of wooden trackways and paths is very labour-intensive. For the hurdle trackway XIV(Bou) it is reasonable to assume that the construction work was done by at least 17 people (3.2.7). For the plank footpath XV(Bou) no such estimation can be made, but it seems plausible to us that for its construction, out of an estimated 15,000 planks of 3 m in length and 6000 pegs (if the path was c. 12 km long), a few dozen people must have been available. In both cases the estimated workforce would appear to be in excess of the number of workers that one or two farmsteads could provide; see also 4.6.

Both trackways showed hardly any or no traces of wear. This cannot simply be taken to mean that they were used hardly or not at all. With a resilient trackway-surface, like that of both the plank footpath XV(Bou) and the hurdle trackway XIV(Bou), the damage caused by use is inflicted not on the wood but on the underlying peat, which becomes crushed flat and thus loses its structure. With trackway XV(Bou) no such disturbed peat has been found under the longitudinal planks; with XIV (Bou) the crushed, amorphous peat concerned was of limited thickness, as is evident from our peat research (Casparie, 1972: figs 78 and 44, respectively). It is now certain that both trackways were used for a short period at most.

As far as the quality of the building material is concerned, the footpath XV(Bou) could undoubtedly have lasted for many years. As a result of the upward growth of the peat-bog surface, however, it could have remained functional for a relatively limited period, possibly 1-2 decades. In this case the bog burst of 500 BC (2.3) put an end to the potential use of the path. We cannot exclude the possibility that this catastrophe occurred while the construction of the path was in progress, and that consequently it could not be completed. The unimpaired state of the longitudinal planks that we excavated (fig. 20) is indicative of (almost) unused wood. When the path was demolished (3.1.7) a number of planks were probably taken back to the settlement to be used there for some other purpose. Where the planks had been driven into the peat, we may literally speak of destruction of the path.

The durability of hurdles of *Salix* is limited to a few years, at least in a damp environment in the open air. Wattlework made out of willow rods loses its strength in a couple of years. In view of the (apparently) unimpaired state of the hurdles that we found (fig. 34), we presume that the trackway very rapidly became overgrown by fresh *Sphagnum* peat, probably within 5 years. Therefore the trackway could only have been in use for a couple of years at most.

4.5. The use of wood and the forest situation on the higher sandy soils

If the footpath XV(Bou) was 12 km long (3.1.8), then the area of forest that was necessary to supply the timber must have been c. 120 ha (3.1.9); the greater part of this would have been present in Southeast Drenthe, close to the western starting-point of the path. We have previously stated (Casparie, 1984) that since c. 1300 BC the entire region, that could have been used for providing the timber, had been reclaimed, with a fairly high degree of deforestation. If we assume that the oaks of c. 40 cm thickness were felled and cut up into planks and pegs in a relatively short time (at most a few years), then we are concerned here with a rather heavy demand on the available wood supply. If the timber had to be transported over a maximum distance of 3 km on the high ground, then 8% of this region must have been forest consisting of oaks about 100 years old (the average age of the trees used). If the transport route was at most 2 km, then c. 20% of the region must have consisted of such forest. It is certain that in such an intensively occupied region (4.6) large stands of forest of such high quality could not have been readily available for a single activity like the construction of this trackway XV(Bou); certainly not close to its starting-point in any case. The transport routes for the timber were undoubtedly very long, possibly more than 3 km (see also further on in this section).

If the hurdle trackway XIV(Bou) was 3 km long, then the willow coppices that provided the willow rods for making the hurdles must have extended over an area of (at least) 18 ha (3.2.7). The total area of land that was available for the cultivation of willow coppice was less than 30 ha. Also in this case we must assume that the building of the trackway made heavy demands on the available wood supply; here it must be borne in mind that this was the yield of one year's harvest of willow rods. In terms of deforestation the building of the hurdle trackway did not play a role of any significance; the annual production of willow rods for wattlework was not disadvantageously affected.

On account of the strict use of timber of *Quercus* for XV(Bou) and *Salix* for the surface-layer of XIV(Bou), these trackways provide little information on the species composition of the forest in the Iron Age. From the substructure of XIV(Bou) it is evident that *Alnus*, *Betula*, *Fraxinus* and *Pinus* were present in the surroundings (3.2.5). From the palynological information (2.1.2; fig. 6) it can be deduced that the forest included also other species of trees, that could have been used for building the wooden trackways. These include *Ulmus*, *Corylus*, and later in the Iron Age also *Fagus* and perhaps *Carpinus*. It is not clear whether *Acer*, *Prunus*, *Taxus* and *Populus* (all of which occur regularly in low values in the pollen diagrams) were also present in exploitable quantities. *Tilia* was hardly present in the Iron Age. In 4.6 we shall devote more attention to the extent of the remaining forest in the region where the road builders lived.

4.6. Aspects of the occupation

In 3.2.7 we stated that the landscape in which the builders of the hurdle trackway XIV(Bou) lived and worked was used intensively, and that forest was only scarcely present. The large concentration of Celtic Fields in this region (fig. 5) and the presence of an extensive area of willow coppice (3.2.7), in a region that had been permanently occupied already since c. 3000 BC (Casparie, 1982; 1984; Casparie & Groenman-van Waateringe, 1980), justify this conclusion. Although the Celtic Fields have no dating, which means that it cannot be said with certainty that they were all in use c. 200 BC, it is nevertheless certain that the land here had already been cleared for a very long time; on this land it would have been hardly or not at all possible for forest to maintain itself for any length of time. Concerning the situation at the time when footpath XV(Bou) was built, c. 500 BC, no such definite assumptions can be made. At that time the occupation was already rather intensive; the extent of the area of arable land was presumably as large as the Celtic Fields indicated by Brongers (1976) and illustrated by us in figure 5. It is known for certain that in the period 700-500 BC a few homesteads with farmhouses and barns within fenced enclosures were present in the southern part of the northern Celtic Fields complex (pers. comm. P.B. Kooi). The fences may have been made of wattlework, though of a type different from that of the hurdles. The marked expansion of *Fagus*, that results in the first *Fagus* maximum of c. 500 BC (Casparie, 1972), indicates the presence of large open spaces where beech was able to become established, presumably abandoned arable land. We consider it very well possible that in the period between c. 600 and 200 BC the higher ground in the vicinity of the starting-points of both wooden trackways, XIV(Bou) and XV(Bou) (fig. 5), consisted for more than 50% of land under tillage. In addition there were the extensive areas of abandoned arable land, where i.a. *Fagus* was able to expand, and the willow coppices. The actual raised-bog region was not exploitable, and the bog-fringing area was too wet to permit reclamation. In depressions on the higher soils of the Hondsrug some peat formation occurred here and there. Consequently the area of long-undisturbed forest that was necessary for providing timber for XV-(Bou) could only have been small.

The bog-trackway research gives no definitive answer as to whether the (relatively) intensive occupation close to the starting-points of the wooden trackways was concentrated in a village, or whether the occupation was more scattered, e.g. in the Celtic Fields. In view of the relatively large number of persons that must have been involved in building each of the trackways (4.4), and in view of the presumed system of organization behind the construction method of XV(Bou), the presence of villages in the Iron Age is very plausible.

4.7. Comparison with other Iron Age bog trackways

In the Dutch peat-bog regions two other Iron Age wooden trackways are well known; both were designed to take wheeled traffic, and measure c. 2.5 m in width or slightly more. The Valtherbrug I(Bou), dated to 345±40 BC (Lanting & Mook, 1977), probably spanned the 12-km-wide Bourtanger Moor (Karsten, 1819; Landweer, 1898; Joosting, 1901; van Zeist, 1958), about 10 km north of the footpath XV(Bou). It consists of a substructure of longitudinals supporting a trackway-surface of transversely laid out planks and roundwood. Pegs were used for fixing the trackway-surface in place.

In the eastern peripheral area of the Smildervenen (i.e. the Smilde bogs) the wooden trackways I and II(Sm) provided passage across two peat-filled depressions, measuring 280 m and 170 m wide respectively, in a route some 9 km long over sand ridges between two Iron Age settlements. I(Sm) has been excavated (Casparie, 1985) and is dated to c. 220 BC. It was made of transversely laid-out roundwood and cleft trunks of *Alnus*, over which a covering layer of sandy heather turves had been placed. A substructure of longitudinals was present in some places; there were no pegs. II(Sm) very probably had the same kind of construction. The Smildervenen are situated about 50 km northwest of the bog region where XIV(Bou) and XV(Bou) are present. The means of construction of I(Bou) and I(Sm) are distinctly different from those of the trackways XIV(Bou) and XV(Bou) dealt with in this article.

In the Somerset Levels in Southwestern England relatively few Iron Age bog trackways are known (Coles, 1975-1982); very little information is available concerning their means of construction. The two known trackways appear to be footpaths, of Early Iron Age date, c. 700-600 BC, consisting of longitudinally laid-out bundles of branches. A few trackways with a surface-covering of wattlework will be discussed in more detail below (4.8); they can be dated in the Neolithic and EBA.

In various bog regions in Northwestern Germany very ingeniously constructed trackways are present, of which a large number have been investigated (Hayen, 1958; 1963a; 1963b; 1965a; 1965b; 1971; 1973; 1975; 1978; 1979; 1984; 1985). These are nearly all trackways for wheeled traffic, measuring c. 2.50 m in width, with a carefully smoothened trackway-surface. The use of planks of *Quercus* is predominant. In a few trackways datable to the Early Iron Age (c. 650 BC) the trackway-surface planks were carefully fixed, by means of substantial pegs driven vertically into the bog, between longitudinally placed wood above and below the trackway surface. This means of construction prevented the planks from shifting and ensured the maximum carrying capacity of the trackway surface. Examples of this type of trackway are IX(Le), XXI(Le), XII(Ip). A few trackways, that can be dated to c. 250 BC or somewhat later, have a more simple means of construction, involving the fixing of the trackway-surface planks to the peat by means of a peg. All of them have carefully smoothened trackway-surfaces. Examples of this type are VI(Pr), XLII(Ip), XXXI(Bou); the Valtherbrug I(Bou) has a similar means of construction. The trackway I(Cl), dated to c. AD 90, has ditches in each verge. The peat that was dug out of these ditches was used as a sub-layer for supporting the trackway-surface planks. In most cases a covering layer of heather turves or sods had been applied, to provide a smooth and comfortable passage.

Also in the Iron Age some trackways were laid out with a surface-layer of roundwood, like I(Le), that can be dated to c. 300 BC and that was in use for a long time. The trackway XLIII(Ip) has a surface-layer of *Alnus* planks supported by a substructure of longitudinally placed *Alnus* trunks; only a few pegs were used in this construction. The trackway, that can be dated to between 200 BC and the beginning of the Christian Era, is partly burnt. When the trackway was restored the partly burnt timbers were laid out longitudinally, in such a way that the trackway became c. 1.30 m wide and thus suitable for wheeled traffic; these planks were fixed by means of pegs. The trackway originally had a covering layer of turves.

Footpaths were built only very seldom in the Iron Age. No true counterparts of XIV(Bou) or XV(Bou) are known from the Iron Age. XV(Bou) fits well in Hayen's typology (1958); in the bogs of Northwestern Germany footpaths occur mainly in the Bronze Age.

4.8. Prehistoric hurdles

The research on wooden trackways in the peat deposits of the Somerset Levels resulted in the discovery of a few trackways with a surface-covering of wattlework mats, that date from the Neolithic and the EBA (Coles et al., 1975-1982). Wattlework was used for various purposes on a wide scale already in Neolithic times and probably even earlier. The use of wattlework mats as a trackway-surface to provide a passage through peat-bog, however, is very limited. Presumably wattlework mats were not suitable for this purpose, almost certainly because the network of wattles was too loose and the mats were too weak to spread the pressure of trampling over a sufficient large area of peat. This applies notably to the Neolithic and EBA hurdles of the Somerset Levels, that differ considerably from the Iron Age hurdles of XIV(Bou) especially in terms of the choice of wood (*Corylus*, *Betula*) and the distance between the stakes or sails – c. 45 cm. The hurdles of XIV(Bou) were made of

Salix, while the distance between the stakes was c. 24 cm. The method of making hurdles of this type had been developed to such a high standard already by c. 170 BC that during the following 2000 years up to now no further improvements were made (3.2.6, 3.2.7). Together with other wattlework articles these hurdles were undoubtedly extremely useful for various purposes, as is still the case today (Westendorp et al., 1985). Yet they were hardly used in the construction of trackways, as can be deduced from the almost total absence of hurdle trackways in peat-bogs.

It is appealing to presume that the Iron Age hurdles of the XIV(Bou) type developed from the less efficient type that has been found in the Somerset Levels. The improvements, that can be dated at some time between c. 1500 and 200 BC, include the shorter distance between the stakes, the use of *Salix* rods and the bending back 180° of the horizontal or longitudinal rods, that ensured the firm fixture of the end-sails in the wattlework. The final product is a very strong, flat mat, that can easily be transported and that has a much greater carrying capacity than the wattlework mats of the type found in the Somerset Levels. Much more force is required for making *Salix* hurdles than for making hurdles of the type in the Somerset Levels, that have been copied experimentally (Coles & Darrah, 1977). The similarities between the two types of hurdle lies especially in the dimensions of the raw material (c. 4-year-old rods with a diameter of 2-4 cm at the base), in the woodland management (Rackham, 1977) and in the harvesting of the raw material: coppicing of the underwood from the stools (3.2.7). The necessary woodland management had already been developed in Neolithic times (Coles, Caseldine & Morgan, 1982). It is not clear why notably in the German bog regions no hurdles were used for constructing prehistoric trackways (pers. comm. H. Hayen).

4.9. Final remarks

In his survey of recent archaeological research in the bogs of Northwestern Germany, Hayen (1985) concludes that the (prehistoric) trackways are parts of a traffic network. The form of this network was determined above all by the situation of different areas of occupation, the nature of the soil and the landscape, and the direction in which more remote destinations lay. It is clear that in the Iron Age this network of trackways, that were designed to take especially wheeled traffic, was of considerable importance. Wide stretches of bog – more than 6 km wide – were spanned, and the technical aspects indicate great insight and skill on the part of the builders of these trackways on a weak substratum. We may assume that the high areas of Southeast Drenthe, with intensive prehistoric occupation, also had such a network of trakways, but that the great width of the Bourtanger Moor made it almost impossible for the bog region to be 'fitted into' this trackway network. Therefore it is much less clear than in the case of many of Hayen's trackways whether and in what way the footpath XV(Bou) and the hurdle trackway XIV(Bou) connected up with this trackway network. An extra complication in our case is that the trackways of Southeast Drenthe were not obviously intended to take wheeled traffic, as the German Iron Age trackways were. The fact that XIV(Bou) and XV(Bou) were in use for only a short time at most means that no supplementary information is available concerning their use, in the form of wear, repairs and lost or discarded objects. From the methods of construction used we can deduce that these trackways were built in a systematic way, involving the effective organization of material, transport and construction activities, in agreement with what Hayen (1985) has formulated; also the presence of well-developed woodland management is certainly in evidence (Rackham, 1977). The skill, demonstrated by Hayen, in constructing an extensive bog trackway was present in this region, as is evident also from the Valtherbrug, dated to 345±55 BC (4.7).

With regard to the hurdle trackway XIV(Bou), it is remarkable that such common articles of use as hurdles, indicative of a long tradition of basketry and wattleworking, were used only exceptionally in building trackways; it is equally remarkable that in this region in the Iron Age only one trackway appears to have been built for the purpose of exploiting bog iron-ore. In the MBA already two of three paths had been laid out for this purpose (Casparie, 1984), and in the Iron Age iron production evidently took place there on a considerable scale (Modderkolk, 1970; Casparie & Smith, 1978; Booij, 1986). There was certainly no tradition here of building trackways on peat-land. Was the raised bog in fact more easily passable than the results of the peat-bog research (Casparie, 1972; 1982; 1984) would suggest? This could only have been the case in dry summer periods at most. The bog burst of 500 BC (2.3) and the demolition of the footpath XV(Bou) (3.1.7) are nevertheless indicative of a great abundance of water, implying that the bog surface was hardly passable.

In the peat-bog of Southeast Drenthe, over a period of about 2000 years (between 2150 and 170 BC) only six wooden trackways were laid out. Partly because they remained in use for only a short period, these trackways played no important role in the development of the prehistoric occupation. Nevertheless the peat-bog trackway research (Casparie, 1982; 1984; this paper) has provided a great deal of information on aspects of building technology and has made important contributions towards our knowledge of the vegetation development and